Power Systems

Andreas A. Neuber (Editor)
Explosively Driven Pulsed Power

Andreas A. Neuber (Editor)

Explosively Driven Pulsed Power

Helical Magnetic Flux Compression Generators

With 203 Figures

Dr. Andreas A. Neuber
Texas Tech University
Center for Pulsed Power & Power Electronics
Dept. of Electrical and Computer Engineering
Mail Stop 3102
Lubbock, TX 79409-3102
USA
Andreas.Neuber@ttu.edu

Library of Congress Control Number: 2005926342

ISBN-10 3-540-26051-X Springer Berlin Heidelberg New York
ISBN-13 978-3-540-26051-6 Springer Berlin Heidelberg New York

This work is subject to copyright. All rights are reserved, whether the whole or part of the material is concerned, specifically the rights of translation, reprinting, reuse of illustrations, recitation, broadcasting, reproduction on microfilm or in other ways, and storage in data banks. Duplication of this publication or parts thereof is permitted only under the provisions of the German Copyright Law of September 9, 1965, in its current version, and permission for use must always be obtained from Springer-Verlag. Violations are liable to prosecution under German Copyright Law.

Springer is a part of Springer Science+Business Media
springeronline.com

© Springer-Verlag Berlin Heidelberg 2005
Printed in Germany

The use of general descriptive names, registered names, trademarks, etc. in this publication does not imply, even in the absence of a specific statement, that such names are exempt from the relevant protective laws and regulations and therefore free for general use.

Typesetting: Dataconversion by author
Final processing by PTP-Berlin Protago-T$_E$X-Production GmbH, Germany
Cover-Design: deblillk, Berlin
Printed on acid-free paper   62/3141/Yu – 5 4 3 2 1 0

Foreword

The operating principles of Magnetic Flux Compression Generators are easy to understand. The details of their construction and performance limits have been mainly described in government reports (many classified). The MegaGauss conference proceedings are also dominated by contributions from government (US and foreign) laboratories. The government work has often been concerned with very large generators with explosive charges which require elaborate facilities and safety arrangements. The work conducted under the MURI program described in this book emphasized small generators (less than 500 grams of high explosives) and the detailed physics understanding, construction details, and parameter variation effects on these generators. The three universities involved (TTU, UMR, and TAMU) worked very well together and each brought different sets of backgrounds, experiences, and facilities to bear on the program.

This MURI project has, for the first time, shown that it is possible for university laboratories and graduate students to be involved in and contribute to this line of research. The book forms an invaluable starting point for other groups getting involved in this research. It proves that small helical flux compression generators (FCG) with high performances can be constructed in a university environment and explains their performance limits and practical uses. In a sense, it can be considered as a handbook for compact helical FCG design. As a personal note; it was a pleasure to be the director/coordinator of this program. The cooperation and synergism between the university partners and the strong encouragement from the technical grant monitors, Dr. R. Barker and Col. Joseph Gregor of AFOSR, were superb and made my role extremely easy.

Texas Tech University *Magne Kristiansen*
March 10, 2005

Preface

This book presents a snapshot in time of the status of university research on explosive-driven pulse power with an emphasis on helical flux compression generators, FCGs, in the United States in collaboration with the U.K. circa 2004. The focus of this book is on the pulsed power output of explosive systems that are comparably small in size and can compete with conventional pulsed power. This book summarizes the research effort started by the 5-year Multi Disciplinary University Research Initiative, MURI, on Explosively Driven Pulsed Power covering the time period from March 1998 to April 2003. Research in the area of explosive pulsed power is at the present time, 2005, ongoing at the MURI partners' laboratories.

Under the auspices of the MURI program, which was managed by Robert J. Barker, scientists at 3 U.S. universities and one collaborating U.K. university have been conducting research that had its strength in the synergism of the wide range of topics needed to tackle the task. Under the consortium leadership by Magne Kristiansen scientists with Electronic & Electrical Engineering, Physics, Mechanical Engineering, and Explosives and Metallurgy background worked closely together to produce the results presented in this book. Funded at a rate of U.S. $1M/year, the participating universities and departments were "The Center for Pulsed Power and Power Electronics" at Texas Tech University (represented by James C. Dickens, Hermann Krompholz, and Andreas A. Neuber), the Mechanical Engineering Department at Texas Tech University (Jahan Rasty), the Explosives Research / School of Mines and Metallurgy at the University of Missouri-Rolla (Jason Baird and Paul Worsey), and Texas A&M University (Bruce L. Freeman). Additionally, the continuing collaboration of Texas Tech University and the University of Loughborough, UK, (Bucur M. Novac and Ivor R. Smith) has resulted in Chap. 6 of this book. Texas Tech as the lead university and UMR were the original proposers, Texas A&M was added to the effort at a later time.

We briefly introduce in Chap. 1 explosive driven pulsed power and discuss basic FCG operation. We will address in Chap. 2 the operating principle of helical FCGs in more depth and present some theoretical aspects of it. Chap. 3 will elucidate why a real FCG is always exhibiting less than ideal performance. In the following Chaps. 4 and 5, we will focus on mechanical aspects of explosively deformed conductors and the physics of FCG operation. Starting with a simple PSpice model, Chap. 6 presents details on state-of-the-art generator performance modeling. We address power conditioning, the link between FCG and load, in Chap. 7 and give an overview of practical seed energy sources in Chap. 8. We finally present an example of a practical explosive Pulsed Power System, PPS, and compare its performance with a conventional PPS in Chap. 9.

Acknowledgements

This book brings together contributions from each of the participants in, and collaborating research groups with, the explosive driven pulsed power MURI program. It is the written contributions of the individual coauthors that have given this book its valuable substance and archival content.

The research leading to this book would not have been possible without the graduate students that were involved in the research over the years. They are in alphabetical order: Daniel J. Dorsey, Troy L. Guy, Tammo Heeren, David J. Hemmert, Thomas A. Holt, Eric Kristiansen, Le Xiaobin, Juan-Carlos Hernandez Llambes, Mark Schmidt, and Teresa E. Tutt.

Finally, special thanks go to Magne Kristiansen and Lynn Hatfield of Texas Tech University for their invaluable input and carefully reading the manuscript.

Texas Tech University *Andreas A. Neuber*
Electrical & Computer Engineering
Center for Pulsed Power and Power Electronics
Lubbock, TX

Contents

Foreword V

Preface VII

Contents IX

1 Introduction 1

2 FCG Overview 7
   2.1 Operating Principle 7
      2.1.1 Simple Generator Circuit 10
      2.1.2 Helical Flux Compression Generator 13
      2.1.3 Performance Figures of Merit 17
      2.1.4 Explosive Considerations and Efficiency 20
   2.2 Applications and Uses for FCG in the Past 22
      2.2.1 Boosting Applications 22
      2.2.2 Limited or No Pulse Forming Network 23
      2.2.3 FCGs with PFNs 26
   2.3 Need for Smaller Helical FCGs 29
      2.3.1 Differences between the Larger Generators and the Smaller Units 29
   References 32

3 Loss Mechanism Basics 35
   3.1 Magnetic Flux Diffusion 35
      3.1.1 Linear Magnetic Flux Diffusion 35
      3.1.2 Magnetic Diffusion and a Hollow Conducting Cylinder 39
      3.1.3 Nonlinear Magnetic Diffusion 41
      3.1.4 Approximate Solution 43
      3.1.5 Practical Implications of Magnetic Diffusion 47
   3.2 Avoidable Flux Losses 49
   References 51

4 Mechanical Aspects 53
4.1 Armature Dynamics 53
4.1.1 Armature Fractures 53
4.1.2 Related Problem 55
4.1.3 Theory 55
4.1.4 Research Postulates 61
4.1.5 Testing the Postulates, Part 1 66
4.1.6 Testing the Postulates, Part 2 71
4.1.7 Simulations and Tests of Multi-Layer Armatures 93
4.1.8 Conclusion Shock Analysis 96
4.2 Analysis of Armature/Stator Contact 97
4.2.1 Introduction 97
4.2.2 Finite Element Model 98
4.2.3 Experimental Verification of FE Model 99
4.2.4 Deformation and Kinematics of Armature/Stator Contact 103
4.2.5 Armature "End-Effect" 106
4.3 Criteria for Prevention of Armature "Turn-Skipping" 109
4.3.6 Introduction 109
4.3.7 Eccentricity Tolerance 109
4.3.8 Armature Wall Thickness Tolerance 112
4.3.9 Combined Eccentricity and Wall Thickness Tolerances 113
4.4 Scaling of the Armature Expansion Angle 114
4.4.1 Finite Element Analysis of the Armature Expansion Angle ... 116
4.4.2 Analysis of Scaling 118
4.4.3 Modification of Gurney Equation 119
4.4.4 Conclusion Expansion Angle 120
4.5 Armature treatment 121
References 122

5 Basic Physics 127
5.1 Shocked Gases within the Flux Compression Generator Volume 127
5.1.1 Generator Model and Shock Tube Design 127
5.1.2 Simulations 128
5.1.3 Diagnostics 128
5.1.4 Synthetic Air Data 130
5.1.5 SF_6 Data 131
5.1.6 Data Overview 133
5.1.7 Shock Tube Summary 133
5.2 Electrical Breakdown of the Generator Volume 134
5.2.1 Resistance of the Generator Volume 134
5.3 Impact of Stator Geometry on Flux Losses 139
5.4 Intrinsic Flux Loss 144
5.4.1 Zipper Contact 146
5.4.2 Magnitude of Ohmic and Intrinsic Losses 149
5.5 Shocked Metal Resistivity 154
References 156

6 Generator Modeling 159
6.1 PSpice Simulation 159
6.1.1 Theoretical Background 159
6.1.2 PSpice® Implementation of a Basic FCG 161
6.1.3 PSpice® Modeling of a Power Conditioning Circuit 165
6.1.4 Combining the Power Conditioning Circuit with the FCG Model ... 171
6.1.5 Including Intrinsic Flux Loss into the FCG Model 173
6.1.6 More accurate PSpice® model for the FCG 174
6.1.7 PSpice® Model Conclusions 180
6.2 Two-Dimensional (2D) Numerical Modeling 181
6.2.1 A Filamentary 2D Model 182
6.2.2 Single-Pitch, Single-Stage Helical FCG 187
6.2.3 Operation of the Simple, Single-Stage FCG 189
6.2.4 Summary 2D Model 197
References 198

7 Power Conditioning 201
7.1 Fuse Opening Switch 201
7.1.1 Historical Overview 201
7.1.2 Fuse Designs 201
7.1.3 Methods to Determine Fuse Parameters 204
7.1.4 Optimal Fuse Cross-Sectional Area 204
7.1.5 Optimal Fuse Length 206
7.1.6 Optimal Fuse Material 209
7.1.7 Fuse Quenching Material 210
7.1.8 Other Performance Influencing Factors 213
7.1.9 Types of Fuse Explosions 213
7.1.10 Equivalent Action Timescales for Capacitive Systems 214
7.1.11 Equivalent Action Timescales for FCG Systems 215
7.1.12 System Equivalency 216
7.1.13 Fuse Resistivity Modeling 217
7.1.14 Exploding Wire Fuse Calculations for a Staged FCG 220
7.1.15 Fuse Summary 222
7.2 Energy Extraction with Transformer 223
7.2.1 Theoretical Background 224
7.2.2 Experimental Work 225
7.2.3 Modeling of an FCG with Energy Extraction by Transformer 229
7.2 Dynamic Transformer 230
References 231

8 Seed Sources 235
8.1 Seed Sources Basics 235
8.2 Capacitor Driven 236
8.3 Explosively Driven Seed Sources 239
8.3.1. Ferroelectric Seed Sources 239
8.3.2. Ferromagnetic Seed Sources 240

    8.3.3. All Explosive FMG-FCG Example System 243
References 245

9 Practical FCG Pulsed Power System **247**
    9.1 Dual Stage Generator 247
        9.1.1 Performance by Stage 247
        9.1.2. Dual Stage Performance 258
    9.2 Inductive Energy Storage 261
        9.2.1 Fuse Opening Switch 261
        9.2.2. Fuse Calculations 264
    9.3 Comparison with traditional PPS 269
        9.2.1. Compact Low Inductance Marx Generator 271
References 275

Index **277**

1 Introduction

Andreas A. Neuber, Bruce L. Freeman

This book summarizes the research effort started by the 5-year Multi Disciplinary University Research Initiative, MURI, on Explosive Driven Pulsed Power[1] covering the time period from March 1998 to April 2003. Research in the area of explosive pulsed power is at the present time, 2005, ongoing at the MURI partners' laboratories. The participating MURI universities and departments were "The Center for Pulsed Power and Power Electronics" at Texas Tech University, the Mechanical Engineering Department at Texas Tech University, the Explosives Research / School of Mines and Metallurgy at the University of Missouri-Rolla, UMR, and Texas A&M University. Additionally, the continuing collaboration of Texas Tech University and the University of Loughborough, UK, has resulted in Chap. 6 of this book. Texas Tech as the lead university and UMR were the original proposers, Texas A&M was added to the effort at a later time.

Magnetic flux compression generators were conceived in the early 50s, one of the fathers was Andrey D. Sacharov, who proposed transforming the energy of explosives into the energy of a magnetic field in January of 1952. His suggestions were followed in the spring of 1952 with the start of early MK-1 experiments at VNIIEF (Russian Federal Nuclear Center - All-Russia Research Institute of Experimental Physics). Max Fowler in the United States fired his first plate generator in 1952 using the pole pieces of a magnetron for the initial magnetic field. Many countries have since then joined in the research on Flux Compression Generators, or short FCGs (also known as Magneto Cumulative Generators, MCGs, in Russian-speaking areas), with the US and Russia starting in the early 1950s. Russian work has been ongoing continually with large staffing since the late 1950's, while the US has had on average much less effort committed during the same time period. Amongst the countries that are currently using or conducting research on FCGs are: Russia, USA, China, Germany, United Kingdom, France, Sweden, Italy, Romania, Poland, and South Africa.

From the beginning, the major interest in the FCGs has been based on the unique properties of the High Explosives, HE, driving these devices. A glance at Fig. 1.1 reveals the uniqueness of HE as far as its specific energy is concerned. However, only the combination with the relatively short "discharge time" makes it

---

[1] This work was primarily funded by the Explosive-Driven Power Generation MURI program funded by the Director of Defense Research & Engineering (DDR&E) and managed by the Air Force Office of Scientific Research (AFOSR).

an excellent choice as part of a Pulsed Power System, PPS. Pulsed Power in general has an extremely wide range of applications in the defense related, medical, and civil world: Radar, tooth/eye/blood vessel treatment, X-ray, oil well exploration, etc. The challenge is always developing "the best" pulsed power system, PPS, for driving a given load. Depending on the set of conditions the PPS and load have to work in, the choice of "best" is not always immediately obvious. However, if a single (one-time) large electrical pulse or a pulse train in a remote location without direct access to external power sources is required, and further, if we wanted to make everything as compact//lightweight as possible, then an explosive PPS based on the FCG is the probable choice for the power source.

A word of caution seems to be appropriate at this point. The energy densities given in Fig. 1.1 do not include the efficiency of converting the available energy into useable output energy. For instance, the capacitors and rotating machinery are very efficient (> 80%), whereas an inductive based system may only be 10 % efficient. The chemical energy stored in HE can indeed be released in microseconds, however due to the small conversion efficiency, it is more realistic to expect that an FCG based PPS has an ~ 10 times higher energy density than a capacitor based PPS. Of course, this number depends on many factors and changes continually as breakthroughs are being made on many frontiers. Nevertheless, a factor 10 provides us with some guidance as we look at many different PPS.

Two important aspects of FCGs are that they are current/energy amplifiers, and require some seed energy. They work best into small inductive loads. Simply speaking, the initially established (seeded) magnetic flux in an FCG is moved into the load, while being conserved.

$$\Phi_0 = L_0 I_0 : \text{Initial Magnetic Flux} \qquad (1.1)$$

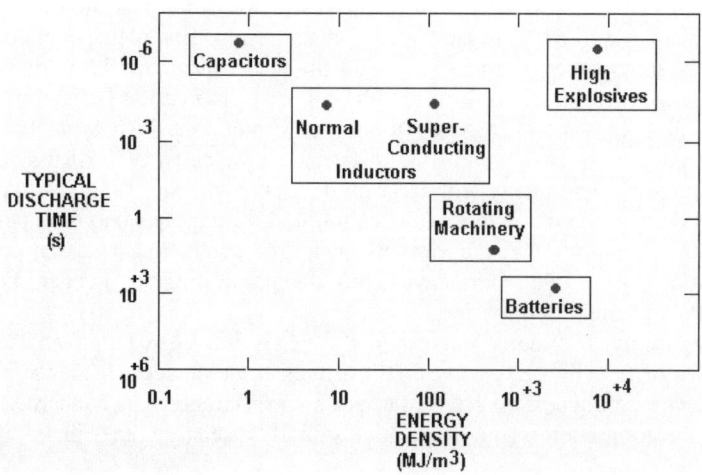

Fig. 1.1. Discharge time and energy density map of energy storage for pulsed power systems, not including conversion efficiencies (some new battery developments allow discharging within milliseconds or less, however, only into a small impedance load).

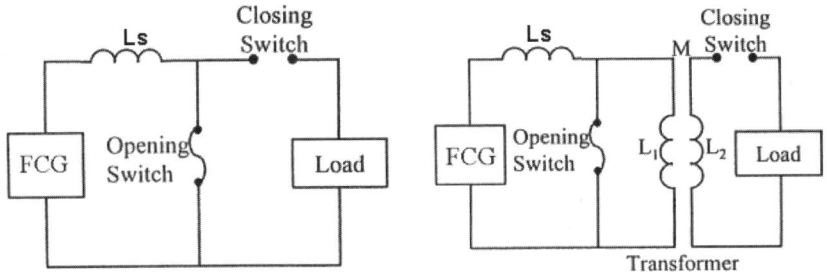

Fig. 1.2. Two basic inductive energy storage systems utilizing the FCG as prime source. L_s is the storage inductor.

In the ideal case of complete flux conservation, we have:

$$\frac{d\Phi}{dt} = 0 \qquad (1.2)$$

Hence,

$$\Phi_0 = \Phi_f \;\rightarrow\; I_f = I_0 \frac{L_0}{L_f}. \qquad (1.3)$$

L_0, L_f (I_0, I_f) are the initial and final inductances (currents) of the FCG-load system. Since typically $L_0 \gg L_f$, L_0 is essentially equal to the FCG inductance, and L_f is equal to the load inductance. This means that higher output currents may be achieved if we reduce the load inductance, cf. Eq. (1.3). As a matter of fact, the current and the energy gain are, in the ideal case, directly proportional to the ratio of FCG inductance to load inductance, L_0/L_f. In this context, the problem is that a small load inductance (to achieve high energy gain) also means a small output voltage.

Most often, the load requires a larger driving voltage so circuits, as indicated in Fig. 1.2, may be employed to match the FCG output to the load. Basically, the FCG amplifies the current in the left loop (FCG and opening switch), while the closing switch is open. At a certain time, the opening switch will open and the associated induced voltage ($V_{ind} = L\, dI/dt$, L being the inductance in the left loop of the circuit) will close the closing switch. This action transfers the inductively stored energy to the load at an increased voltage level and during a shorter time period. However, in many applications such an approach is not needed, as the load inductance may be inherently small. For instance, imploding liner experiments, high magnetic field experiments, etc. are driven by FCGs and pushed into regimes currently not achievable with traditional PPS, e.g. with capacitive energy storage. The diameter and length of such generators is typically of the order of several tens of centimeters to several meters, respectively. Refer to Chap. 7 for a more in-depth treatment of energy storage and power conditioning.

Another method of increasing the output current of an FCG is increasing the FCG's initial inductance itself, L_0 in Eq. (1.3). The FCG with the largest L_0 for a given volume utilizes a current carrying helix, see Fig. 1.3. The seed current is applied to the FCG such that it reaches its maximum at the moment of crowbar, which is electrically connected to the helix, cf. Fig. 1.3b. Typically, the FCG reaches the state indicated in Fig. 1.3e a few 10s of μsec after initiation. We note that the FCG completely disintegrates on a millisecond time scale.

Our research effort focused exclusively on helical FCGs having a diameter of no more than twice that of a soda can. The main thrust has been developing an improved picture of the physics of magnetic flux compression generators. However, the design and construction of even simple FCGs for studying the underlying physics made it necessary to acquire and hone a variety of engineering skills, which we then have used to develop a more applied, multi stage FCG, cf. Chap. 9.

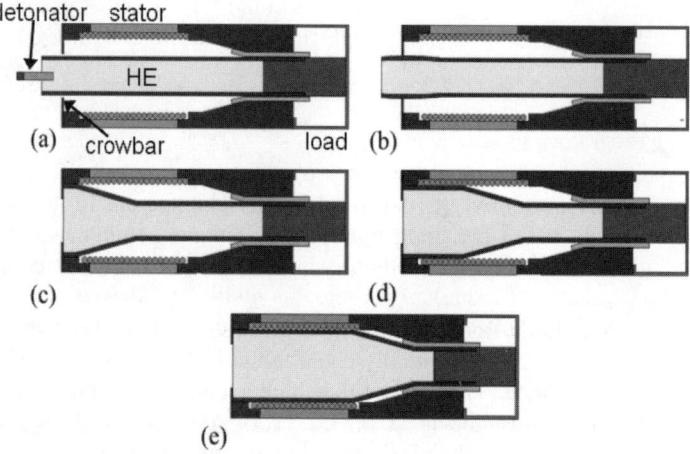

Fig. 1.3. Operation of the helical FCG consisting of an HE filled armature (metal tube), the stator (wire helix), and the low inductance load (simple current return): (a) before initiation of detonation, (b) few μs after initiation (moment of crowbar), (c) armature has expanded and makes contact with the first helix turn, (d) initial inductance, L_0, roughly reduced by a factor 4, (e) most of the remaining magnetic flux is pushed into the load.

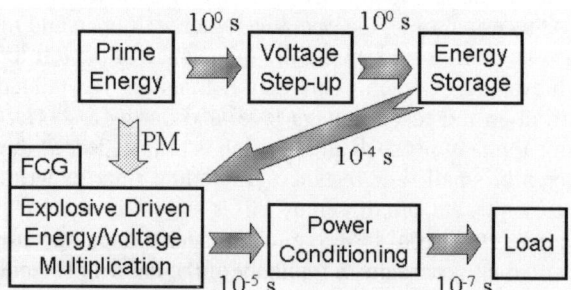

Fig. 1.4. Energy flow diagram of an explosive driven pulsed power system. "PM" indicates direct seeding with permanent magnets without the need for voltage step-up and energy storage. Timescales for energy transfer from one to the next stage are shown as well.

Overall, we have been less interested in producing very high magnetic fields or currents, but rather in a scheme as it is shown in Fig. 1.4. Starting with some prime energy (battery, AC power, etc.) the voltage is stepped-up and the energy is capacitively stored at a level of several kilovolt. Closing a switch dumps the stored energy within $\sim 10^{-4}$ s into the FCG and provides the needed seed energy. The output energy of the FCG is increased considerably over the seed energy and the output power is conditioned, cf. Fig. 1.2, to match the high impedance (tens of Ohms) load. Alternatively, the seed energy for the FCG may also be provided by the magnetic field energy of permanent magnets, PM, or similar. In this case, the energy flow is directly from prime energy to FCG, as indicated by the vertical arrow in Fig. 1.4, and voltage step-up or energy storage are not needed.

We will address in Chap. 2 the basic operating principle of helical FCGs in more depth and present some theoretical aspects of it. Chap. 3 will elucidate why a real FCG always exhibits less than ideal performance. In Chaps. 4 and 5, we will focus on mechanical aspects of explosively deformed conductors and the physics of FCG operation. Starting with a simple PSPICE model, Chap. 6 presents details on state-of-the-art generator performance modeling. We address power conditioning, the link between FCG and load, in Chap. 7 and give an overview of practical seed energy sources in Chap. 8. We finally present an example of a practical explosive Pulsed Power System, PPS, and compare it with a conventional PPS in Chap. 9.

2 FCG Overview

Bruce L. Freeman and Andreas A. Neuber

2.1 Operating Principle

If magnetic field lines are cut or the magnetic field itself is compressed, work has been done on the system, and energy will be transferred from the mechanical system to the electrical system. In particular, the compression of the magnetic field by the action of an explosively driven piston is generally referred to as explosively driven magnetic flux compression. If the device is able to use this magnetic flux either directly or as electrical output to drive a remote load, then we have an explosive-driven magnetic flux compression generator or FCG, see Fig. 2.1 for a cylindrical FCG, which has no stator helix. In this instance, the chemical energy of the explosive is transferred into the kinetic energy of the armature, and subsequently into electrical energy. In extreme cases, the output of the FCG is limited by the kinetic energy available to transfer into electrical energy.

As a practical matter, the assessment of the generator performance based on the efficiency of converting explosive energy into electrical energy is neither a desirable viewpoint nor a useful way to design these devices.

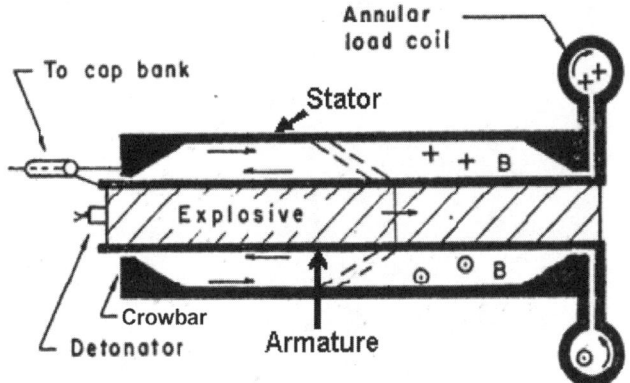

Fig. 2.1. The cross section of a cylindrical generator is shown before the initiation of the explosive and at a later stage with the armature expanded [Fow75].

The reasons for this are that there is generally more explosive energy available than can be productively used, and stopping the moving element, or armature, in flight leads to at least two negative aspects. These are enhanced non-linear magnetic flux loss and a slowing of the output pulse peak.

The expansion of explosively driven armatures may be studied utilizing fast optical imaging, see Fig. 2.2, and x-ray imaging, Fig. 2.3 and Fig. 2.4. Since the detonation front moves with about 8 mm/µs, a temporal resolution of ~ 15 ns or better is needed to freeze the movement within ~ 0.1 mm. As we will see later in more detail, the observed expansion angle or "Gurney" angle of the end-initiated armature depends on the mass of the high explosives, HE, and the armature material. For instance, the 3 mm wall thickness aluminum armature has approx. the same Gurney angle of ~ 15 degree (half angle of expanding cone) as the 1 mm wall copper armature with the same outer diameter, see Fig. 2.2 and Fig. 2.3, respectively. In the US, the armature material has been historically aluminum, whereas in Russia, copper has been the material of choice. One can find many arguments why copper or aluminum should be better. Typically, the choice of armature material is driven by whether or not the final armature velocity requires optimization. Thus, if higher armature velocities are required for the application, then aluminum is the material of choice. However, if armature velocity is not a critical factor, then often copper is used. In almost all cases, the outer current carrying conductors (helical turns for the helical FCG) have almost exclusively been made from copper due in part to its higher cold conductivity.

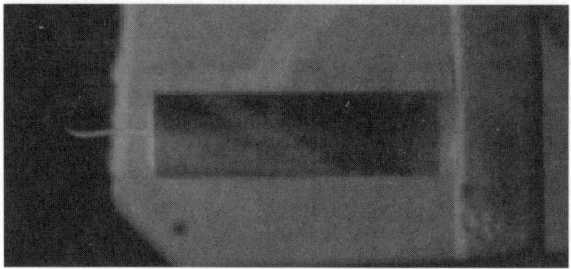

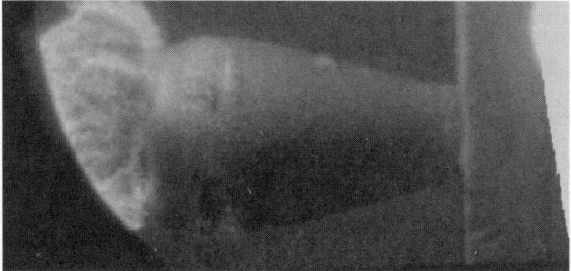

Fig. 2.2. Flash photograph of a 38.1 mm diameter aluminum armature (3 mm wall thickness) filled with HE expanding freely in air before and 18 µs after initiation of detonation (exposure time 14 ns).

2.1 Operating Principle   9

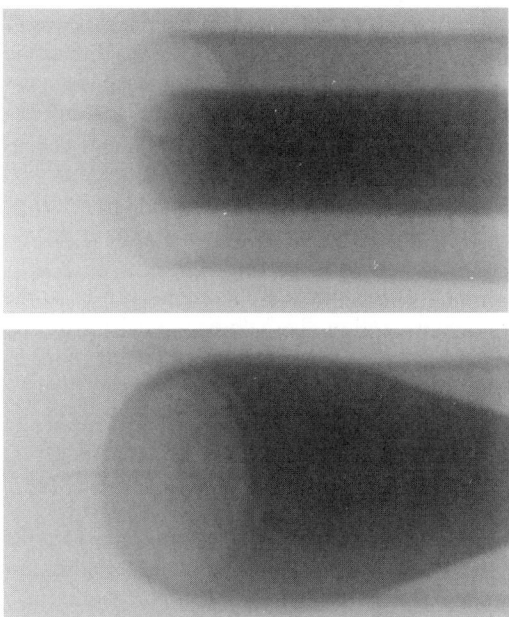

Fig. 2.3. Flash X-ray photograph of a 38.1 mm diameter copper armature (1 mm wall) inside a 3mm wall Lexan tube before and 16 μs after initiation of detonation (15 ns exposure time). The tube virtually stops the expanding armature on the microsecond time scale.

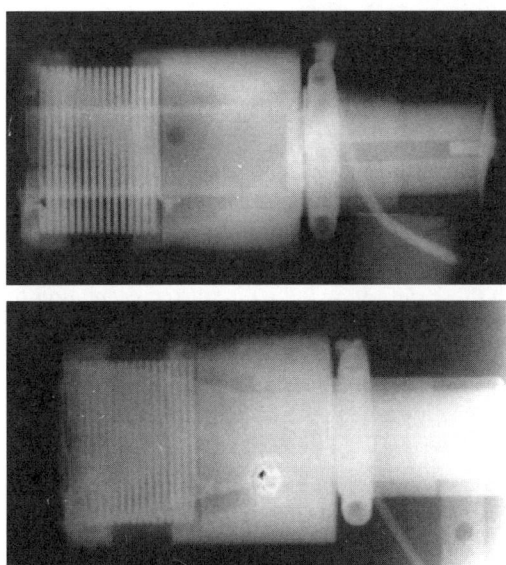

Fig. 2.4. Flash X-ray photograph of a single stage helical generator with 16 turns before and 7 μs after initiation of detonation (15 ns exposure time).

Only seamless metal pipes should be used as armatures in FCGs. Having seams leads to non-uniformity in mass and properties (temper) of the material, causing uneven expansion of the armature and heavy flux losses. We have used seamless tubes, as received, for some of our generators and have not observed any deterioration in performance as long as the manufacturing tolerances were tight. It is generally recommended to check the concentricity of inner and outer diameter as well as for any sagging along the tube. In larger generators where the sound transit time through the armature material is significant, the armature metal should be annealed to be in a completely soft state of temper or hardness.

For a properly aligned helical generator, the idea of operation is to have the leading edge of the armature/helix – contact follow the path of the helix wire. As we will see in later chapters, the quality of this spirally moving contact edge is most important for the generator performance.

2.1.1 Simple Generator Circuit

To begin to understand the process of magnetic flux compression, consider the section of a cylindrical geometry generator shown in Fig. 2.1 for times before initiation and at a later stage for the armature expansion. The FCG is essentially an inductor that is "collapsed" by the action of the explosive. The static portion of the device, for time scales of interest, is normally called the stator. The portion that is moved by the explosive is typically referred to as the armature. For the present discussion, the initial current for the magnetic field is supplied from a fast current source, such as a capacitor. In the case of the cylindrical generator portrayed in the Fig. 2.1, the armature moves outward under the influence of the detonated explosive pressure and closes off the interior volume of the unit, and therefore, reduces the inductance. Thus, the magnetic flux is forced out of the generator volume and into the load, presumably a smaller inductance. The initial inductance of the generator is L_G, or a time dependent, $L_G(t)$. The connection between the generator volume and its load will have an inductance that is associated with the transmission line, L_T. The static load inductance for this discussion is represented as L_L. A static resistance is represented as R, but in practice, this element of the circuit may be a static or dynamic resistance or the diffusion of magnetic flux into the conductors. With these definitions, the circuit representation of a simple FCG system is shown in Fig. 2.5.

The equation for the performance of this circuit (upper mesh only after switch S_1 is closed) is given with Kirchhoff's voltage law by

$$\frac{d}{dt}[(L_G(t) + L_T + L_L)I(t)] + I(t)R(t) = 0, \qquad (2.1)$$

with the initial condition, $I(t = 0) = I_0$.

Time is measured from the instant that the armature closes the input gap from the capacitor feed, typically referred to as crowbar for the generator.

2.1 Operating Principle

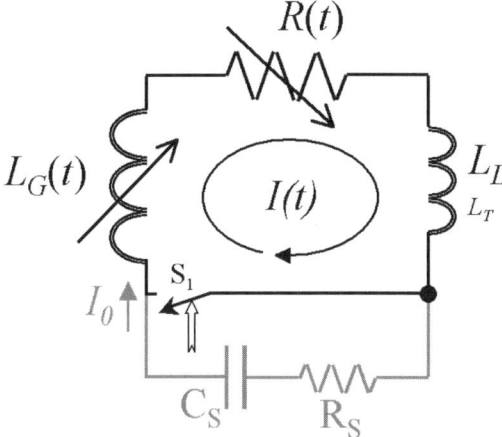

Fig. 2.5. The simplest FCG circuit. $L_G(t)$, $R(t)$ –generator inductance or resistance (both decrease during generator operation). L_L, L_T – constant load inductance and parasitic inductance (accounting for the connection between generator and load). The energy storage capacitor, C_S, for seeding the generator with the current I_0 is also shown along with the finite resistance in the seed circuit, R_S.

If the resistance for this circuit is set to zero, then the solution for the circuit equation may be expressed strictly in terms of inductances and currents. The flux conserved solution is:

$$I(t) = \left[\frac{L_G + L_T + L_L}{L_G(t) + L_T + L_L}\right]I_0, \text{ or simply } I(t) = \left(\frac{L_0}{L(t)}\right)I_0, \quad (2.2)$$

with the lumped inductance, $L(t)$:

$$L(t) \equiv L_G + L_T + L_L, \quad (2.3)$$

its initial value being $L_0 = L(t=0)$.

This expression is readily recognized as the flux conserved generator equation that has been derived and published in many places. For the energy gain of a generator circuit, we find that the initial energy is

$$E_0 = \frac{1}{2}(L_G + L_T + L_L)I_0^2, \text{ or simply } E_0 = \frac{1}{2}L_0 I_0^2, \quad (2.4)$$

and the energy at any time through the generator run is

$$E(t) = \frac{1}{2}(L_G(t) + L_T + L_L)I(t)^2, \text{ or simply } E(t) = \frac{1}{2}L(t)I(t)^2. \quad (2.5)$$

Then, the lossless energy gain or current gain for the simple FCG circuit is

$$G_E \equiv \frac{E(t)}{E_0} = \frac{L_0}{L(t)} \text{ or } G_I \equiv \frac{I(t)}{I_0} = \frac{L_0}{L(t)}, \tag{2.6}$$

At the end of the generator run, or burnout, the lossless case results in $L_G(t)$ approaching zero. Thus, the final current, I_F, and final energy, E_F, as a result of the FCG action is given by;

$$I_F = \left[\frac{L_G + L_T + L_L}{L_T + L_L}\right] I_0, \text{ and} \tag{2.7}$$

$$E_F = \left[\frac{L_G + L_T + L_L}{L_T + L_L}\right] E_0. \tag{2.8}$$

These equations provide an ideal version of FCG performance and are useful when comparing actual performance to the ideal. However, the resistance in our circuit is never zero, so these equations must be modified to include the circuit resistance. Typically, the resistance of a flux compression generator can be rather complex and several different phenomena have to be included. Most importantly, the current in the generator will flow mostly on the surface of the conductors that make up the generator. In general, a skin-depth is defined that gives a measure for how deep the current has diffused into the conductors, typically only a fraction of a millimeter in FCGs, cf. Eq. (3.28) and Fig. 3.5. The skin depth itself is a function of time, see Chap. 3, which makes it rather difficult to calculate the time dependent $R(t)$. For a helical generator (cf. Fig. 2.4) using an average skin depth and linear reduction of the helix wire length, as it is reduced by the expanding armature, provides a first order approximation for $R(t)$, see Fig. 2.6.

Other factors such as nonlinear magnetic diffusion due to a temperature rise of the conductors, see Chap. 6.1.6, the path length of the current along those conductors, and shock heating of the armature from the explosive will also tend to change the effective resistance within time scales of interest.

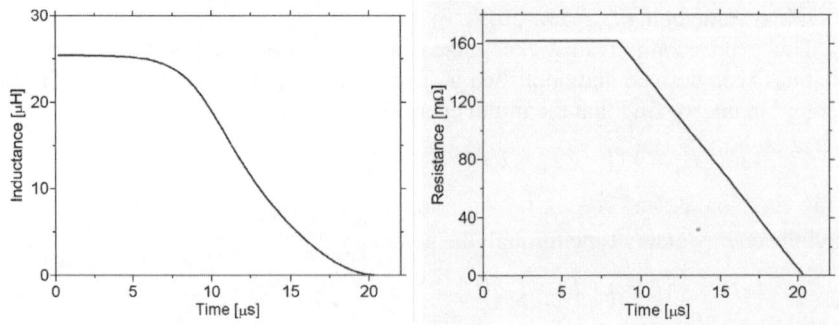

Fig. 2.6. Time dependent resistance and inductance for a single stage, single pitch helical FCG, TTU I, cf. Fig. 2.4.

However, for rather small maximum current amplitudes (< 10 kA for AWG 12 wire) and short burn times (< 10 µs), both of which limit the heating of the conductors, our first order $R(t)$ will suffice. Solving Eq. (2.1) for the current yields

$$I(t) = I_0 \frac{L_0}{L(t)} \exp\left[-\int_0^t \frac{R(t')}{L(t')} dt'\right]. \tag{2.9}$$

Similarly, the energy in the system as a function of time may be expressed as;

$$E(t) = E_0 \frac{L_0}{L(t)} \exp\left[-2\int_0^t \frac{R(t')}{L(t')} dt'\right]. \tag{2.10}$$

In any case, it should be noted that the current and energy gain, G_I and G_E, increase with an increasing ratio of initial to final inductance, L_0/L_f. However, making the final inductance, $L_f = L_T + L_L$, as small as possible for the benefit of an increased energy gain is not always feasible if the load requirements are taken into account. For instance, an inductive energy storage system that is to be energized by an FCG requires that L_L is of the order of several µH. Since the initial inductance of the cylindrical FCG is low to begin with, the helical flux compression generator with an inherently larger L_0 is widely used. But even for the helical FCG, the energy gain of a single stage FCG is typically too small to drive large inductive loads, as we cannot make the initial inductance too large due to the increasing overall resistance, $R(t)$, of the FCG in Eqs. (2.9) and (2.10). Of course, we could use larger wires to keep the resistance low, which, however will lead to an increased helix pitch and thus lower the initial inductance. Nevertheless, as we will discuss, large helical FCGs generally outperform their smaller counterparts.

Increasing the dL/dt of a helical FCG is an approach to increasing the effective output impedance of the FCG. This is usually accomplished by using a closing phase velocity to reduce the time for the majority of the inductance of the generator to be eliminated. For example, the armature of a helical FCG may be tapered to an angle slightly larger than the opening angle of the armature. This geometrical modification will lead in limits to a shorter pulse length, a higher output impedance, and a higher output voltage. For a more in-depth theoretical treatment of the generator performance see Chaps. 3 and 6.

2.1.2 Helical Flux Compression Generator

Thus far, we have mostly used the cylindrical generator as a model for our considerations of FCG performance because of the geometrical simplicity of this style generator, cf. Fig. 2.1. However, the generator type that is of most interest generally is the spiral or helical flux compression generator, cf. Fig. 2.4. The reason is that the helical FCG may have a very large initial inductance, so the ideal gain from one of these units may be very large. Conceptually, a helical flux compression generator is shown in Fig. 2.7 a.) before the explosive is initiated and in Fig. 2.7 b.) after the explosive has been detonated. The spiral winding of the stator provides inductance proportional to the number of turns squared. While there are several formulations for the inductance of a helical FCG, a very useable one is presented by Fowler [Fow75] as;

$$L_{Helical} = 3.95 \times 10^{-6} \frac{N^2 (R_{outer}^2 - R_{inner}^2)}{l + 0.9(R_{outer} - R_{inner})} \text{ Henry,} \quad (2.11)$$

where, N is the number of turns in the solenoid of the stator, R_{outer} is the inner radius of the stator, R_{inner} is the outer radius of the armature, l is the length of the generator, and the term $[0.9(R_{outer}-R_{inner})]$ is a correction term for the aspect ratio of the generator since it has a finite length. Both l and R are in units of meter.

Originally, this formula is attributable to Terman [Ter43] in its more generalized form that does not include the presence of the armature. This formula assumes a uniform current density for its application. Multiple turn stators are appropriately calculated when the stator length is broken into calculational segments, cf. Fig. 6.31. Wide turns in the stator will lead to higher current densities at the edges of the turn. This can lead to a calculated inductance being larger by 10 to 20% than the physical value. Similar inductance calculations have been reported by Wheeler [Whe28], Grover [Gro73], and Kalantarov [Kal58].

These generators generally operate over longer time scales, several 10s to 100s μsec, so even though the ideal predicted gains can be enormous, there are numerous phenomena that serve to limit the practical performance of helical FCGs. These range from physical tolerances for machining and assembly to avoid flux pocketing and turn skipping to internal voltage breakdown problems between turns.

The longer times are particularly important with respect to magnetic flux diffusion into the conductors. This magnetic flux loss generally limits helical FCGs to linear current densities of less than ~ 1 MA/cm of conductor width. In addition to these issues, there are long-standing unresolved questions concerning the rotation of the magnetic field as it exits the generator, eddy losses that may add additional flux losses, and possible armature cracking along the generator axis. The result of these various losses is that good generator design generally means relatively widely spaced turns using wide output conductors with tight machining tolerances on both the stator and armature.

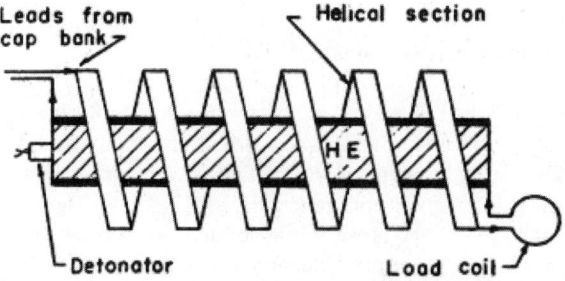

Fig. 2.7. a) The helical FCG, before HE initiation, has a cylindrical armature and a solenoidal stator, which provides much larger initial inductance [Fow75].

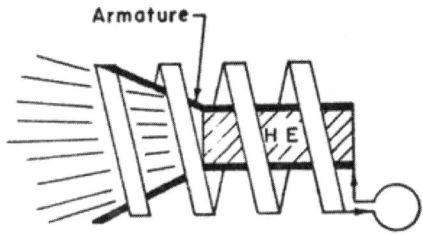

b) After explosive initiation, the armature progressively expands to contact the stator winding in a spiral-motion, sweeping contact that reduces the initial inductance of the FCG. The picture is incomplete with respect to the fact that helix and armature will still exist to the left of the contact edge on the time scales of FCG operation (microseconds).

2.1.2.1 Helical FCG, Basic Theoretical Treatment

Simply differentiating the first term of Eq. (2.1) leads to

$$I\frac{dL}{dt} + L\frac{dI}{dt} + IR(t) = 0, \qquad (2.12)$$

where we used the definition for the lumped inductance, $L(t)$, of Eq. (2.3).

This is pretty much the standard equation of an RL circuit, however, with the additional term dL/dt. Obviously the current in a simple RL circuit merely decreases over time, thus the dL/dt term is identified as the driving term, mathematically responsible for any increase in current over time.

Solving Eq. (2.12) is straightforward and leads to the result given in Eq. (2.9). However, no matter how sophisticated or accurate the time varying resistance, $R(t)$, is modeled, the calculated performance from Eq. (2.12) always overestimates the experimentally observed performance of the helical FCG. Hence, an intrinsic loss is introduced, scaling the driving term:

$$I \cdot \alpha \cdot \frac{dL}{dt} + L\frac{dI}{dt} + IR(t) = 0 \qquad (2.13)$$

We reasonably assume that the intrinsic flux loss parameter, α, varies between 0 and 1. Obviously, with $\alpha = 0$, Eq. (2.13) reduces to the standard passive RL circuit, whereas with $\alpha = 1$, the generator behaves ideal with respect to its intrinsic flux loss. We will discuss the physics behind the intrinsic flux loss in a later chapter, cf. Chap. 5.4, and merely note for the moment that experimentally observed values of α lie in the range from 0.7 to ~ 0.95 depending primarily on the generator size and to some lesser extend also on the geometry.

Eq. (2.13) can be analytically solved for the current, assuming that α is time independent:

$$I(t) = I_o \cdot \left(\frac{L_o}{L(t)}\right)^\alpha \exp\left[-\int_0^t \frac{R(t')}{L(t')} dt'\right] \tag{2.14}$$

It should be noted that the magnetic flux in the generator remains constant over time if $R(t) = 0$ and $\alpha = 1$, or expressed mathematically:

$$\frac{d}{dt}(\Phi) = \frac{d}{dt}(L \cdot I) = 0 \Rightarrow I_o \cdot L_o = I_f \cdot L_f \tag{2.15}$$

Hence, the name flux compression generator and hence the intrinsic flux loss parameter, describing additional losses beyond the ohmic losses.

Using Eq. (2.14) we calculated the output current for one of our single pitch generators, see Fig. 2.8, Fig. 2.9. It should be noted at this point that we used this simple FCG type primarily for the study of FCG basics (The design and operation of a more applied FCG will be discussed in Chap. 9). Since α is, *a priori*, unknown, no true prediction could be made, α is adjusted so that the calculation fits the experiment. Although this may sound unsophisticated, all helical FCG models, currently known to the authors, use at least one parameter to account for the a priori unknown intrinsic flux loss. Although we have made progress on quantifying the intrinsic loss, it is still not entirely understood, as discussed in Chap. 5.4.

Eq. (2.13) can be used to derive a few FCG design rules. Simply rewriting yields

$$\frac{dI}{dt} = -\frac{I}{L}\left(\alpha \frac{dL}{dt} + R\right), \tag{2.16}$$

which should be > 0 during the generator burn time. Obviously, dL/dt is negative so that we find

$$-\alpha \frac{dL}{dt} > R, \tag{2.17}$$

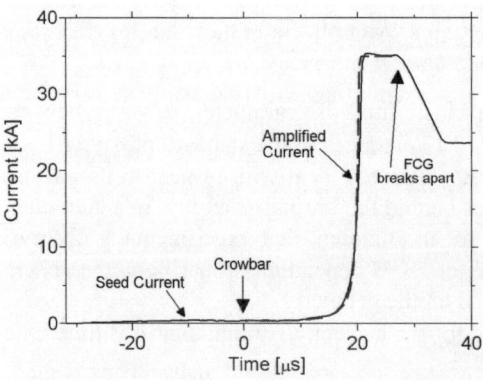

Fig. 2.8. Calculated and experimental current waveforms for the single pitch generator shown in Fig. 2.9.

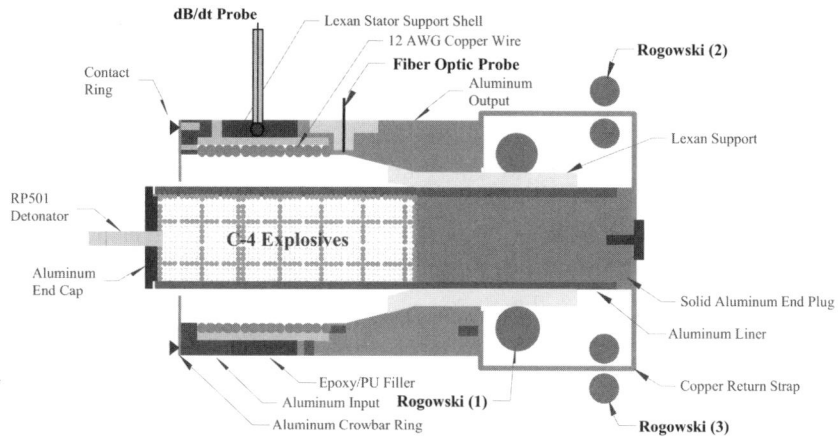

Fig. 2.9. Simple single-stage generator (38.1 mm armature diameter, approx. 20 cm long) with aluminum end cap on detonator side and various diagnostic probes indicated. The output copper straps are about 2.5 cm wide and serve as current return path/load. The standard generator has 16 helix turns as shown. The initial generator inductance is 20.2 µH with 32 turns, the final load inductance is 46 nH.

as a condition for positive current gain, $G_I > 0$.

In a similar fashion, we derive as condition for positive energy gain, $G_E > 0$,

$$(1-2\alpha)\frac{dL}{dt} > 2R, \qquad (2.18)$$

which is an even more demanding criterion. We note that dL/dt itself is negative since $L(t)$ is a constantly decreasing function over time.

Clearly, for end-initiated FCGs both criteria demand a small pitch at the beginning to overcome the large initial resistance, whereas the helix pitch towards the end should be somewhat relaxed. A larger pitch towards the end will help to keep the maximum induced voltage, $d(LI)/dt$ in the generator below the electrical breakdown threshold of the insulation in the system. Not to mention that the larger conductor surface has a larger current handling capability.

Some advantage can be gained from double end initiated versions of helical FCGs, where the output current tap is in the axial center of the FCG [Che94]. Such geometry requires careful design of the collision plane, where the armature expanding from both ends meets the current tap.

2.1.3 Performance Figures of Merit

As noted, the various phenomena that may adversely impact the performance of FCGs in general, and helical FCG in particular, range from physical tolerances to rather complex considerations of magnetic flux diffusion. However, a shorthand approach has generally been adopted by the flux compression community to account for generalized performance losses, without having to get deeply into the details of why the losses have occurred. This approach was first reported by Shearer, et. al. [She68]. We define the figure of merit, β, by the equation,

$$\left[\frac{I_f}{I_0}\right]_{experimental} = \left[\frac{I_f}{I_0}\right]_{ideal}^{\beta}. \tag{2.19}$$

If we use the simple equation for the lossless case, Eq.(2.2), developed earlier, then the figure of merit can be determined from

$$G_I = \left[\frac{I_f}{I_0}\right]_{experimental} = \left[\frac{L_0}{L_f}\right]^{\beta}. \tag{2.20}$$

Similarly, the energy ratio may be expressed in terms of a generalized figure of merit, β, as,

$$G_E = \frac{E_f}{E_0} = \frac{\frac{1}{2}L_f I_f^2}{\frac{1}{2}L_0 I_0^2} = \left[\frac{L_0}{L_f}\right]^{2\beta-1}, \tag{2.21}$$

Lossless performance is obtained when β is equal to 1. From Eq. (2.21), there is no energy gain when

$$\beta \leq 0.5. \tag{2.22}$$

Generally, larger values of β will be achieved for larger diameter generators with relatively widely spaced turns, lower current densities, and smaller theoretical current gains, cf. Fig. 2.10.

We found it very instructive to look at the figure of merit as a function of the angular frequency of the leading edge of the contact between armature and stator (FCG helix), see Fig. 2.7. b.) for several generators whose performance has been detailed in the open literature. Although many of these generators vary widely in pitch and size, most of them have armatures whose diameters remain constant along their axes. Adopting the simple equation for the inductance of a long solenoid with closely spaced turns (A: cross sectional area of solenoid)

$$L_0 = \mu_0 \cdot N^2 \cdot \frac{A}{l}, \tag{2.23}$$

(which is probably accurate within 10 – 20 % for most of the generators in this study), we establish the following set of equations

$$L_0 = \mu_0 \cdot \frac{\pi \cdot (x^2 - 1)}{4} \cdot N^2 \cdot \frac{d^2}{l}, \tag{2.24}$$

$$\omega = v_0 \cdot \frac{N}{l} \cdot 2\pi, \tag{2.25}$$

$$V = \frac{\pi \cdot (x^2 - 1)}{4} \cdot d^2 \cdot l, \tag{2.26}$$

with L_0 the initial FCG inductance, x the expansion ratio, i. e. ratio of stator diameter to armature diameter (typically $x \approx 2$), N the number of helix turns, d the armature diameter, l the length, v_0 the detonation wave velocity (approx 8 mm/µs for C-4), ω the angular frequency of the azimuthally moving contact point, and V the compressed volume between armature and stator. A few simple substitutions result in

$$\omega = \frac{2\pi v_0}{\sqrt{\mu_0}} \cdot \sqrt{\frac{L_0}{V}}, \tag{2.27}$$

an expression that simply relates the three basic quantities ω, L, and V (v_0 is thought as constant). Although Eq. (2.27) disregards any varying pitch, it can be used to calculate an average ω for any helical FCG just by using the initial inductance, L_0, and the compressed (active) volume, V. As a rule of thumb, smaller sized FCGs have a larger angular frequency, ω, of the contact point.

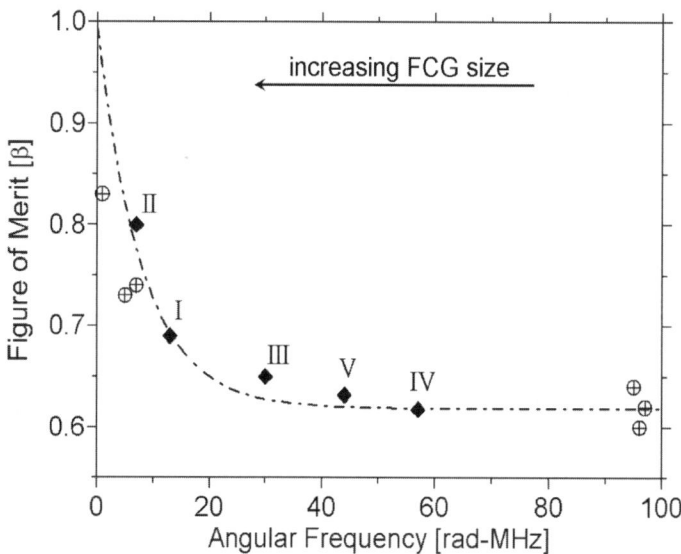

Fig. 2.10. Dependence of the figure of merit on the contact point's angular frequency for FCGs with constant diameter armature and single stage design. Data points are calculated from experimental data in references [Gov79], [Nov95], [Fow89], [Jon79], [Fre79], circle symbols, and own tests, diamond symbols. The solid line serves as guide for the eye only. Indicated are the authors' [Neu01] generators: TTU I – V, cf. Table 5.4.

We have used the trend line in Fig. 2.10 to predict the performance of some of our generators by just simply substituting the initial inductance, L_0, and the compressed volume, V into Eq. (2.27). Generally, we were surprised how close all generators came to this trend line. Generators with avoidable flux losses, - turn skipping, breakdown and such, cf. Chap. 3.2 -, are excluded from Fig. 2.10, as their figure of merit would be quite below the trend line for "flawless" FCGs. It should be noted that a figure of merit, $\beta = 0.6$, typically means an energy gain, G_E, of ~ 2. Hence, the very small generators with an angular frequency, ω, of more than 30 rad-MHz (corresponding to a helix diameter below a few tens of millimeter) are unsuitable for most pulsed power applications, which would like to see an energy gain, G_E, in excess of ~ 10.

2.1.4 Explosive Considerations and Efficiency

In magnetic flux compression generator design, one rarely encounters fundamental limitations on the generator performance due to the quantity of explosive present in the FCG. Part of the reasoning for this is that only a comparatively small layer near the armature actually efficiently couples to the armature kinetic energy. Another important issue is that the kinetic energy of the armature must be absorbed by the magnetic field for the efficiency to be maximized. Thus, near the end of the generator run, the armature would have to be stopped by the back pressure exerted by the magnetic field for there to be significant efficient energy transfer between the kinetic energy and the field. This second point means that the initial magnetic field inside the FCG must be high enough for the final currents to produce sufficiently intense magnetic fields to slow or stop the armature in flight. There are only a few instances where this has been achieved.

For completeness, the Gurney model for coupling explosive to moving metal liners should be considered [Gur43]. The two principal hypotheses for the Gurney model are:
(a) The Gurney specific energy, E, represents only a part of the chemical energy available per unit mass of the explosive charge, ΔH_{det}, which may be about 70%.
(b) The velocity distribution within the detonation products is assumed to be linear.

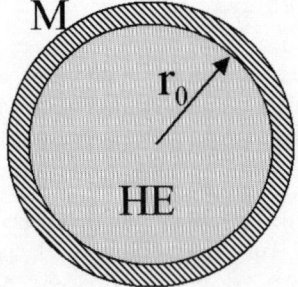

Fig. 2.11. Explosive driven metal liner (armature in the FCG) in cylindrical geometry.

Now, if M is the armature mass per unit length and r_0 is the initial charge radius, see Fig. 2.11, then at a later time and a radius, r, the gas velocity, $v(r)$, within the detonation products is

$$v(r) = V_a \frac{r}{r_0}, \qquad (2.28)$$

where V_a is the final armature velocity. If $C = \pi r^2 \rho$ is the explosive mass per unit length and W_k is the detonation product kinetic energy per unit length, energy conservation is now

$$CE = \frac{MV_a^2}{2} + W_k, \qquad (2.29)$$

where a cylindrical charge and liner geometry has been assumed. By definition, W_k is

$$W_k = \int \frac{1}{2} mv^2 dx = \frac{CV_a^2}{4}. \qquad (2.30)$$

To get the familiar Gurney result, Eq. (2.30) is substituted into Eq. (2.29) to give

$$\frac{V_a}{\sqrt{2E}} = \frac{1}{\sqrt{\frac{M}{C} + \frac{1}{2}}}. \qquad (2.31)$$

The value of $\sqrt{(2E)}$ can be estimated from the explosion velocity of the explosive, cf. Eq. (4.60). It should be noted that Eq. (2.31) is derived for the case of a simultaneous centerline on-axis detonation, see Fig. 2.12. Still, using this formulation, one may derive reasonably accurate approximations for the armature velocity and expansion angle for end-initiated FCGs. As a matter of fact, most FCGs, specifically the smaller ones, will be end-initiated, as it is not straightforward to simultaneously initiate detonation with the required accuracy along the centerline.

We refer to Chap. 4 for a more in-depth treatment of armature expansion.

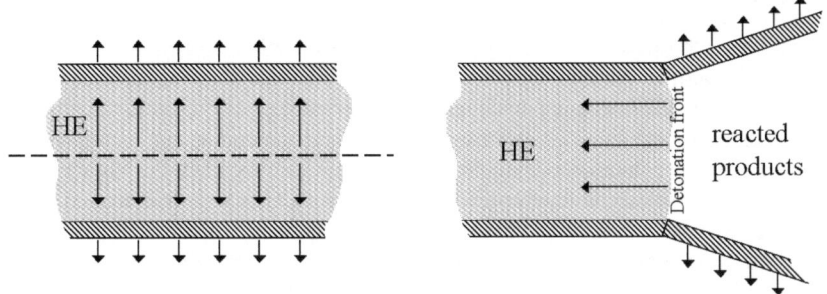

Fig. 2.12. Cross section of expanding metal armatures. Left: simultaneously centerline initiated; Right: end-initiated.

2.2 Applications and Uses for FCG in the Past

The uses for various types of FCGs have ranged from the direct use of the magnetic field for high field studies to powering rather complex plasma physics systems as loads. In general, there are three overall reasons to use an explosive power supply. These are;
- Feasibility studies
- Large energy and power capability, and
- Portability.

Historically, the first two of these requirements have dominated the applications of FCGs and FCG-driven power supplies. One reasonable approach to powering a given load is to ask if the job can be done with conventional pulsed power. If the answer is affirmative, then this is likely to be the better course of action. However, there are numerous instances where conventional pulsed power simply does not work for reasons of total power or energy, cost, expendability, portability, etc. Several examples for earlier applications for flux compression generators will be given, though no attempt will be made to be comprehensive. Also, this discussion will not be limited to helical FCG. Though the division will be somewhat artificial, the various uses for FCGs will be divided into three categories; (1) boosting, (2) limited or no PFN (Pulse Forming Network), and (3) FCGs with PFNs. As one may understand, many times one will find that the application of a specific power supply will result in some governmental sensitivity. Thus, only those applications actually published in the open literature can be accessed in this context. Nevertheless, the list is still reasonably long.

2.2.1 Boosting Applications

Boosting applications have used helical FCGs almost exclusively. The reason for this emphasis is the potentially large gains that can be realized with these units because of their large initial inductances. Possibly the most spectacular of these systems is the 1 J to 100 MJ explosive system that was fired in Sarov. There were six booster helical flux compression generators in this experiment, including a C-640 unit. There are two primary methods of coupling the flux from one generator to the next. The first coupling technique is simply running the current from the booster FCG through the next generator. However, this procedure requires the second generator to contain significantly less inductance than the booster unit, or there will be little or no gain in either current or energy. Often, this technique is used when a helical FCG is boosting a cylindrical generator, for example. The second coupling technique is inductively coupling the flux from the booster to the next stage. Essentially, a load coil for the booster FCG is wrapped around part or all of the stator winding of the next stage, cf. Chap. 9.1. The advantages are that the booster can work much more efficiently into a load tailored to maximize its gain, and the boosted FCG has no current acting on its conductors until it is crowbarred. A disadvantage to this approach is that the coupling between the booster load coil and the boosted FCG acts as a step-up, air-core transformer. Thus, the second generator must, before crowbarring, effectively stand off the voltage of this transformer coupling. There are several variations for the actual implementation of each of these coupling approaches, but the result is the same in all cases, transfer of magnetic flux from one system to the next.

2.2.2 Limited or No Pulse Forming Network

This category of application is meant to indicate that these applications utilize essentially the direct output of the flux compression generator. If there is a PFN present, then it is typically in the form of a closing switch. With this introduction, it is reasonable to find possibly the widest variety of generator styles being used in this division. The reason for this is that several styles of FCGs have evolved to either directly address a particular application or the experimental application requires a faster performance than is normally achieved with an end-initiated explosive system. The primary advantage of designing the FCG specifically for the task to be done is that the energy efficiency is typically maximized. The disadvantage is that one never has an "off the shelf" FCG to perform all jobs presented. Of course, such a situation is unreasonable because the effective impedance of an FCG is typically relatively low, usually measured in milliohms to 10's of milliohms. Thus, the load and its FCG power source are usually very tightly coupled.

2.2.2.1 High Magnetic Field Experiments

Perhaps the first use of magnetic flux compression was the generation of very high magnetic fields for scientific study that were clearly beyond the capabilities of other methods for producing these fields. The simplest and most efficient version of this type of experiment is the imploding liner on an initial magnetic field. In the United States, the highest magnetic fields obtained have been about 14 Mgauss (1,400 Tesla) using an implosion system. This result was achieved by Fowler's team at the Los Alamos Scientific Laboratory in 1959 [Fow60]. A plot of the experiment is shown in Fig. 2.13.

The system that produced this field used a split liner to enable injection of the initial magnetic field. Once the ring charge of Composition B explosive began to drive the liner, the slot closed to form the required flux trap to contain the field. In addition to B-dot probes, Zeeman splitting and Faraday measurements were used to assure that the measured magnetic field data were consistent and correct.

Another method for fielding high field experiments is to use a coil block as a single-turn load for an FCG. The advantage of this approach is that the working volume for the experiment is well defined, but the ultimate magnetic field achievable is much lower than for an implosion system. An example of this style of generator is depicted in Fig. 2.14. This system was dormant for about 15 years and was redeveloped in the early 1990's to provide an experimental capability to measure the critical magnetic field in the high-temperature superconductor materials, specifically YBCO. During one series of experiments, the single-stage strip generator produced a maximum magnetic field of 1.26±0.02 Mgauss, or 126±2 Tesla [Goe94].

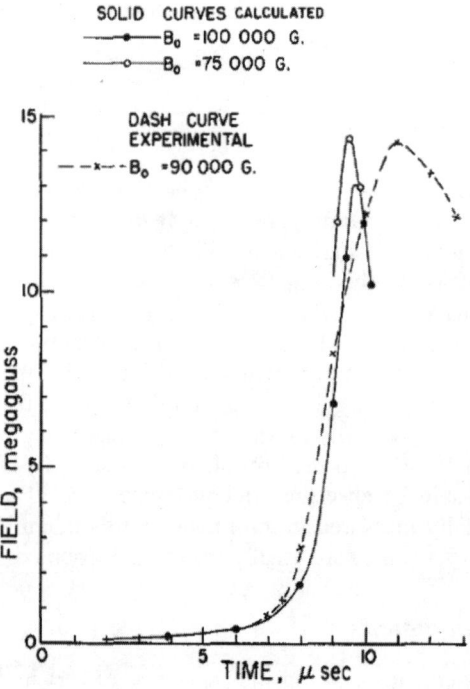

Fig. 2.13. Measured magnetic field as a function of time for a high magnetic field experiment that achieved a maximum field of ~14 MegaGauss [Fow60].

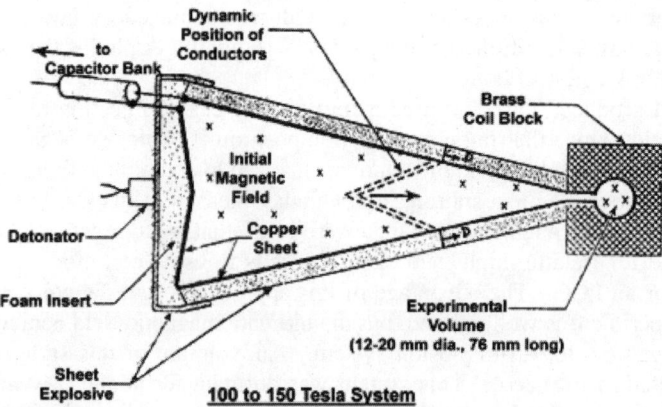

Fig. 2.14. Sketch of a single-stage strip generator. The initial field is provided from a capacitor bank, and the explosive is PETN Detasheet [Goe94].

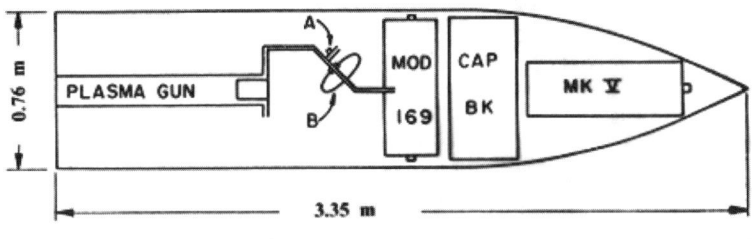

Fig. 2.15. The schematic layout of the Birdseed payload shows the positioning of the capacitor bank, the Mark V FCG, the Model 169 FCG, and the plasma gun [Fow73].

2.2.2.2 Birdseed Experiments

In the early 1970's, a series of experiments were performed to inject a high-energy neon plasma into the ionosphere [Fow73]. The experiments may be the best known example of a fully contained, flight-worthy, magnetic flux compression power supply powering an active load. In this case, the load was a Marshall gun. The power supply started with charged capacitors, a booster generator, and a final stage FCG to power the plasma gun. The weight budget for the entire experiment on the Strype rocket was 220 kg. One of the significant challenges of this project was that the Strype rotated around its long axis for spin stabilization. Therefore, the capacitor bank had to be certified to be able to hold a charge during launch and flight until the system was initiated.

The helical FCGs used for these experiments were a Mark V FCG developed at Los Alamos specifically for this effort and the Sandia Model 169 FCG. The Model 169 generator was selected for this program because it was a faster, double end-initiated, center output helical FCG. From Fig. 2.15, the only pulse forming consisted of a ballast inductor and a closing switch. Essentially, the dummy load or ballast inductor is where the current flows until the intended load is switched in parallel by the detonator switch. The capacitor bank was positioned within the payload volume to provide blast protection for the Model 169 from the Mark V generator. These experiments injected a maximum of ~300 kJ of neon plasma into the ionosphere.

2.2.2.3 Plasma Focus Experiments

Between 1977 and 1981, a series of experiments were fielded to power a plasma focus device to produce intense bursts of neutrons using explosive power supplies [Fre83]. The generator used in these experiments was the "plate" style FCG. This power supply uses areal initiation of the PBX-9501 HE to drive the nearly parallel plates of the generator together simultaneously for a relatively fast pulse output.

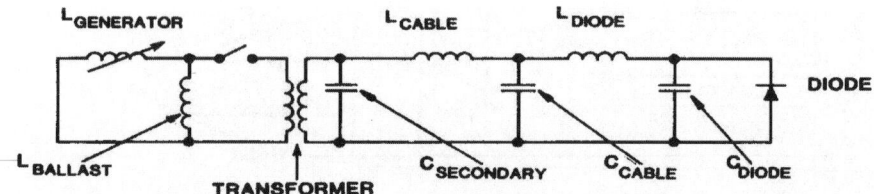

Fig. 2.16. Schematic of the circuit used in the 1983 experiments at Los Alamos [Fre86].

To address the few microsecond needs of the plasma focus device as the generator load, the plate were separated by 7.62 cm at the input end and 12.7 cm at the output end. The plate generator powered a shunt or ballast inductance until the proper time in the pulse when the plasma focus was switched into the circuit as a parallel load. The plasma focus was successfully driven with currents ranging from 1.1 MA to 2.4 MA.

2.2.3 FCGs with PFNs

In this area of application, the more widely accepted power flow model is intended. In this model the power flow chain consists of a generator, a Pulse Forming Network (PFN), and its load. The pulse-forming network may be comprised of closing and/or opening switches, transformers, and transmission lines. Much effort has been invested in this generic area for applications, due in part to desires to be able to standardize generator designs and subsequently tailor their output to be appropriate for intended loads.

2.2.3.1 Electron Beam Experiments

In 1983, a series of three experiments were performed to demonstrate the practicality of powering electron beam devices using a standard design plate-type FCG [Fre86]. The pulsed power system consisted of a 13.2 cm X 52.8 cm, two-sided plate generator with its ballast inductor, a closing switch to connect the primary in parallel with the ballast inductor, a 1:33, air-core transformer, a 10-m length of Sieverts high-voltage cable, and a vacuum diode, Fig. 2.16. The pulsed power system functioned well for all three experiments, achieving peak voltages of about 520 kV and currents of ~135 kA, Fig. 2.17. The electron beam pulse width was ~70 ns.

2.2.3.2 RF Production Experiments

FCG-powered vircator experiments were performed in 1986 to produce RF radiation using explosive power supplies at the Los Alamos National Laboratory [Fre87]. The power flow arrangement for the 1986 test succeeded in powering a large-area vircator diode which produced RF radiation for a 1 μs pulse duration. The large plate FCG with its ballast inductor was again used as the energy source.

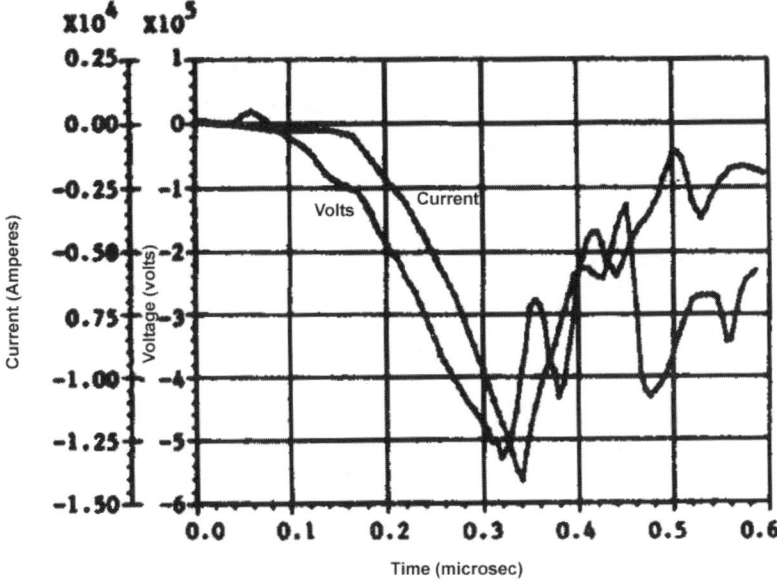

Fig. 2.17. Voltage and current waveforms recorded for one of the three electron beam experiments conducted in 1983 [Fre86].

The air-core transformer was connected to the vacuum diode with a relatively short, silicone fluid-filled coaxial transmission line. The diameter of the cathode for this test was 69.8 cm.

In a more recent report, A. I. Pavlovskii, et. al. reported RF production experiments that were driven by a significantly more sophisticated circuit than the Los Alamos tests [Pav94]. The RF generator in this case was a Cherenkov relativistic oscillator. The FCG power supply was an EMG-160, containing 7.5 kg of TG 50/50 explosive, 50% TNT and 50% RDX. The power supply was divided into six sections. The first section included the EMG-160 flux compression generator and a ~1:1037 step-up, cable transformer. The second stage included a 25 µH solenoidal storage inductor and a ~20 m long twin lead transmission line. The total inductance of this section was ~45 µH. A sixth section contained the solenoidal coil to produce the axial, 2 T magnetic field for the electron diode, where the combined solenoid and transmission line inductance was <45 µH. The third section contained the opening switch that used five copper 0.72-mm diameter, 4.5-m, polyethylene insulated long wires. By winding the copper wires on helical forms, the total length of the fuse package was 1.1-m long. Sulfur hexafluoride gas was used to increase the electrical voltage breakdown of the unit. The fourth section was a two-electrode, overvoltage switch using air as the working gas. The breakdown voltage of this switch was set to be 0.8-0.9 MV. The fifth section was the electron beam diode that produced a 35-cm diameter cylindrical beam into the slow-wave structure. This system produced an RF pulse of 3-cm radiation with a power level of 100 MW and a duration of 0.8 µs.

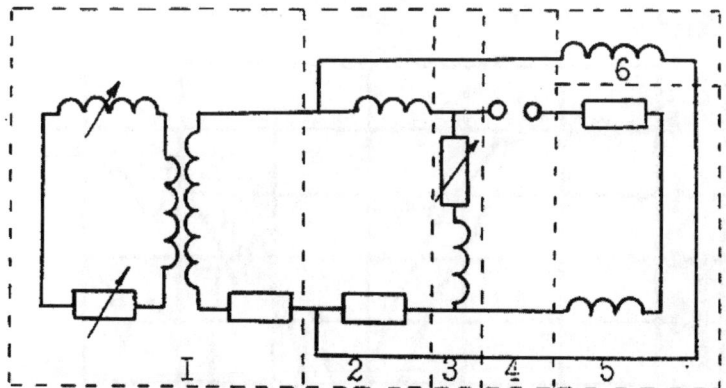

Fig. 2.18. The schematic depicts the circuit used by Pavlovskii, et. al. to produce a 600 kV, 20 kA, 1 μs electron beam [Pav94]. 1-transformer, 2-intermediate inductive storage, 3-opening switch unit, 4-sharpening gap, 5-high current diode, 6-solenoid for magnetic field.

We will discuss in Chap. 9 a similar approach on a more compact scale.

2.2.3.3 Imploded Foil Experiments

Beginning in 1981 and continuing into the early 1990s, a series of experiments were performed to implode thin metal foils using both plate and helical generators with pulse forming circuits. The initial experiments were performed using a plate generator switched to the foil at late times. The Procyon experiments employed a LANL Mark IX helical generator with an explosively formed fuse opening switch, detonator actuated closing switches, and a copper wire array as a plasma flow switch to connect the implosion load to the generator [Gof94]. The circuit used for these experiments is shown in Fig. 2.19. These tests delivered ~15 MA to the plasma flow switch and ~1 MJ kinetic energy to a z-pinch implosion.

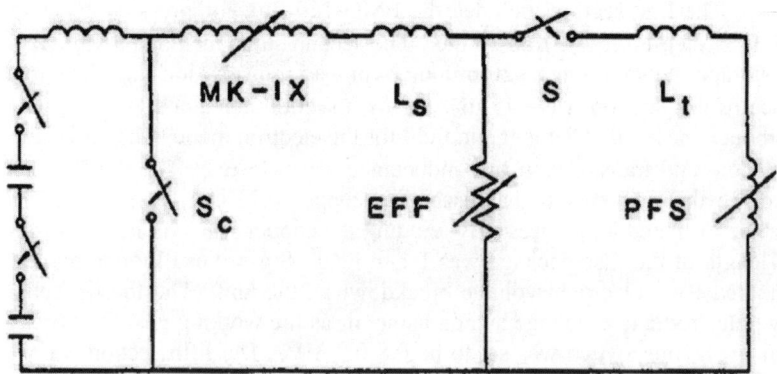

Fig. 2.19. The circuit schematic is shown for the Procyon tests reported at the 1993 Megagauss Conference [Gof94].

2.3 Need for Smaller Helical FCGs

Given the growing need for very small but still very energetic power supplies, the size of flux compression generators must be reduced to serve the needs where conventional pulsed power cannot be productively used. Thus, newer applications are placing a premium on the "portability" aspect of the explosive power supplies. A seripus problem is, however, that the efficiency of helical FCGs is limited, especially if their size is reduced, cf. Fig. 2.10. Nevertheless, given a charge mass of ~ 200 grams of C-4 high explosives and assuming an upper limit of 10% conversion efficiency, one still has about 80 kJ of useable energy for an application. Therefore, there will be significant efforts made to use the smaller FCGs in newer applications.

2.3.1 Differences between the Larger Generators and the Smaller Units

In the studies that have been pursued under this MURI, we have observed several differences that exist between the larger FCGs and the smaller units that have been developed and studied. Important points must begin with the behavior observed in the small flux compression generators that we studied. Our findings are summarized by the following points [Fre03];
- Raising the wire size increases the figures of merit, β.
- Round magnet wire results in better performance than square magnet wire.
- Lowering ideal gain on small FCGs tends to increase both β and the energy gain.
- Raising the initial current (initial field) tends, in limits, to raise β.
- Tapered stator or armature generators provide good performance and figures of merit, cf. Fig. 5.20.
- It has been difficult so far to exceed a current gain of 10:1 in the very small FCGs.

Some of these results are somewhat contrary to the historical experience with large helical FCGs. However, it should be noted that we have studied very small FCGs relative to the historical efforts. Our armature diameters have ranged from 0.95-cm to 3.8-cm diameters. Typical armature diameters had been at least 5 cm in the past.

The least surprising finding in these studies is that increasing the wire size increases the figure of merit. The effect may be attributable to two causes. First, with a larger ware size there is more conductor surface available to carry the currents within the generator. This reduces the current density on any given conductor and lowers the magnetic field for a given current. Thus, the magnetic flux diffusion loss is reduced with the larger conductor. Also, the guidance for helical design has been that the more widely space the turns in the stator winding will result in lower flux losses and better performance. By increasing the wire size this spacing is increased, so for these two reasons the performance should improve.

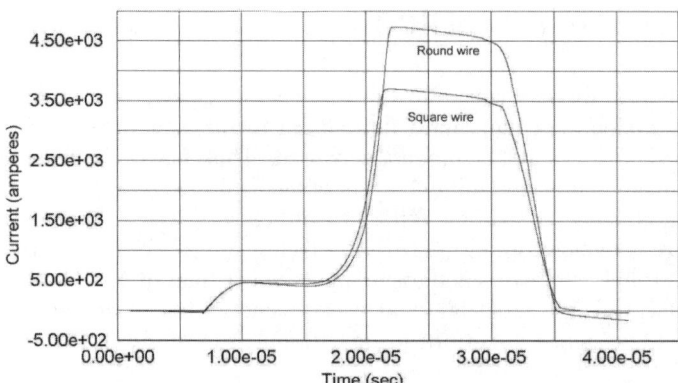

Fig. 2.20. Performance is shown for two generators with nearly identical characteristics except one used round wire and the other used square cross section wire.

Another less controversial finding is the comparison between round and square wire wound stators. Though it is counter-intuitive, both large and small FCGs perform better using round wire in the stator. In the larger units, it was assumed that voltage breakdown is the cause for this difference, but with the smaller systems, voltage breakdown between turns appears to be a very unlikely explanation. With external voltages of only 1-2 kV, the internal voltages would be about twice this value. However, the 2-4 kV would be divided by the number of turns on the stator to approximate a voltage per turn. For example, for a 30 turn stator, the voltage per turn would be about 100 volts, which is a much smaller value than the expected capability of the insulation on the stator wire. One example for this result is shown in Fig. 2.20. The two FCGs were configured as nearly identical as possible where the only difference was that one unit was fabricated with 12 gauge round wire and one used 12 gauge square wire. The initial currents in both tests were ~ 500 A, so the comparison is very straight forward.

Also, lowering the ideal gain on small FCGs tends to increase both β and the energy gain, cf. Eq. (2.21) (an ideal generator has $\beta = 1$). Thus, if a larger passive load inductance is used relative to a smaller inductance, the figures of merit are increased for both large and small generators. For example, the Los Alamos Mark IX helical generator has a measured β of 0.85 with a 140 nH load versus an β of 0.79 with a 35 nH load [Fow89]. We have observed the same behavior in the very small generators.

The next three observations are distinctly different from observations of larger flux compression generators. Raising the initial current or initial magnetic field results in improved β for the generator is contrary to what is expected, based on the historical data. A "small signal" experiment on a new, larger, helical FCG design will generally provide the largest measured gain with respect to its ideal gain. The β is expected to slowly decrease as the initial current is increased until the final current approaches the region of non-linear magnetic flux loss. At that level of current density, the figure of merit will rapidly decrease with increases in the initial currents. This progressive increase in loss is essentially the result of magnetic diffusion with increasing magnetic field strength. An improvement in this figure of merit is not expected with increasing initial currents. Nevertheless, this result has been obtained with comparative experimental tests of the smallest FCGs, particularly ones with armature diameters of 0.95 cm. The observed behavior might be due to the available kinetic energy of the expanding armature being smaller than or comparable to the energy needed for adequate collisional contact connection between the armature and the stator. In this case, a higher induced voltage between armature and stator would aid in breaking down the stator insulation, thus ensuring better and faster contact and hence better performance for higher FCG current amplitudes.

In nearly any style of FCG, changing the parameters to provide a faster output pulse usually gives in a decrease in gain and the associated figures of merit. In other words, as one shortens the pulse and increases the voltage on most generator designs, they tend to perform more as pulse-forming networks with diminished gains. However, in our 2.54-cm diameter FCGs, we observed just the opposite effect. By tapering the stator of a small helical FCG at 12°, both the current and energy gains were improved, cf. Chap. 5. In the larger generators a decrease in gain performance is expected because the internal voltages are increased while producing a more extensive shock ionization length within the generator volume due to the smaller closing angle between the armature and stator.

The smaller generators simply have not produced current gains much over about ten to one, so ideal gains in excess of this figure result in lower β's. Contrary to this result, large helical FCGs have been routinely designed to have gains of several hundred to one, depending on intended loads and load history. In fact, it is precisely these relatively large gains, compared to cylindrical generators, for example, that have lead to widespread interest in developing helical FCGs. Thus, the much smaller gains obtained so far with the very small FCGs is both a puzzle and an issue that requires resolution if these generators are to be used in the roles presently being proposed. Or differently put, a helical FCG with only a few tens of mm in stator diameter tends to be at present of limited use due to its intrinisically smaller energy gain.

Thus, a summary of the differences between the small and large helical flux compression generators has three points. These are;
- Raising the initial current on large helical FCGs has a weak but negative effect on β, within reasonable limits, while the very small FCGs tend to improve in their performance with increasing current, within the limits of current saturation.
- Larger, pulse shaped FCGs tend to perform more poorly on the figures of merit while providing PFN-like pulse tailoring. The small FCGs have performed better in these configurations.
- Large helical FCGs easily produce current gains of >100:1, while the very small generators have not proven able to provide current gains of more than a few 10:1 with an even smaller energy gain, typically close to 1. Thus, typical expected β's for larger helical generators usually range from 0.7 to > 0.8, not the ≤ 0.6 seen in the small systems.

From this viewpoint, the physical limitations of the smaller generators are more severe than for the larger units. While there is a reasonable overall understanding for the processes that may lead to these differences, some of the details are require further study.

Overall, several performance issues have been revealed in these MURI studies that require resolution. The most critical of these are the small current and energy gains that have been achieved using small helical FCGs. From the discussion, it is clear that a β of 0.5 is required to achieve a unity energy gain. Values of β's of ~ 0.6 provide energy gains of a few, but still not a gain of 10:1, for example. While these small FCGs can indeed be very useful in the role of strictly current multipliers for limited volume and mass application, their real utility will only be realized if their energy gains can be improved to provide ~ 10:1.

References

[Che94] Chernyshev VK, Demidov VA, Skokov VI (1994) Helical explosive magnetic generator, having double-sided HE charge initiation. In: Cowan, M, Spielman RB (eds) Megagauss magnetic field generation and pulsed power applications, part 1. Nova Science Publishers, Commack, NY, pp 533-537

[Fow60] Fowler CM, Garn WB, Caird RS (1960) Production of very high magnetic field by implosion. J. Appl. Phys. 31: 588-594

[Fow73] Fowler CM, Thomson DB, Garn WB, Caird RS (1973) LASL Group M-6 Summary Report. The Birdseed Program. (Los Alamos Scientific Laboratory Report LA-5141-MS)

[Fow75] Fowler CM, Caird RS, Garn WB (1975) An introduction to explosive magnetic flux compression generators. (Los Alamos Scientific Laboratory Report LA-5890-MS)

[Fow89] Fowler CM, Caird RS (1989) The Mark IX Generator. In: White R, Bernstain BH (eds) Digest of Technical Papers of the 7th IEEE Pulsed Power Conf., IEEE press, New York NY, pp. 475-478

[Fre79] Freeman JR, McGlaun JM, Thompson SL, Cnare EC (1980) Numerical Studies of Helical CMF Generators. In: Turchi PJ (ed) Megagauss physics and technology. Plenum Press, New York, pp. 205-218

[Fre83] Freeman BL, Caird RS, Erickson DJ, Fowler CM, Garn WB, Kruse HW, King JC, Bartram DE, Kruse PJ (1984) "Plasma focus experiments powered by explosive generators. In: Titov VM, Shvetsov GA (eds) Ultrahigh magnetic fields. Moscow, Nauka, pp. 136-144

[Fre86] Freeman BL, Erickson DJ, Fowler CM, Hoeberling RF, King JC, Kruse PJ, Peratt AL, Rickel DG, Thode LE, Toevs JW, Williams AV (1987) Magnetic flux compression generator powered electron beam experiments. In: Fowler CM, Caird RS, Erickson DJ (eds) Megagauss technology and pulsed power applications. Plenum Press, New York, pp. 729-737

[Fre87] Freeman BL, Brownell JH, Davis HA, Rickel DG, Sheppard MG, Stokes JL, Toevs JW (1987) Microwaves from an explosively-driven virtual cathode oscillator (U). In: Sixth Biennial Nuclear Explosives Design Physics Conference, Los Alamos National Laboratory, October 19-23, 1987, LA-11185-C, Vol. 1, p. 198, (SECRET).

[Fre03] Freeman BL, Altgilbers LL, Fowler CM, Luginbill AD (2003) Similarities and differences between small FCGs and larger FCGs. J. of Electromagnetic Phenomena 3: 467-476
[Goe94] Goettee JD, Brooks JS, Skocpol WJ, Smith JL, Rickel DG, Freeman BL, Fowler CM, Mankiewich PM, DeObaldia EI, O'Malley ML (1994) Megagauss exploration of HC2 and vortex dynamics in $YBa_2Cu_3O_{7-x}$ thin films. Physica B 1805: 194-195
[Gof94] Goforth JH, Oona H, Fowler CM, Greene AE, Herrera DH, King JC, Martinez EC, Parker JV, Seitz GJ (1993) Explosively driven opening switch for the 20-Megampere Procyon experiments. Cowan, M, Spielman RB (eds) Megagauss magnetic field generatoion and pulsed power applications, part 1. Nova Science Publishers, Commack, NY, pp 849-855
[Gov79] Gover JE, Stuetzer OM, Johnson JL (1979) Small helical flux compression amplifiers. In: Turchi PJ (ed) Megagauss physics and technology. Plenum Press, New York, pp 163-180
[Gro73] Grover FW (1973) Inductance Calculations. Dover, New York
[Gur43] Gurney RW (1943) The initial velocities of fragments from bombs, shells, and grenades. (Army Ballistic Research Laboratory report BRL-405)
[Jon79] Jones M (1979) An equivalent circuit model of a solenoidal compressed magnetic field generator. In: Turchi PJ (ed) Megagauss physics and technology. Plenum Press, New York, pp 249-264
[Kal58] Kalantarov PLL, Teitlin LA (1958) Inductance Calculations. Technica, Bucharest
[Leh99] Lehr M, Bamert L, Bell K, Cavazos T, Chama D, Coffey S, Degnan J, Gale D, Kiuttu G, Pellitier P, Sommars W (1999) Helical explosive flux compression generator research at the Air Force Research Laboratory. In: Cooperstein G, Vitkovitsky I (eds) Digest of technical papers of the 12th IEEE International Pulsed Power Conference, IEEE press, Piscataway NJ,, pp 339-342
[Neu01] Neuber A, Dickens J, Cornette JB, Jamison K, Parkinson R, Giesselmann M, Worsey P, Baird J, Schmidt M, Kristiansen M (2001) Electrical behavior of a simple helical flux compression generator for code benchmarking. IEEE Trans. on Plasma Science 29: 573-581
[Nov95] Novac BM, Smith IR, Stewardson HR, Senior P, Vadher VV, Enache MC (1995) Design, construction and testing of explosive-driven helical generators," J. Phys. D.: Appl. Phys. 28: 807-823
[Pav94] Pavlovskii AI, Kravchenko AS, Selemir VD, Brodskii AYa, Bragin YuB, Ivanov VV, Konovalov IV, Suvorov VG, Shibalko KV, Chernyshev VV, Cherepenin VA, Vdovin VA, Korzhenevskii AV, Sokolov SA (1993) EMG magnetic energy for superpower electromagnetic microwave pulse generation. In: Cowan M, Spielman RB (eds) Megagauss magnetic field generation, part 2. Nova Science Publishers, Commack, NY, pp 961-968
[She68] Shearer JW, Abraham FF, Aplin CM, Benham BP, Faulkner JE, Ford FC, Hill MM, McDonald CA, Stephens WH, Steinberg DJ, Wilson JR (1968) Explosive-driven magnetic-field compression generators. J. Appl. Phys. 39: 2102-2116
[Ter43] Terman FE, in Radio Engineers Handbook, McGraw-Hill, p. 55, 1943.
[Whe28] Wheeler HA (1928) Simple Inductance Calculations for Radio Coils. Proceedings of the Institute of Radio Engineers 16, pp 1398-1400

3 Loss Mechanism Basics

Bruce L. Freeman and Andreas A. Neuber

3.1 Magnetic Flux Diffusion

Magnetic flux diffusion into the metal conductors of a magnetic flux compression generator is the most basic loss of magnetic flux in these systems. As long as the magnetic diffusion is such that the resistivity of the conductors may be treated as a constant for times of interest, the diffusion is considered linear. If the diffusion becomes large enough to affect the conductivity of the metals during time scales of interest (ohmic heating → temperature increase → increased resistivity → increased diffusion), the diffusion is considered non-linear in nature.

3.1.1 Linear Magnetic Flux Diffusion

For deriving the magnetic flux diffusion into a conductor as a function of time, we begin with the general Maxwell equations:

$$\nabla \bullet \overline{B} = 0 \qquad (3.1)$$

$$\nabla \bullet \overline{D} = \rho_e \qquad (3.2)$$

$$\nabla \times \overline{H} = \overline{j} + \frac{\partial \overline{D}}{\partial t}, \qquad (3.3)$$

$$\nabla \times \overline{E} = -\frac{\partial \overline{B}}{\partial t}, \qquad (3.4)$$

where

$$\overline{B} = \mu \overline{H}, \qquad (3.5)$$

with

$$\mu = \mu_0 \mu_R, \tag{3.6}$$

where $\mu_0 = 4\pi \times 10^{-7}$ Henry/meter, and

$$\overline{D} = \varepsilon \overline{E}, \tag{3.7}$$

with

$$\varepsilon = \varepsilon_0 \varepsilon_R, \tag{3.8}$$

where $\varepsilon_0 = 8.85 \times 10^{-12}$ Farads/meter.

To complete this description, Ohm's Law is

$$\overline{j} = \sigma \overline{E} = \frac{1}{\rho} \overline{E}, \tag{3.9}$$

where j is the current density, σ is the conductivity, and ρ is the resistivity, the reciprocal of the conductivity. The resistivity is introduced and will be carried in the derivation to emphasize the more prevalent usage of the resistivity over the conductivity in the laboratory environment.

Now, we want to obtain an equation that describes the generalized propagation of the magnetic field into a conductor. By substituting (3.9) into (3.3) and taking the *Curl*,

$$\nabla \times \nabla \times \overline{H} = \frac{1}{\rho_0}(\nabla \times \overline{E}) + \varepsilon \frac{\partial}{\partial t}(\nabla \times \overline{E}), \tag{3.10}$$

and recognizing the vector identity;

$$\nabla \times \nabla \times \overline{\Psi} = \nabla(\nabla \bullet \overline{\Psi}) - \nabla^2 \overline{\Psi}, \tag{3.11}$$

we obtain the general wave equation. The form of this equation is

$$\Delta \overline{H} = \frac{1}{\rho_0}\frac{\partial \overline{B}}{\partial t} + \varepsilon \frac{\partial^2 \overline{B}}{\partial t^2}, \tag{3.12}$$

where the Laplacian is defined as:

$$\Delta \overline{\Psi} \equiv \nabla^2 \overline{\Psi} = \nabla \bullet (\nabla \Phi). \tag{3.13}$$

In a non-conducting medium ($\rho_0 = \infty$), Eq. (3.12) becomes the vacuum wave equation. In a conducting medium, displacement currents may be neglected since conduction currents dominate. Thus, the last term of Eq. (3.12) is negligible, and since it derives from the displacement term in (3.3), it can be omitted. The resulting equation is the magnetic diffusion equation;

$$\Delta \overline{H} - \frac{1}{\kappa_0} \frac{\partial \overline{H}}{\partial t} = 0, \qquad (3.14)$$

where

$$\kappa_0 = \frac{\rho_0}{\mu} \qquad (3.15)$$

is the magnetic diffusivity. By taking the Curl of (3.14), substituting (3.3), and again neglecting the displacement currents, the companion equation for the current density, j, is obtained,

$$\Delta \overline{j} - \frac{1}{\kappa_0} \frac{\partial \overline{j}}{\partial t} = 0. \qquad (3.16)$$

Both (3.14) and (3.16) are very reminiscent of the thermal diffusion equation.

To simplify the analysis, we limit the problem under consideration to a one-dimensional diffusion of the magnetic field outside a conducting half space into the conductor as a function of time, Fig. 3.1. Though the geometry is idealized, it offers a practical way to approach the diffusion of the magnetic field into a conductor with a fixed resistivity, ρ_0. We will use a coordinate system with the x-axis pointing into the conductor. Thus, the vector quantities of H, E, and j become $H(0,0,H_z)$, $E(0,E_y,0)$, and $j(0,j_y,0)$. With these definitions, the Eqs. (3.3) and (3.4) may be written as

$$\frac{\partial H_z}{\partial x} = -j_y \qquad (3.17)$$

$$\frac{\partial E_y}{\partial x} = -\mu \frac{\partial H_z}{\partial t}. \qquad (3.18)$$

For the case of ρ_0 and μ being constant, the magnetic diffusion equation in one dimension becomes

$$\frac{\partial^2 H_z}{\partial x^2} - \frac{1}{\kappa_0} \frac{\partial H_z}{\partial t} = 0. \qquad (3.19)$$

Before solving Eq. (3.19) for a specific boundary condition, we introduce two useful quantities. The first, the magnetic skin depth, δ, is defined as the depth in the conductor where the magnetic field is a value of 1/e of the magnetic field outside the conductor at some time. It can be shown that for a sinusoidal boundary condition, the classical skin depth is [Kno70]

$$\delta = \sqrt{\frac{2\kappa_0}{\omega}} = \sqrt{\frac{\rho_0 T}{\mu \pi}}, \qquad \begin{matrix}(3.20)\\ \text{a)}\end{matrix}$$

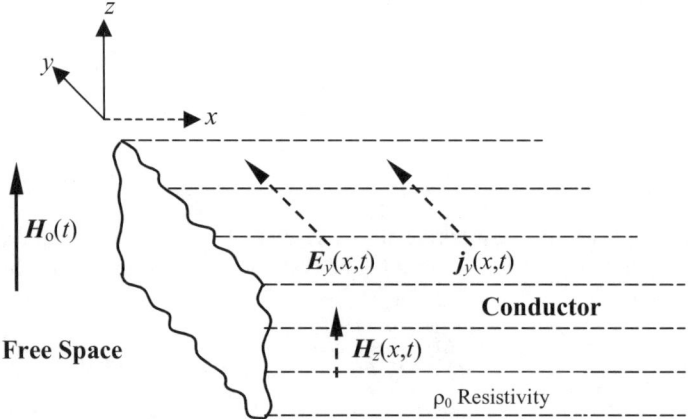

Fig. 3.1. The geometry for the one dimensional solution is shown as a conducting half space that extends in the positive x direction.

with ω and T being the angular frequency and the period of the sinusoid, respectively.

Or similarly, for a step function $H_z(0,t) = H_0$ for $t > 0$ and 0 otherwise, we get:

$$\delta = 2\sqrt{\kappa_0 t} = 2\sqrt{\frac{\rho_0 t}{\mu}}. \tag{3.20}$$
b)

The second quantity that is useful is the magnetic flux skin depth, δ_ϕ, which is the total diffused flux, $\phi(t)$, as a function of the surface flux density, $B_z(0,t)$ over the skin depth. This is expressed by

$$B_z(0,t)\delta_\phi = \int_0^\infty B_z(x,t)dx. \tag{3.21}$$

Taking the permeability as space and time independent, this definition may also be expressed in terms of the magnetic field,

$$H_z(0,t)\delta_\phi = \int_0^\infty H_z(x,t)dx. \tag{3.22}$$

One very instructive boundary condition for solving Eq. (3.19) is the exponential condition,

$$H_z(0,t) = H_0 e^{t/\tau}, \text{ for } -\infty < t < \infty. \tag{3.23}$$

This boundary condition approximates the nearly exponential increase in field observed in magnetic flux compression generators. The solution is found by seeking a solution to Eq. (3.19) in the form $e^{t/\tau}f(x)$. From Eq. (3.19), the differential equation for $f(x)$ is

$$\frac{\partial^2 f}{\partial x^2} = \frac{f}{\kappa_0 \tau}. \tag{3.24}$$

Hence, the general solution for Eq. (3.19) is

$$H_z(x,t) = H_0 e^{t/\tau}\left(A_1 e^{-x/x_0} + A_2 e^{+x/x_0}\right), \tag{3.25}$$

where A_1 and A_2 are constants. The term, x_0, is

$$x_0 = \sqrt{\kappa_0 \tau}. \tag{3.26}$$

The only physically reasonable solution for this problem is

$$H_z(x,t) = H_0 \exp\left[\frac{t}{\tau} - \frac{x}{x_0}\right]. \tag{3.27}$$

One can now identify the characteristic magnetic flux skin depth in Eq. (3.22) as

$$\delta_\phi = x_0 = \sqrt{\kappa_0 \tau} = \sqrt{\frac{\rho_0}{\mu}\tau}. \tag{3.28}$$

If the parameter, τ, is time dependent the equation for $f(x)$ becomes,

$$\frac{\partial^2 f}{\partial x^2} = \frac{f}{\kappa_0 \tau}\left[1 - \frac{t}{\tau}\frac{d\tau}{dt}\right], \tag{3.29}$$

where the form of τ is

$$\tau = \frac{\overline{H}}{d\overline{H}/dt}. \tag{3.30}$$

These relations are valid as long as

$$\frac{d\tau}{dt} \ll \frac{t}{\tau}. \tag{3.31}$$

The skin depth derived for the various boundary conditions, cf. (3.20)a, (3.20)b, and (3.28), differ at most by a factor 2.

3.1.2 Magnetic Diffusion and a Hollow Conducting Cylinder

The diffusion of magnetic field into a conducting, hollow cylinder is a problem that is a common consideration for the magnetic flux compression generator applications. Specifically, we will address a conducting tube with an inner radius, a, an outer radius, b, and a wall thickness, d, see Fig. 3.2. The coordinate system is transferred from Cartesian to cylindrical, but the problem remains an essentially one-dimensional consideration. The known boundary condition is:

$$H_z(b,t) = H_z(t), \text{ for } 0 \leq t < \infty, \tag{3.32}$$

By using Faraday's induction law, the second boundary condition may be obtained. We will assume that the external magnetic field is a constant outside the tube. Further, the permeability $\mu = \mu_0$ within the conductor. If one considers a closed contour resulting from the intersection of a normal plane with the inner surface and the area, A, bounded by it, the Faraday law is

$$-A\mu_0 \frac{dH_z(a,t)}{dt} = \oint_C E \bullet dl \cdot \tag{3.33}$$

Within the interior of the tube, we have

$$\nabla \times \overline{H} = \frac{1}{\rho_0} \overline{E} \cdot \tag{3.34}$$

The tangential component of E must be continuous at the surface. Thus, the second boundary condition is

$$\frac{dH_z(a,t)}{dt} = \frac{\rho_0}{A\mu_0} \oint_C [\nabla \times \overline{H}] \bullet d\overline{l} \cdot \tag{3.35}$$

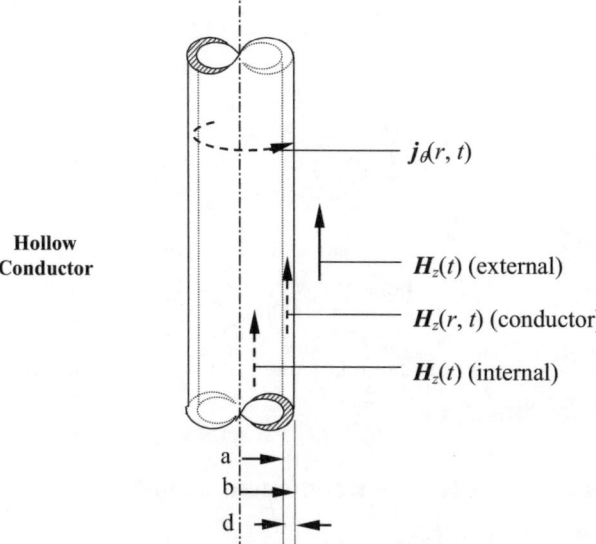

Hollow Conductor

Fig. 3.2. The tubular conductor is shown with the dimensions and quantities of interest.

Thus, for a tubular cylinder, Eq. (3.35) reduces to

$$\frac{dH(a,t)}{dt} = 2\frac{\kappa_1}{a}\left[\frac{\partial H_z(r,t)}{\partial r}\right]_{r=a}, \qquad (3.36)$$

where $\mu_R = 1$ and the magnetic diffusivity,

$$\kappa_1 = \frac{\rho_0}{\mu_0}. \qquad (3.37)$$

is used. For a tube, as illustrated, and a uniform magnetic field imposed on the outside, the companion boundary condition to Eq. (3.32) is

$$H_z(r,t) = 0, \qquad \text{for } 0 \le r < b. \qquad (3.38)$$

The magnetic field that has penetrated the tube of Fig. 3.5 can be solved [Jae40], assuming that $\mu_R = 1$ in the form

$$H_z(t) = H_0 - 4H_0 \sum_{n=1}^{\infty} e^{-\kappa_1 \alpha_n^2 t} \frac{J_2(a\alpha_n)J_0(b\alpha_n)}{(a\alpha_n)^2\left[J_0^2(b\alpha_n) - J_2^2(a\alpha_n)\right]}, \qquad (3.39)$$

where the α_n are the roots of

$$J_0(b\alpha)K_2(a\alpha) - K_0(b\alpha)J_2(a\alpha) = 0. \qquad (3.40)$$

The functions, J and K, are Bessel functions of the first and second kind, of order $n = 0, 2$.

3.1.3 Nonlinear Magnetic Diffusion

The most serious limitation of the linear theory of magnetic flux diffusion into a conducting medium with finite resistivity is that we have assumed that the resistivity of the material is a constant. In reality, the current carrying layer in the conductor heats up with an accompanying increase in the local resistivity, or decrease in the conductivity, as a function of temperature. If one considers the linear current density, current per unit width of conductor carrying the current, some intuitive idea for the practical limits for current to be carried on a conductor may be obtained, for the time scale of interest. For low linear current densities and relatively short times, the conductor heating is typically small enough that the approximation of a constant resistivity during the current pulse is acceptable. However, when the current densities reach values of 750 kA/cm to 1 MA/cm, the conductor heating becomes large enough that the resistivity will increase significantly during a pulse lasting 10's of microseconds. The associated magnetic field with the larger value is approaching 1 Mgauss or 100 Tesla. In such a situation, the resistivity becomes a time dependent variable for the equations describing the system under study.

While we will show the general equations for the nonlinear magnetic diffusion into a conductor, a detailed solution for specific problems is typically a problem best solved using numerical techniques. Since the magnetic fields accompanying the high current densities are very large, it is reasonable to assume that the relative magnetic permeability is equal to one, or $\mu \equiv \mu_0$. To a first approximation, we can take the resistivity in a relatively simple form, namely,

$$\rho = \rho_0(1+\beta Q), \qquad (3.41)$$

where ρ_0 is the resistivity at 0 °C, β is the heat factor, and Q is the heat content or internal energy density relative to 0 °C. In the solid phase, Q is related to the temperature, T, by

$$Q = c_v T, \qquad (3.42)$$

where c_v is the specific heat per unit volume, at constant volume. It can be shown that this model of the resistivity applies reasonably well until the temperature of vaporization. In general, the increase in Q in a conductor carrying a current density, j, is given by

$$\frac{\partial Q}{\partial t} = \rho j^2 + \lambda \Delta T, \qquad (3.43)$$

where λ is the thermal conductivity and the temperature, T, is a function of x and t. Physically, the first term on the right side of the equation represents heat source, or Joule's dissipation, and the second term is the heat conduction in the conductor. The heat conduction term is often neglected in magnetic field diffusion problems because the magnetic skin depth in conductors is a factor of 10 or more larger than the equivalent thermal skin depth. For a one-dimensional problem, Eq. (3.43) can be written as

$$\frac{\partial Q(x,t)}{\partial t} = \rho j_y^2. \qquad (3.44)$$

Substituting the temperature-dependent resistivity (3.41) in (3.44), we have the heat equation,

$$\frac{\partial Q(x,t)}{\partial t} = (1+\beta Q)\rho_0 j_y^2. \qquad (3.45)$$

The integral form of this equation is

$$(1+\beta Q) = \exp\left[\rho\beta \int_0^t j_y^2 dt\right]. \qquad (3.46)$$

We note that this integral is the action integral that is usually encountered in fuse considerations, cf. Chap. 7, Eq. (7.8). More specifically, the action or current integral is given as

$$J = \int_0^t j_y^2 dt, \quad (3.47)$$

where J is dependent only on the conductor properties. By integrating (3.44), the resulting equation is

$$\int_0^t j_y^2 dt = \int_{Q_0}^{Q_f} \frac{1}{\rho} dQ = \int_{Q_0}^{Q_f} \sigma dQ. \quad (3.48)$$

In this case, both the resistivity and conductivity forms of the integral are provided. The expression shows that J is dependent only on the conductor properties of either resistivity or conductivity as a function of Q and the beginning and ending energy densities, Q_0 and Q_f. If one considers Eq. (3.42), then J is related to the beginning and ending temperatures, T_0 and T_f.

Now, referring to Fig. 3.1 for our geometry, the diffusion of magnetic field into a plane, incompressible conductor, as a one dimensional problem, can be written as a complete set of equations. This is done using the assumptions applied in this section concerning the permeability and resistivity. In this set of equations, the current density, j_y, has been eliminated and the permeability is assumed to be that of free space. Thus, the set of equations for non-linear magnetic diffusion is:

$$\frac{\partial H_z}{\partial x} = -\frac{1}{\rho(1+\beta Q)} E_y \quad (3.49)$$

$$\frac{\partial E_y}{\partial x} = -\mu_0 \frac{\partial H_z}{\partial t} \quad (3.50)$$

$$\frac{\partial Q}{\partial t} = \rho_0(1+\beta Q)\left(\frac{\partial H_z}{\partial x}\right)^2. \quad (3.51)$$

In addition to these three, coupled equations, the complete description of the equation set requires the boundary and initial conditions. In the half space of the conductor, these conditions are

$$x = 0: \quad H_z(0,t) = H_0(t), \quad (t \geq 0); \quad (3.52)$$

$$t = 0: \quad H_z(x,0) = 0, \quad (x \geq 0). \quad (3.53)$$

3.1.4 Approximate Solution

One feature of non-linear magnetic diffusion stands out from all other issues, and that is the rapidity that a field of sufficient strength can penetrate a conductor. The approximated solution for the Eqs. (3.49), (3.50), (3.51) for the hyperbolic boundary condition,

$$H(0,t) = H_0(t) = h_c \sqrt{\frac{t}{t_0}}, \qquad (3.54)$$

provides the appropriate information to illustrate the field penetration indicated. For convenience, the magnetic field has been split into a characteristic field, h_c, which has the form

$$h_c = \sqrt{\frac{2}{\mu_0 \beta}}, \qquad (3.55)$$

and a free parameter, t_0, which defines the time scale of the boundary condition. This pulsed field boundary condition provides a reasonable approximation for the first quarter period of a transient, sinusoid field pulse.

In the limit that $\beta Q >> 1$, the approximated solution for the set of Eqs. (3.49), (3.50), (3.51), is

$$\frac{H_z(x,t)}{h_c} \approx \sqrt{\left(\frac{t}{t_0} - \frac{x}{s_0}\right)}, \qquad (3.56)$$

$$\beta Q(x,t) \approx \left(\frac{t}{t_0} - \frac{x}{s_0}\right), \qquad (3.57)$$

where

$$s_0 = \sqrt{\frac{1}{2}\kappa_0 t_0} \qquad (3.58)$$

and

$$\kappa_0 = \frac{\rho_0}{\mu_0}. \qquad (3.59)$$

From Eqs. (3.56) and (3.57), the current density and effective energy flow are

$$j_y(x,t) = \frac{h_c}{2s_0}\sqrt{\left(\frac{t}{t_0} - \frac{x}{x_0}\right)} \qquad (3.60)$$

and

$$\beta Q(x,t) \approx \left[\frac{H_z(x,t)}{h_c}\right]^2 = \frac{1}{2}\beta\mu_0 H_z^2(x,t). \qquad (3.61)$$

Recall that this solution is only valid for $\beta Q >> 1$, or that the energy flow into the conductor is significant. This means that the magnetic fields must be large,

$$H_z(x,t) \gg h_c, \qquad (3.62)$$

for the solution to be valid. Alternatively, the boundary condition (3.54) shows that the solution is valid for a time, t, where

$$t \gg t_0. \qquad (3.63)$$

In addition, the solution is only valid for a limited depth into the conductor, defined by

$$x \ll h_c \frac{t}{t_0}. \qquad (3.64)$$

Given the solution (3.61), the physical meaning of critical field, h_c, is defined. By letting $H=h_c$, we see that h_c is the field at which

$$\beta Q = \frac{1}{2}\beta\mu_0 H^2 = 1. \qquad (3.65)$$

Thus, this is the field that leads to a doubling of the resistivity, from Eq. (3.41). Therefore, this is the transition magnetic field above which, heating effects become dominate in the magnetic diffusion process.

The flux skin depth, s_ϕ, may be obtained using Eqs. (3.56) and (3.60). The result is that

$$\frac{s_\phi}{s_0} \approx \frac{2}{3}\frac{t}{t_0} = \frac{2H_0(t)}{3h_c}\sqrt{\frac{t}{t_0}}. \qquad (3.66)$$

To better understand the various relationships in energy diffusion, let us define two more parameters. These are the surface energy factor, $X(t)$, and the energy skin depth, $s_e(t)$. In general, the surface temperature, $T(0,t)$, may be expressed in the general form,

$$c_v T(0,t) = \frac{1}{2}\mu_0 H_z^2(0,t)X(t), \qquad (3.67)$$

where the surface energy factor, $X(t)$, depends on the pulse shape and duration. However, this factor is always of the order of unity in magnitude. In a similar manner, the total dissipated energy, W_T, is simply

$$W_T = \frac{1}{2}\mu_0 H_z^2(0,t)s_e(t). \qquad (3.68)$$

With these definitions, the energy skin depth and surface energy factor may be related to our nonlinear diffusion example. The energy skin depth is related through the expression,

$$\frac{S_e}{S_0} \approx \frac{t}{t_0} = \frac{H_0(t)}{h_c}\sqrt{\frac{t}{t_0}} = \frac{3s_\phi}{2s_0}, \qquad (3.69)$$

and the surface energy factor is essentially unity, $X \sim 1$.

For the linear diffusion case, $H \ll h_c$, one set of solution equations is given by

$$\frac{H_z}{h_c} = \sqrt{\frac{t}{t_0}}\left\{e^{-\varphi} - \sqrt{\pi}\varphi[1 - erf(\varphi)]\right\}, \qquad (3.70)$$

and

$$\frac{j}{h_c/s_0} = \frac{1}{4}\sqrt{\pi}[1 - erf(\varphi)]. \qquad (3.71)$$

The associated skin depths and energy factors are

$$\frac{s_\phi}{s_0} = \sqrt{\left(\frac{1}{8}\pi\frac{t}{t_0}\right)}, \qquad (3.72)$$

$$\frac{s_e}{s_0} = \frac{1}{3}\sqrt{\left(2\pi\frac{t}{t_0}\right)} = \frac{4s_\phi}{3s_0}, \qquad (3.73)$$

and

$$x = \frac{1}{2}\pi, \qquad (3.74)$$

where the dimensionless parameter, φ, has been defined as

$$\varphi = \frac{x}{2\sqrt{2s_0}}. \qquad (3.75)$$

From the division in the solutions, the magnetic field diffused into a conductor may be divided into two regions (Fig. 3.3a). The characteristic field provides the dividing the line between the two solutions, i.e. $H_0/h_c \sim 1$. This condition corresponds to the depth

$$\frac{x_0}{s_0} = \frac{t}{t_0} - 1. \qquad (3.76)$$

The initial magnetic flux diffusion is given by the classical, linear diffusion in Eq. (3.70). This holds until one reaches the condition where the diffused field and the characteristic field are about equal. At this point, the nonlinear diffusion Eq. (3.56) applies. The corresponding current density in Fig. 3.3b is peaked around the critical field value. The initial current density is governed by Eq. (3.71), and the nonlinear portion of the current density is approximated by Eq. (3.60).

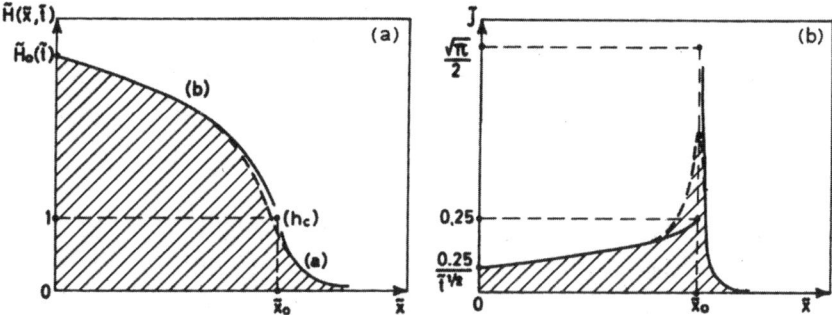

Fig. 3.3. (a) Plot showing the magnetic field inside the conductor as a function of depth in the conductor, and (b) Plot shows the current density as a function of depth in the conductor. [Kno70].

It is important to note that the general behavior of the current density is similar for several boundary fields. The characteristic that is unique about nonlinear flux diffusion is the current density concentration that propagates its way through the conductor [Kid59]. This has the effect of appreciably increasing the amount of flux diffused into the conductor, and thereby lost to a flux conserving system. This peaking in the current density, derived using a step function boundary condition, is depicted in Fig. 3.4.

3.1.5 Practical Implications of Magnetic Diffusion

The first important point is that the definition of h_c in Eqs. (3.64) and (3.65) means that when the imposed magnetic field exceeds h_c for the system, the magnetic diffusion becomes non-linear in nature. The threshold for this is when the resistivity in the material is doubled. By non-linear, we mean that the material resistivity is changing significantly within time scales of interest.

Now, we have seen that the change of resistivity of a material, the history of the pulse, and the magnetic field pulse amplitude all impact the onset of non-linear magnetic field diffusion. Further, the threshold values and subsequent history of the pulse after the threshold is reached affect the progression of the non-linear diffusion into or through real conductor materials. We introduce the concept of linear current density as the current flowing through a conductor per unit width. As a rule of thumb, for a pulse rising in 10 to 10's of microseconds in copper or aluminum, the limiting linear current density is ≤ 1 MA/cm. It is worth noting that a current density of this magnitude will produce an associated magnetic field approaching 100 Tesla or 1 Megagauss. For longer rise time pulses, the limiting linear current density is somewhat lower, perhaps in the range of 500-750 kA/cm. At the other extreme in time, a pulse rise time of 1 microsecond or less results in significantly higher limiting current densities. The considerations for the limiting current densities assume room temperature resistivities for the conductor under consideration.

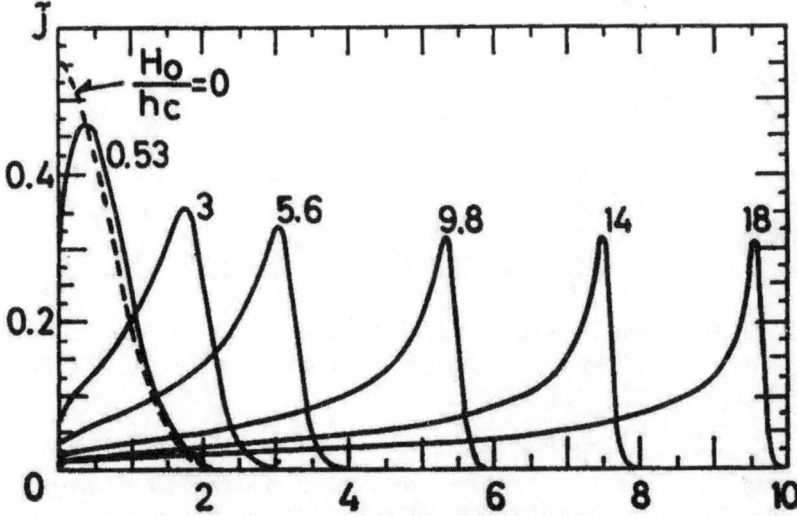

Fig. 3.4. Dimensionless current density, $j(\varphi)\sqrt{j(\kappa_0 t/H_0)}$ for a step function boundary field, $H(t)=H_0$, where $\varphi = x/(s_0 2\sqrt{j2})$. [Kid59]

From Chap. 5.5, we see that the propagation of a shock through the armature material in an explosive driven generator leads initially to a drop in the material's resistivity on the 100 ns time scale, while deformation and heat transfer from the explosive products to the metal cause an increase in the temperature of the metal and an associated increase in the material's resistivity on the microsecond timescale. Typically, in the action of an FCG, this jump in resistivity happens before the current densities become significant. Thus, the conductor resistivity has often increased, doubled or more, before the current rises to levels that would typically drive the onset of nonlinear magnetic diffusion. The practical implication is that the limiting current of an FCG may be more dependent on the conductor state after the explosive shock has passed through the armature. Since the better conductors such as copper and aluminum typically have larger rises in resistivity with temperature than others, small modifications in these materials are usually not seen in practical experiments. However, methods that serve to lower the initial temperature or the temperature rise from the explosive drive can have significant impact on a generator's ultimate performance.

The conductor thickness in a flux compression generator should be chosen such that the skin depth is smaller than the conductor dimension. Since the exponential boundary condition is most relevant to magnetic flux compression, we used Eq. (3.28) to calculate δ_φ for copper and aluminum, the two materials that have found widespread use in flux compression generators, see Fig. 3.5.

Virtually all flux compression generators are indeed designed such that the skin depth is smaller than the thickness of the armature enclosing the magnetic flux in order to avoid excessive loss of flux during compression. We can use Fig. 3.5 as a simple design rule.

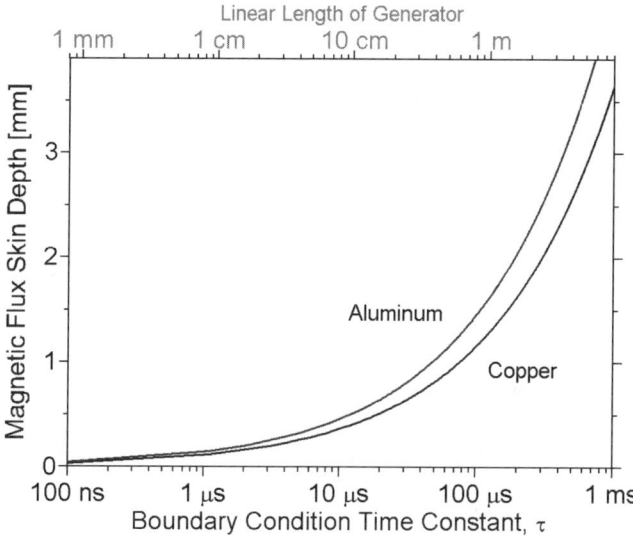

Fig. 3.5. Magnetic flux skin depth for copper ($\rho = 1.68 \times 10^{-8}$ Ohm-m) and aluminum ($\rho = 2.65 \times 10^{-8}$ Ohm-m). Assuming an HE detonation velocity of 8 mm/µs, the time constant, τ, (bottom abscissa) is translated into the linear length of the generator (top abscissa).

If we assume that we want to build a generator with a linear length of 20 cm, then we read from Fig. 3.5. that the thickness of the metal boundaries should be at least 0.5 mm. If we choose > 1 mm wall thickness for our design, we have made sure that only very little flux will diffuse through the conductors. It should be noted that the armature expansion causes thinning of the armature wall thickness, which needs to be accounted for as well.

3.2 Avoidable Flux Losses

All previously discussed flux loss mechanisms contribute to the physical limits of flux compression with respect to the performance and efficiency of practical FCGs. While these limits are physics based, the situation can even be worsened by flaws in the FCG design. As an example, a non-uniformly expanding armature will cause the contact between armature and helix to jump in the worst case from turn to turn. Such turn-skipping or partial turn-skipping is easily observed in experimental waveforms of the current's time derivative. Distinct jumps will in the extreme case dominate an otherwise expected smoothly running waveform. We refer for more details about armature expansions or misalignment to Chap. 4 and for a waveform exhibiting partial turn skipping to Fig. 5.15.

We will briefly discuss a critical moment in FCG operation, the moment of crowbar, where the seed current circuit is separated from the FCG, see Fig. 1.3b.

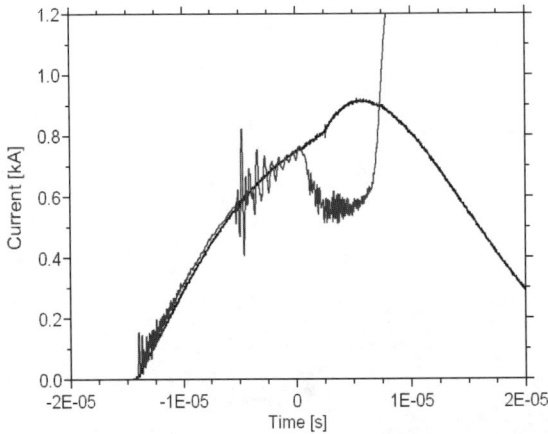

Fig. 3.6. Seed current (current in lower branch, C_S and R_S, of Fig. 2.5) and FCG current. Seed and FCG current are identical until crowbar (switch S1 is closed in Fig. 2.5). The moment of crowbar coincides with $t \sim 0$ s. The FCG current goes eventually off scale.

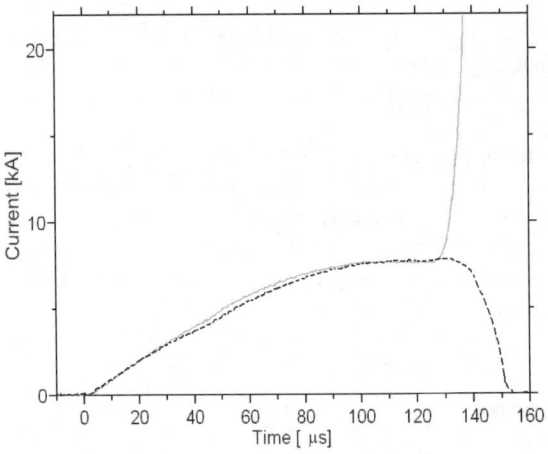

Fig. 3.7. Seed current, dashed waveform, and FCG current, solid waveform for crowbar disk with insulation.

If the crowbar is not carefully designed, a loss in current will be experienced, cf. Fig. 3.6. The crowbar switch needs to switch fast such that little energy is dissipated during the switching process. Further, once contact between expanding armature and crowbar has been established, it has to be ensured that the contact is never lost during the armature's expansion to the full helix diameter.

Several crowbar geometries can be found in the literature, ranging from a simple radially inward directed round metal bar to an angled glideplane, cf. Fig. 2.1, that is carefully designed such that the armature will not loose contact while expanding. We have primarily used a simple, not-angled, brass disk as crowbar, cf.

Fig. 1.3. Our initial attempts with this approach produced a significant loss in current amplitude during the crowbar process, see Fig. 3.6. The FCG current visibly exhibits a dip in current amplitude before going off-scale. However, adding a thin layer of insulation to the crowbar and the armature at the crowbar position resulted in flawless operation, see Fig. 3.7. Without putting much effort into investigating the physical reason for the difference in the observed behavior, we believe that the initial contact in the case of the bare crowbar disk is established through an air arc while the armature is still a distance away from the crowbar. Thus, some of the FCG's seed energy is consumed by the air arc, whereas the insulation hinders the formation of an arc while ensuring a fast switching process due to mechanical and electrical breakdown of the thin insulation layer. However, in larger FCG designs, the loss of armature contact with the glideplane has demonstrated significantly larger losses. This is another example of differences between the smaller FCG designs and the larger units.

Although the current amplitudes in the two example FCG shots are quite different, we have this behavior observed in a wide range of seed current amplitudes. The advantage of using a simple crowbar disk is clear: The cost and time consuming machining of a glide plane becomes unnecessary, the crowbar disk can simply be stamped out of sheet metal.

References

[Jae40] Jaeger JC (1940) Magnetic screening by hollow circular cylinders. Phil. Mag. 29: 18-31.

[Kid59] Kidder RE (1959) Nonlinear diffusion of strong magnetic fields into a conducting half-space. (Lawrence Livermore Laboratory Report UCRL-5467)

[Kno70] Knoepfel H (1970) Pulsed High Magnetic Fields. North Holland Publishing Company, Amsterdam, London and American Elsevier Publishing Company Inc., New York, pp. 92 ff.

4 Mechanical Aspects

Paul N. Worsey, Jason Baird, and Jahan Rasty

4.1 Armature Dynamics

The armature is part of the electric circuit within the generator, and as the flux within the generator is compressed, electric currents flow on the outer surface of the armature in a helical direction due to the magnetic field orientation. The flow of these currents must not be disturbed. If their flow direction is altered, the magnetic field direction will be affected. If features on the surface of the armature retard the current flow an arc can form between the armature and the stator or simply the armature/helix contact cannot progress smoothly, cf. Chap. 5.4.1. In the extreme case, the arc will create a hot plasma, which can cause the stator insulation to break down before the contact edge reaches that location. Since arcing causes current flowing from the armature to the stator to jump ahead of the contact edge and the contact is no longer part of the current path, compressed magnetic flux is trapped in the region between the sliding contact and the arc, thus lost for further compression [Kno70].

4.1.1 Armature Fractures

Longitudinal cracking in the surface of the expanding armature interferes with electrical currents generated on the surface of the armature, affects the generation of plasma in the sliding contact area, and causes magnetic flux to be trapped within the sliding contact area. Trapping of the flux within the contact area, called "flux cut-off" in the idiom of flux-compression generators, does not allow the flux to be forced into the stator coil, keeping the flux from being compressed. As the contact edge moves down the length of the armature, the trapped flux is lost to the compression process [Kno70]. All engineering materials begin to break when stressed beyond their strength limitations, and when the metal in the armature is expanded beyond a certain point it begins to crack. Originally, this limit was thought to be reached when the armature is expanded to more than twice its original diameter. Cracking at that point has no effect on generator performance, as the armature would have already expanded through the stator and the time for flux compression would be past.

In a typical armature test, the explosive detonation is initiated at one point in the center of one end of the explosive charge. As the detonation wave proceeds through the length of the explosive charge within a typical armature, the armature begins to expand into a truncated conical shape and the cone moves down the cylinder in the same direction as the explosive detonation wave. This expanding, moving cone forms the contact with the stator coil in an assembled generator. After expanding to about two times its original diameter, the cylinder begins to break apart, with high-speed photography showing detonation products visibly escaping through fractures in the metal.

The fracturing, however, occurs much sooner in the expansion process, and does not extend along the expanding armature as would be expected if the fracturing was purely a result of explosive expansion of the armatures.

This type of longitudinal fracturing occurs within two armature diameters of the initiation point within the armature explosive charge, as shown by high-speed photography of the expansion of explosive-loaded armatures during flux-compression generator component tests. As will be explained later, several clues lead researchers to suspect that shock dynamics, rather than explosive pressurization, was the root cause of the fracturing.

Longitudinal fractures that happen early in the armature expansion are a likely cause of magnetic flux losses within helical flux-compression generators due to the reasons cited above. Flux cut-off in these generators is a source of inefficiency that can result in the loss of a major portion of their electrical pulse generation capability. If shock dynamics were the cause of fracturing, armature design could be modified to prevent or minimize this sort of flux loss inefficiency. In addition, other types of explosive-driven flux-compression generators may suffer from similar inefficiencies as a result of surface fractures in current-carrying components. It is hoped that discovery of the previously undetected fracturing process leads to helical generator design changes that minimize effects of fracturing on generator performance.

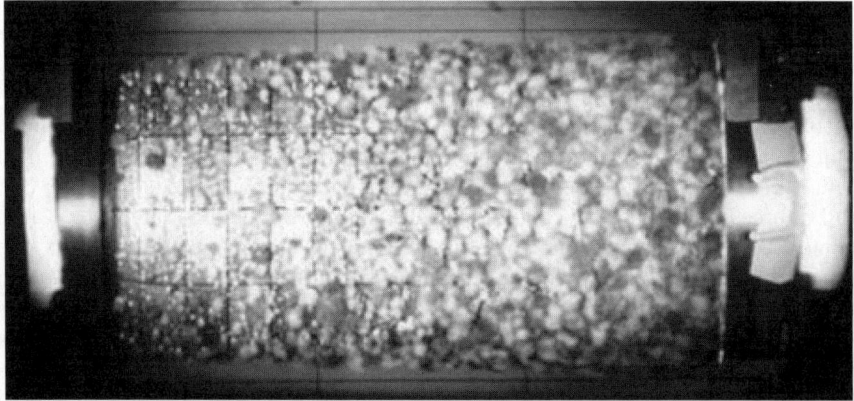

Fig. 4.1. Los Alamos Mark101 armature [Fre02].

4.1.2 Related Problem

Study of the longitudinal cracking of armatures evident in our tests was seen as a starting point for the explanation of other phenomena, such as those seen in the testing of foam staging in the armature of a Los Alamos flux-compression generator called the Mark 101. The origin of Mark 101 phenomena and our observed armature cracking appeared to be related, with both resulting from similar causes at the explosive/armature interface.

The Los Alamos Mark 101 armature in question was constructed by pressing beads of PBX 9501 (95% HMX, 2.5% estane, 2.5% BDNPA-F) explosive into an aluminum armature that was lined with thin, very low-density foam [Fre02]. The explosive charge was initiated along its central axis, in contrast with the end-point initiation of the FCG armature. Fig. 4.1 is a photograph of the Mark 101 armature during its expansion, immediately after its explosive was detonated. The explosive initiation on the right-hand side of the photo was slightly ahead of the left-hand side. The smooth original surface of the armature was colored to appear cross-hatched in a checkerboard pattern, alternating an aluminum silver-grey color with a burnished copper color.

The photo raised some questions among the researchers at Los Alamos [Fre00]. Why was the armature breaking up, instead of smoothly expanding? Why did it break-up in that sort of pattern? The researchers inferred that the white and brown splotches were blobs of explosive detonation products combined with material from the foam, and that the pattern was due to gaseous jetting through the metal of the armature in an arrangement that resulted from a combination of foam cell size and explosive effective bead size which generated Mach stems within the detonation products. Mach stems are also called Mach wave interactions, and they begin near an interface between two media of different density. The combination of shock waves that are impinging on the interface, and shock waves that have reflected from it, results in an amalgamated waveform that has the general shape of a "Y", see Fig. 4.2. The bottom of the "Y" tends to travel through the medium at Mach (sonic) speed; hence the name Mach stem. For subsequent shots, the foam composition and thickness was changed and the effect seen in Fig. 4.1 did not reoccur [Fre00].

4.1.3 Theory

It is necessary at this point to give a short primer in shock dynamics and in the thermodynamic theory of detonations. The normal laws of thermodynamics and the physical conservation laws are the bases for all theoretical work in thermodynamics, and this introduction will start with the laws of conservation of mass and conservation of momentum in a situation where the flow is one-dimensional, steady, with a discontinuous detonation wave (the chemical reaction is assumed to take place instantaneously). Of course, none of these assumptions are totally accurate in real reactions, but these generalities are sufficiently accurate for this purpose, and they greatly simplify matters.

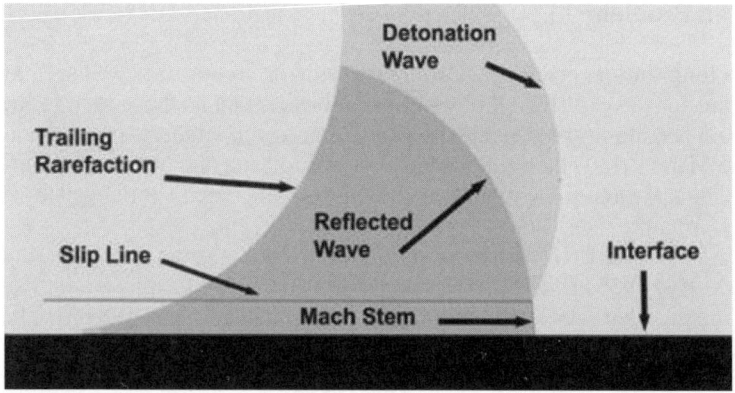

Fig. 4.2. Mach stem.

4.1.3.1 Shocks

Solids under explosive stresses can be considered to be inviscid fluids under hydrostatic stresses. A shock wave traveling through the solid has a thickness approximately equivalent to the mean free path of a molecule of the fluid, so shocks are extremely thin and may be considered discontinuities in fluid properties such as velocity, pressure, and density. Drawing a control volume with a width approaching zero around the shock allows the assumption that there is no heat transfer (i.e., the process is adiabatic) across the control volume because the time for the fluid to transit the control volume is approximately zero (Fig. 4.3).

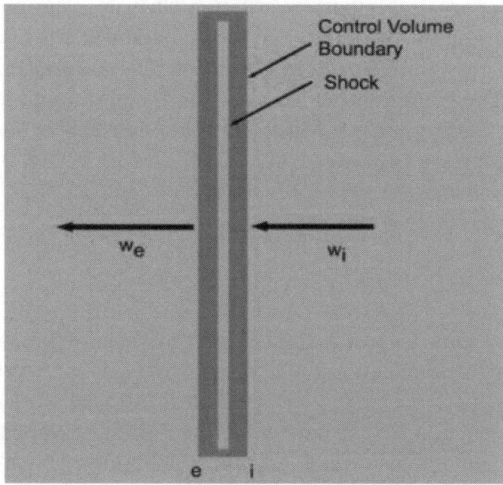

Fig. 4.3. Control volume around shock.

According to the Second Law of Thermodynamics, entropy may not decrease during an adiabatic change, so pressure must always increase as a fluid crosses a shock wave. Therefore, rarefaction discontinuities (shock waves) are not possible; rarefactions become less steep as they propagate, so no discontinuities are present and rarefactions exist as gradual pressure losses. As such, they are sometimes referred to as rarefaction fans [Sha53].

4.1.3.2 Shock Mechanics Based on the Momentum Theorem

Utilizing the shock's frame of reference in the material, as the fluid passes from the upstream side of the shock to the downstream side, the conservation of momentum theorem can be written as:

$$\sum F_z = \frac{d}{dt}(mw_z) \qquad (4.1)$$

In other words, the net force acting instantaneously on the material within the control volume is equal to the time rate of change of the momentum within the control volume, plus the excess of outgoing momentum flux as compared to incoming momentum flux. This is the analogous case to a normal shock in supersonic fluid dynamics, as the fluid does not change its vector direction of travel as it passes through the shock, see Fig. 4.4.

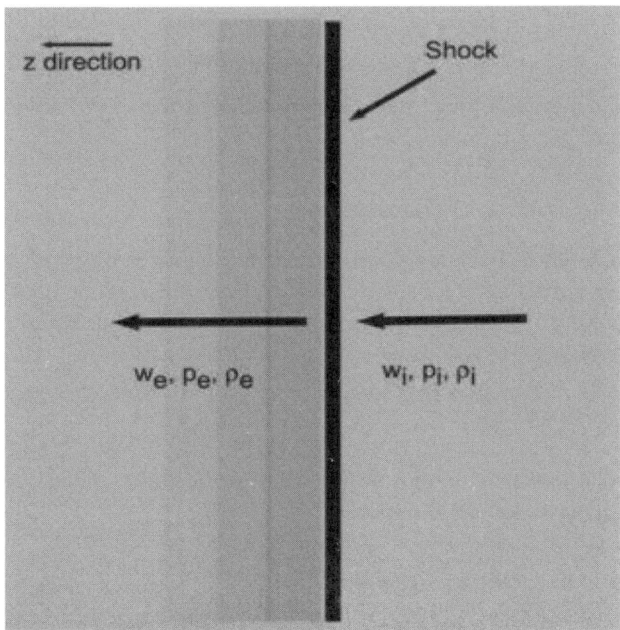

Fig. 4.4. Shock frame of reference.

Therefore, steady flow exists across the shock, because at each point there is no variation of any property with respect to time. Also, since there is no heat transfer across the shock, the steady-flow energy equation can be used to describe the enthalpy change across the control volume

$$h_i + \frac{w_i^2}{2} = h_e + \frac{w_e^2}{2} = h_o, \qquad (4.2)$$

where h_o is the stagnation enthalpy per unit mass.

Since the cross-sectional area on the upstream side of the control volume is equal to the cross-sectional area on the downstream side, the continuity equation can be written as

$$\frac{W}{A} = \rho_i w_i = \rho_e w_e, \qquad (4.3)$$

where W is the mass flow rate across the control volume.

Applying the momentum theorem to the flow across the shock (through the control volume) generates

$$p_i - p_e = \frac{W}{A}(w_e - w_i), \qquad (4.4)$$

and combining Eqs. (4.2) and (4.3) results in

$$p_i + \rho_i w_i^2 = p_e + \rho_e w_e^2, \qquad (4.5)$$

an equation helpful in understanding what state changes occur as the shock moves through the material [Fic79].

4.1.3.3 Thermodynamic Theory of Detonations

Immediately after the detonation of an explosive, a detonation wave propagates into the unreacted mass of explosive at a velocity, D. This gives the reaction products of the detonation wave a velocity, w, immediately behind the wave. Mass is conserved across the detonation wave, so Eq. (4.3) can be rewritten as

$$\rho_0 D = \rho(D - w) \qquad (4.6)$$

with ρ_0 the density of the unreacted explosive (in front of the wave) and ρ the density of the reacted explosive (behind the wave).

The conservation of momentum Eq. (4.4) across the wave is represented by

$$p - p_0 = \rho_0 w D, \qquad (4.7)$$

where p_0 and p are the pressures in front of and behind the wave, respectively. When the velocity, w, is eliminated from these equations by substitution,

$$\rho_0 D^2 - \frac{\rho_0^2 D^2}{\rho} - (p - p_0) = 0 \tag{4.8}$$

results. This is the equation of a straight line, called the Rayleigh line for a detonation. It can be plotted on a pressure versus volume plane when the density terms are replaced by specific volume terms using the relation $\rho = 1/v$.

Shock Hugoniots are curves of pressure versus compressibility (v/v_0, where the subscripted 0 represents the initial condition), and are also plotted on a pressure versus volume plane. This plane represents all of the states in a particular reaction. The equations that represent Hugoniot curves are obtained by combining the conservation of mass, momentum, and energy conditions for specific circumstances. As a result, the general form of a Hugoniot is

$$E - E_0 = \frac{1}{2}(p + p_0)(w - w_0) \tag{4.9}$$

Use of Hugoniots and Rayleigh lines allows the determination of pressure, volume, and density of matter involved in different temporal parts of the explosive event. It also allows the monitoring of changes in those values in a particular place or places over time. This information is essential to determine the state of matter at a particular time or place in a detonation. Calculation of a Hugoniot curve from Eq. (4.9) for a detonation requires knowledge of the equation of state of the detonation products of the explosion. An equation of state defines the state of matter at a particular time and place within a detonation reaction. But because such a reaction produces very extreme conditions, there is little direct experimental data or self-contained theory sufficient to describe matter under those ephemeral conditions. The best that can be done is to choose an empirical form for the equation, and then to perform a curve-fit of the available data to produce the information necessary to make the equation specific to the experiment. The general form of an equation of state is

$$E = f(p, w) \tag{4.10}$$

When a good empirical fit for the equation of state is found, E is expressed as a function of p and w that can be substituted into Eq. (4.9) to allow derivation of the Hugoniot curve for the explosion products [Mad98].

Curve-fitting an empirical solution to experimental data is laborious, and so it is typically done using computer programs.

4.1.3.4 Numerical Solutions of Reactive Flow

Computer codes allow the simulation of explosive events to varying degrees of accuracy. The solutions typically utilize either Lagrangian or Eulerian equations of motion, or they may use the Method of Characteristics because the flow is supersonic (the shock velocity is greater than the acoustic velocities of the media through which the shock is passing).

Lagrangian equations of motion use a coordinate system attached to a laboratory reference, with the fluid being studied moving through the coordinate system; the Lagrangian equations are often used in structural analysis (note that the use of the term "fluid" includes matter that behaves as a fluid under the circumstances; it does not require that matter be in its the liquid phase). There are several analogs of the Lagrangian equations of motion of a compressible fluid that have been written in finite difference computer code.

Eulerian equations of motion use a coordinate system that moves with a point attached to the fluid. Eulerian equations are often used in fluid dynamics and are more useful for the numerical solution of highly distorted flows than are Lagrangian equations. Lagrangian solutions are better at accounting for viscous effects, and are more accurate than Eulerian solutions [Mad98]. A Lagrangian solution, therefore, was used to assist in finding the source of cracking in flux-compression generator armatures in this research.

Use of a Method of Characteristics solution allows maximum accuracy when compared to finite difference solutions. Applying the Method of Characteristics to front- tracking solutions, where material interfaces (such as shock waves) become part of the numerical representation of the flow, allows one to follow each separate shock wave explicitly, and to determine the effects of each on the overall flow. In the case of supersonic hydrodynamics, the movement of a wave in one direction with no change of shape can be described by a partial differential equation, the so-called first order wave equation

$$c\frac{\partial w}{\partial x} + \frac{\partial w}{\partial t} = 0 \qquad (4.11)$$

The Method of Characteristics reduces such a partial differential equation to an ordinary differential equation that can be solved by numerical methods [Sha53], [Hoo01], [Gro99]. In contrast, finite difference solutions do not explicitly describe the shock, because mathematically a shock is a discontinuity that cannot be pinpointed on the finite difference net. How precisely a shock can be located within a finite difference solution depends on the net density, and the denser the net, the more computationally intense the problem.

Originally, to aid in simulating the processes within the FCG armature during testing, the researchers obtained two hydrocodes for comparative use in the flux-compression generator research, TDL and CHARADE.

A Lagrangian finite-difference code, TDL (an acronym for two-dimensional Lagrangian) was compiled for use on IBM-PC computers and distributed by Mader Consulting Company [Mad98]. TDL has evolved into an advanced hydrocode developed at the Los Alamos National Laboratory that includes elastic-plastic flow, real viscosity, heat conduction, and gravity in its calculations.

The other hydrocode is called CHARADE, a Method of Characteristics - based on code also developed at Los Alamos. It was the only Method of Characteristics code the author was able to obtain, because public domain Method of Characteristics hydrocodes for detonation physics are no longer being maintained at the National Laboratories in the US or the UK. ([Mad01], [Ham01], [Joh01]). The code listing for CHARADE was presented in *CHARADE: A Characteristic Code for*

Calculating Rate-Dependent Shock-Wave Response (LA-11993-MS, 1991). Unfortunately, CHARADE had been written for use on mainframe computers, and as such, it was not suitable for desktop workstations.

TDL is capable of producing a model detailed and accurate enough for this research study, so porting CHARADE to Visual FORTRAN for use in desktop workstations was not necessary to this research. However, CHARADE will be of interest in future research of this sort, and effort will be required to make it useable on desktop workstations.

4.1.4 Research Postulates

Discovery of premature fracturing of armatures during explosive expansion tests prompted this investigation into the causes of the fracturing. Unfortunately, the fractures were occurring at much lower armature diameter expansion ratios than was expected. In theory, an armature should expand to twice its original diameter before breaking up [Kno70]. The immediate and easy assumption was that armature expansion due to explosive gas pressurization was the cause of early fracturing. There are two problems with this assumption, and each concerns timing. Chemical reactions during detonation of the explosive progress relatively slowly as compared to when the fractures occurred. This implies that gas expansion happens too late to cause the fractures, but the implication cannot be dismissed out of hand; it must be tested. Also, the strains due to explosive expansion probably do not become intense enough to fragment the armatures until much later in each test. This conjecture can also be tested.

So, if explosive expansion is not a probable cause of premature fracturing, what other, less obvious cause might there be? What other forces act on the armature during testing? Aside from atmospheric pressure, gravity, and explosive expansion, the only other forces acting on the armatures are shock loads from detonation of the explosive charge. So, strains within the armature that cause premature cracking might be shock-induced, as they are probably not due to explosive expansion, gravity, or atmospheric pressure.

How might these concepts about the cause(s) of premature cracking be tested? First, one must look at how the armature loading can be analyzed, using the engineering tools available. Afterwards, certain analysis tools may be used to confirm or deny the applicability of the concepts.

4.1.4.1 Helmholtz Free Energy

To understand what was causing the premature cracking of armatures during test firings, an attempt was made to go back to first principles to derive the state function of the armature expansion process. Regrettably, as will be shown, this was unsuccessful.

To derive the state function of the expansion process, one must examine the energy fluxes involved in armature expansion, so examination of the thermodynamic principles surrounding entropy was a good starting point.

Entropy is a measure of the disorder of a system; i.e. it is the amount of "useless energy" in the system. In the case of an explosive-armature system, molecular motion, dislocations within crystal structures, and fractures within the materials are the major contributors to entropy because the energy required to produce them cannot be utilized to do useful work [Cer66]. During detonation of the explosive part of the system, thermal stresses and stresses due to gas expansion also cause an increase in entropy. Moreover, as intense pressure discontinuities (shocks) pass through the metal structure after the explosive detonates the resulting dislocations, slippages, fractures, and crystal plane shears increase entropy.

The thermodynamic work function used by some explosives researchers [Coo96] which incorporates energy, temperature, and entropy is the Helmholtz free energy, and is defined by

$$F = E_n - TS , \qquad (4.12)$$

where E_n is energy, T is temperature, and S is entropy

When the system changes its thermodynamic state (during explosive expansion, for example), the change in Helmholtz free energy is given by

$$dF = dE - TdS - SdT \qquad (4.13)$$

The closed-form solution of the Helmholtz free energy equation is performed using a simplifying assumption that the system is at thermal equilibrium and that the reaction is taking place at a constant volume, so as to produce a minima in (4.13). In other words, the Helmholtz free energy is at its minimum at thermal equilibrium [Wei00].

Unfortunately, in the armature expansion problem the control volume is the armature itself, and the reaction within the control volume does not take place at constant volume; also, thermal equilibrium is not achieved within the original control volume. In addition, the solution to Eq. (4.13) considers only the starting and ending points of the detonation process. Lastly, the energy and entropy of the detonation products and of the remnants of the armature are not easily calculated. Therefore, the Helmholtz free energy method of analysis was not useful in analyzing the energy used to do work on the armature.

4.1.4.2 Strain Energy Method

Another, more complex method of analysis is to look at work done within the system by forces (shock forces, gas pressure, gravity, etc), so as to derive the strain energy within the system during expansion. Then, the strain energy is analyzed in terms of stresses. A comparison of stresses to material strength and toughness properties might then provide insight into the origin of premature longitudinal cracking in our armatures.

Strain energy is the work done by external forces in causing deformation within a body or a structure. In this case, the structure is the armature tube. The strain energy is stored in the structure until it fails, or until it releases the energy through noise, heat transfer, etc.

If an external force is directed onto a unit surface of the structure, d_y, it produces a stress, σ_x within the structure. The resulting strain, dU is

$$dU = \int_0^{E_x} \sigma_x d\left(\frac{\partial u}{\partial x} dx\right) dydz = \int_0^{E_x} \sigma_x dE_x (dxdydz), \qquad (4.14)$$

where $\frac{\partial u}{\partial x} = E_x$ and σ_x is normal stress.

The total stored energy, U, in a structure due to all external forces is

$$U = \int_V U_0 dV = \iiint_{xyz} U_0 dxdydz, \qquad (4.15)$$

where U_o is the total strain energy per unit volume and U is a nonlinear function of the load or of the deformation produced by the external forces. Therefore, because of its nonlinearity, the mechanical engineering Principle of Superposition is not valid. This means that a breakout of plain stress and shear stress (or hydrostatic and point loads) cannot be solved separately and then summed, if the purpose is to equate total stored energy to energy due to loading [Ugu79].

Once again, as with the Helmholtz free energy method of analysis, the basic strain energy method cannot be used to discover the source of work that is causing the premature armature cracking. The analytic method of simplifying the problem by solving for stresses individually and then combining the results cannot be used, so approximate numerical methods are the only ones available to evaluate the total energy stored in the armature. This is unfortunate, but it suggests another way of looking at the problem through use of the total strain energy per unit volume of the structure.

4.1.4.3 Total Strain Energy per Unit Volume Method

The total strain energy per unit volume due to normal forces is represented by

$$U_0 = \frac{1}{2} E \varepsilon_i^2, \qquad (4.16)$$

where E is Young's Modulus and ε is plane strain.
Since

$$E = \frac{\sigma_i}{\varepsilon_i} \text{ and } \varepsilon_i = \frac{\sigma_i}{E}, \qquad (4.17)$$

Eq. (4.16) can be converted to constitute stress only, through substitution. This results in

$$U_0 = \frac{1}{2} E \left(\frac{\sigma_i}{E}\right)^2 = \frac{1}{2} \frac{\sigma_i^2}{E} = \frac{1}{2E} \sigma_i^2 = \frac{1}{2E}\left(\sigma_x^2 + \sigma_y^2 + \sigma_z^2\right), \qquad (4.18)$$

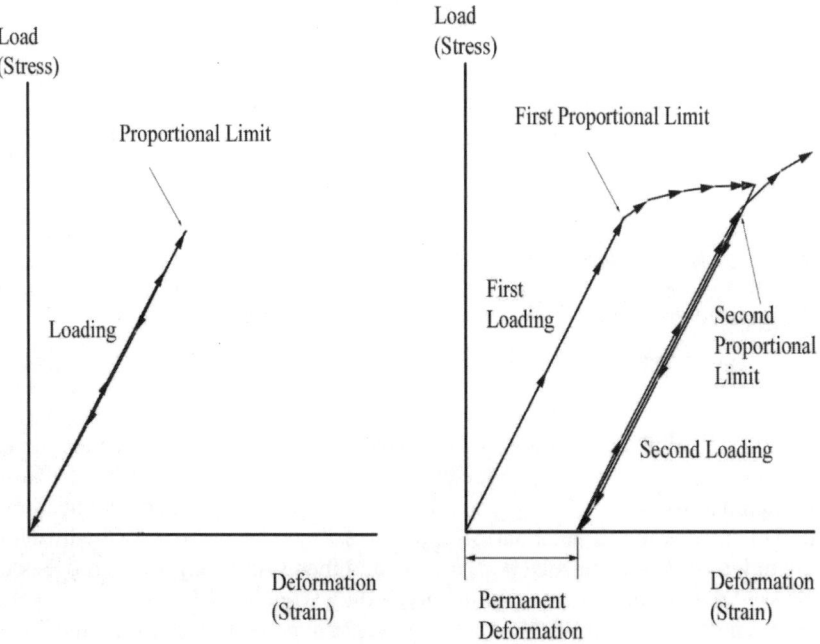

Fig. 4.5. Elastic deformation with strain energy recovery (left), and permanent set with strain energy lost (right). Arrows denote paths during loading and unloading.

Next, the effect of shear is accounted for. The total strain energy per unit volume due to shear forces is [Ugu79]

$$U_0 = \frac{1}{2G}\tau_{ij}^2 \text{ and } U_0 = \frac{1}{2G}\left(\tau_{xy}^2 + \tau_{yz}^2 + \tau_{xz}^2\right), \tag{4.19}$$

where G is shear modulus and τ is shear stress.

In strength of materials terminology, an elastic material is one that returns to its original, unloaded shape after external forces are removed. There is, however, in all elastic materials a point beyond which the material cannot be loaded and expected to return to its original shape. That point, the end of linear elasticity, is termed the material's proportional limit, see Fig. 4.5. In situations where the strain in the armature is below the proportional limit of its constituent metal, the strain energy stored in the metal is recoverable, and it is equal to the work done on the armature by external forces.

Therefore, to estimate if a set of forces is the cause of armature structural failure, one can use the following method: find the amount of external work done on the armature, equate external work to strain energy, calculate stress from strain energy and compare that amount against handbook data to see if the stress is high enough to fracture the metal.

What might be some other contributing factors to the premature cracking of armatures? The magnitude of the external load is not the only important factor in the failure of armatures during explosive testing. The fractures of the sort seen in photographs of our armature tests are identical to those seen in impact loading of ductile materials used in armatures. During impact loading of ductile materials, the enormous structural loading rate (producing strain rates $> 10^5$ cm/cm/second) causes brittle failure [Ugu79, Zuk92]. So, it is probable that premature, longitudinal cracking of armatures is some sort of brittle failure.

Interestingly, research performed by others has pointed out that this type of brittle failure of ductile material is primarily due to shear stresses rather than tensile stresses [Ugu79]. Ductile materials under impact loads yield due to slip (shear) along crystal planes. Therefore, any failure criteria used to determine if a set of external forces is the cause of structural failure should relate to shear rather than tensile stresses.

The octahedral shear stress theory of ductile failure typically provides the best match to experimental results in the failure of ductile materials, with $\tau_{yp} = 0.5770 \times \sigma_{yp}$ (The yp subscript denotes the yield point, typically at a 2% to 5% stress offset from the origin of the stress-strain plot in a uniaxial tension test of the material) [Ugu79]. As a check, this agrees favorably with handbook results for bulk properties of both the 6061-T6 aluminum and OFHC copper used as armature materials. For example, the tensile yield strength of 6061-T6 aluminum was reported to be 400 MPa, and the shear strength was reported to be 234 MPa, which results in a ratio of shear yield strength to tensile yield strength of 0.585 [Cer66].

There is another simplifying assumption that can be used when dealing with the structural properties of the armatures. Instead of considering individual metal crystals or grains, one may assume isotropy and homogeneity of the material properties. This is because of the large number of crystals and randomly arranged flaws (failure initiation loci) in the macro structure. Therefore, Hooke's Law can be used to evaluate stress/strain relationships within the macro isotropic and homogeneous structure of a thick-walled cylinder [Ugu79]. Using this assumption and Hooke's Law, the following are true:

$$\varepsilon_x = \frac{1}{E}\left[\sigma_x - v_P(\sigma_y + \sigma_z)\right]$$

$$\gamma_{xy} = \frac{\tau_{xy}}{G}$$

$$\varepsilon_y = \frac{1}{E}\left[\sigma_y - v_P(\sigma_x + \sigma_z)\right], \quad (4.20)$$

$$\gamma_{yz} = \frac{\tau_{yz}}{G}$$

$$\varepsilon_z = \frac{1}{E}\left[\sigma_z - v_P(\sigma_x + \sigma_y)\right]$$

$$\gamma_{xz} = \frac{\tau_{xz}}{G}$$

where v_P is Poisson's Ratio and γ is the shear strain.

The relationships in Eq. (4.20) will be used in a following section to estimate the loads that might cause armatures to fracture prematurely. Fractures of the sort seen in the armature photographs, if caused by stresses in the metal tube, are produced by tensile loads in the circumferential direction.

Temperature, as an indication of internal energy, is the most influential condition determining the point during loading history at which localized yielding in a structure becomes generalized yielding [Cer66]. At that point, material flaws coalesce throughout the structure into macro cracks. Did heat from explosion gas influence armature failure?

One might expect the armatures to gain internal energy due to contact with the hot gases resulting from the detonation of their explosive charges. As it turns out, they do not. In *Shaped Charge Temperature Measurement* [Von76], the temperatures of metals in contact with detonating charges were reported. There was no evidence that detonation products heated the metals. Similarly, our armature failure occurs over an extremely short time interval; too short for appreciable heat transfer from explosion gases to the metal, so elevated metal temperatures due to the detonation of the armatures' explosive charges did not affect armature failure.

Other effects that might influence armature cracking are stress distributions within the armatures as they are explosively expanded, as well as their material properties (whether soft or hard, brittle or tough, etc). These effects are analyzed in turn, and the results are presented, below.

Consider the stress in an armature cross-section (a ring, basically) due to shock wave passage and subsequent explosive pressurization. Since the armature thickness, 3 mm, was more than 10% of the armature inner radius, 16 mm, the armature is considered a thick-walled rather than a thin-walled cylinder for stress calculations. The 10% condition is a standard one in this type of investigation, and is due to the variation of stress through the wall thickness. The variation of stress through the wall thickness causes more than just a "membrane" stress, so a plane stress analysis cannot be used. Unfortunately, this means that the cylinder is in plane strain rather than plane stress, which is an important distinction for structural analysis. Membrane stresses in thin-walled cylinders allow analysis under simpler plane stress conditions than the analysis of thick-walled cylinders [Ugu79].

Therefore, the calculations are more complex, as they have to include the variation in stress through the thickness of the cylinder wall.

4.1.5 Testing the Postulates, Part 1

The idea that explosive expansion due to gas pressurization is the cause of the cracking can now be tested. After performing a plane strain analysis on the expansion process, the calculated results will be compared to experimental results.

If the loads developed by explosive expansion are enough to produce premature armature fracturing, and if the stresses are distributed in such a way as to produce the fracturing as seen in the photographs, and if the stress versus time history matches the onset of fracturing, then the culprit is explosive expansion.

First, some simplifying assumptions are made, which do not change the outcome of the analysis:
• the armature is assumed to be a thick-walled cylinder, so it was in a condition of plane strain.
• within a cross-section of the armature cylinder that is away from end effects (effects due to the open ends of the tube), it is assumed that there is uniform internal pressure due to explosive expansion.

Symmetry within such a cross-section dictates that shear stresses are zero ($\tau_{r\theta} = 0$, where r is in the radial direction and θ is in the circumferential, or hoop, direction), otherwise the cross-section will undergo rotation about one of its axes. This appears contrary to the concept that shear stresses cause brittle fractures; it must be remembered, however, that there is zero shear stress because the analysis is being applied to a non-rotating, cross-sectional surface of the armature. The stress derived from these computations must have the $\tau_{yp} = 0.5770 \times \sigma_{yp}$ reduction factor applied when failure of the armature structure is examined.

As far as the external forces are concerned, there is no loading along the long axis of the armature, as the cross-section is selected away from end effects. There are no rotational body forces, so the polar equations of equilibrium [Ugu79] reduce to

$$\frac{d\sigma_r}{dr} + \frac{\sigma_r - \sigma_\theta}{r} = 0, \qquad (4.21)$$

where r is the inner radius of the cross-section

Due to symmetry within the cross-section, there are no tangential displacements, so the strains are:

$$\varepsilon_r = \frac{du}{dr}$$
$$\varepsilon_\theta = \frac{u}{r} = \frac{(r+u)d\theta - rd\theta}{rd\theta}, \qquad (4.22)$$
$$\gamma_{r\theta} = 0$$

where u is the radial displacement due to pressurization of the cross-section. Since $u = r\, \varepsilon_\theta$, it follows that

$$r\frac{d\varepsilon_\theta}{dr} + \varepsilon_\theta - \varepsilon_r = 0. \qquad (4.23)$$

Following the Ugural and Fenster [Ugu79] treatment for internal pressurization of cylinders in plane strain, Lamé's equations reduce to

$$\sigma_r = \frac{a^2 p_i}{b^2 - a^2}\left(1 - \frac{b^2}{r^2}\right)$$
$$\sigma_\theta = \frac{a^2 p_i}{b^2 - a^2}\left(1 + \frac{b^2}{r^2}\right) \qquad (4.24)$$

$$u = \frac{a^2 p_i r}{E(b^2 - a^2)} \left[(1 - v_P) + (1 + v_P) \frac{b^2}{r^2} \right],$$

where p_i is internal pressure, a is the inner radius of the cross-section, b is its outer radius, and v_P is Poisson's Ratio.

Since

$$\frac{b^2}{r^2} \geq 1, \tag{4.25}$$

σ_r is compressive (the sign is negative) everywhere except for on the outer surface of the armature (at $b = r$), and $\sigma_r = 0$ at that location. In addition, the maximum compressive (negative) radial stress is at $r = a$ (at the inner surface of the armature tube), and σ_θ is tensile (positive) everywhere and is maximal at $r = a$.

Therefore, under this line of reasoning any stress developed by explosive pressurization loads which would produce the premature armature fracturing are tensile circumferential, or hoop, stresses that are at their greatest on the inside surface of the armature. So, utilizing this analysis, what is the actual stress magnitude in armatures under explosive expansion?

Using the theoretical Lamé's relationships for radial and circumferential stresses, see Eq. (4.24), and inserting the inner and outer radius values ($a = 16$ mm, $b = 19$ mm) for cross-sections of both our aluminum and copper armatures, results in

$$\sigma_r = 2.44 p_i \left(1 - \frac{361 mm^2}{r^2} \right).$$
$$\sigma_\theta = 2.44 p_i \left(1 + \frac{361 mm^2}{r^2} \right) \tag{4.26}$$

The Lamé's radial displacement relationship for a cross-section of a 6061 - T6 aluminum armature, with $E = 69$ GPa, and $n = 0.33$ [Mat01] becomes

$$u = 3.53 \times 10^{-11} p_i \left[0.67 r + \frac{480.13}{r} \right]. \tag{4.27}$$

The Lamé's radial displacement relationship, cf. Eq. (4.24), for a cross-section of an OFHC copper armature, with $E = 117$ GPa, and $n = 0.31$ [Mat01] becomes

$$u = 2.08 \times 10^{-11} p_i \left[0.69 r + \frac{472.19}{r} \right]. \tag{4.28}$$

It now remains to obtain p_i for use in Eqs. (4.27) and (4.28), and to decide upon a radius within the armature wall at which flaws coalesce to initiate cracks. A theoretical quantity called CJ pressure will be used as p_i for this calculation.

4.1.5.1 Determination of Internal Pressure, p_i

In detonation theory, the detonation region is broken into several sections: a leading shock immediately in front of the reaction zone, the reaction zone, the Chapman-Jouguet (CJ) plane at the immediate rear of the reaction zone, and the Taylor wave which makes up the expansion flow following the CJ plane. The Taylor wave is a rarefaction, and pressure within it is lower than that in the undisturbed explosive. Theoretically, the greatest pressure in the detonation region is at the very front of the shock, and the pressure decreases continuously across the reaction zone until thermodynamic equilibrium is reached at the CJ plane [Coo58]. Since there are two similar, but slightly different main charges used in this research, different CJ pressures are experienced by our armatures when they are tested.

Published data for the explosive used in our early experiments, military standard C-4, showed the CJ pressure to be 25.7 GPa, as calculated by the TIGER hydrodynamic-thermodynamic computer code [Dob85] when the explosive was at its theoretical maximum density (TMD) of 1.67 g/cm^3. Our later shots utilized C-4 obtained from Accurate Arms, Inc. This explosive was produced as the precursor for a molding process, and its density was less than that of military C-4 because it was a crumbly solid instead of a firm, plastic brick (the usual configuration of military C-4). In addition, the plasticizer used by Accurate is slightly different than the plasticizer used in the handbook C-4 (Accurate uses dioctyl adiapate, versus di [2-ethylhexyl] sebacate in the military explosive).

Density is an important factor in the CJ calculation; the higher the density, the greater the CJ pressure. Our armatures are loaded by hand, with small portions of the explosive charge rolled into balls and then rammed into place in the armatures with a wooden dowel. The dowel is rotated slightly upon contact with the explosive, to create a shearing force in addition to compression. According to an Accurate Arms company representative [Ric01], the highest density for C-4 that Accurate achieves when it molds the precursor explosive into linear shaped charges is 1.60 g/cm3. A check on the density of explosives within our FCG armatures loaded with military standard C-4, and of those loaded with Accurate C-4, resulted in identical average densities (within the accuracy limits of the density measurements). This check was done by water-displacement density calculations, and based on three explosive samples cut from each hand-loaded armature. The calculations showed an average density of 1.58 g/cm^3 for both types of C-4. Therefore, the maximum explosive density obtainable in this manner was a function of the size of the explosive balls and of the loading method consistency, and it was not a function of the relative firmness of the portions of explosive being packed into the armatures.

The CHEETAH hydrodynamic-thermodynamic computer code [Fri98] is an improvement over the TIGER code that was used at the Lawrence Livermore National Laboratory to generate the military C-4 explosive performance figures cited above. We ran CHEETAH to check on the performance of both types of C-4 in the armatures, with inputs values incorporating a range of explosive densities from 1.58 g/cm^3 to the CHEETAH-derived TMD for Accurate C-4 of 1.6509 g/cm^3. At a density of 1.58 g/cm³, the CHEETAH-derived CJ pressure is 22.36 GPa, and at TMD the CJ pressure is 24.91 GPa. The corresponding CJ pressures for military standard C-4 are 22.55 GPa and 25.09 GPa, respectively.

Another method of detonation pressure derivation is the one proposed by Cooper, which in this case gave a p_i (recall that CJ pressure is being used as p_i) 10% higher than CHEETAH (24.81 GPa for Cooper's method, using a detonation velocity of 7.565 mm/μsec) [Coo96]. This 10% difference may be a bit disconcerting, but published measurements of CJ pressures from experimental data are known to have a 10% error overall, and published values of CJ pressures based on a JWL equation of state (such as those in the Lawrence Livermore National Laboratory Explosives Handbook) are even more uncertain because they are extrapolated from data at lower pressures [Fri98]. Therefore, the 10% difference is within the expected.

In the interest of conservatism, the pressure on the internal surface of the armature tube (p_i) is taken to be the lowest of the calculated pressures, which is the CHEETAH-derived Accurate Arms C-4 pressure of 22.36 GPa at a density of 1.58 g/cm³.

Accordingly, using the CJ pressure of 22.36 GPa in Eqs. (4.27) and (4.28) produces radial displacements (at the inner surfaces of armature cross-sections) of 32.15 mm for the aluminum armature, and 18.88 mm for the copper. So, in the amount of time it takes for the CJ pressure to develop within a cross-section (on the order of a few microseconds), the ring of armature wall material is displaced to a radius of more than twice the original radius (for copper), or slightly more than three times the original radius (for aluminum). Even though the armatures are made of ductile metals, dynamic loading at these extreme rates definitely causes brittle failure in the armatures.

Now that we have an idea of the amount of armature expansion produced by detonation pressurization, the resultant stresses and their effects on the armature tube will be examined.

4.1.5.2 Cracking Due to Pressurization

Since the maximum tensile circumferential stress is at the armature's inside surface, the stress magnitude due to detonation pressure will be determined on the inside surface.

According to Eq. (4.26), the circumferential stress during explosive expansion in an armature cross-section is greater than the radial stress. This is expected, as reason shows that circumferential and not radial stress is the stress that creates longitudinal cracks in the armature tubes. Using the same p_i in Eq. (4.26) for circumferential stress as was used for radial displacement in Eqs. (4.27) and (4.28) results in a value of = 131.5 GPa at the inner surface for both aluminum and copper armatures. This stress, according to the octahedral shear stress theory, should be reduced to $\tau_{yp} = 0.5770 \times \sigma_{yp}$, or 75.9 GPa for the armature as a structure.

How do these stresses affect the armature structure? Examination of the handbook material properties for these two metals, and of the armature expansion photographs, reveals that this sort of stress does not cause premature longitudinal cracking in the our armatures. There are several reasons for this statement.

In the first place, the ultimate tensile strength of OFHC copper (soft temper) is 235 MPa, and the same property of 6061-T6 aluminum is 310 MPa [Mat01]. Each of these is at least two orders of magnitude less than the circumferential tensile stress generated by the expansion of explosive gases, 75.9 GPa. Therefore, if the cracking seen in the armature photographs is due to explosive pressure, the cracking should initiate at the inner surface of the armature. It does not; the longitudinal cracks are external. They do not begin as cracks on the inner surface of the armature tube, because there is no photographic evidence of detonation products escaping through them, see Fig. 4.6 and Fig. 4.7; the black smoke seen escaping through cracks in the armature in Fig. 4.7 is characteristic of RDX-based explosives such as C-4, and the bright line along the armature is flash lamp reflection). The cracks originate on the external surfaces of the armatures, where the circumferential stress due to explosive pressurization is the lowest. So, since the cracks do not originate on the internal surfaces of the armatures, where the stress due to explosive expansion is the highest, they are not a result of explosive expansion. (N. B. - It is reasonable to expect the longitudinal cracking in the armatures to initiate slightly below the outer skin of the armature rather than on the surface itself, as the finished outer surface has fewer microflaws.)

Secondly, for explosive expansion to be the culprit the timing is off, as the cracking begins prior to explosive expansion. Proof of this statement follows.

4.1.6 Testing the Postulates, Part 2

If premature longitudinal cracking of the armatures is not caused by the rapid, violent expansion of the armatures due to explosive expansion, what other cause might there be? The only other forces acting on the armatures are gravity and hydrodynamic forces that result from the passage of shock waves through the armatures when the explosive charge is detonated.

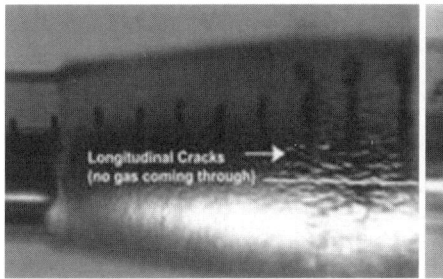 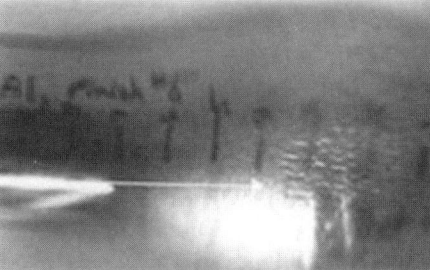

Fig. 4.6. Longitudinal cracking (left - copper, right – aluminum). Absence of gas escaping from cracks shows that they have not penetrated into the inner volume of the armature.

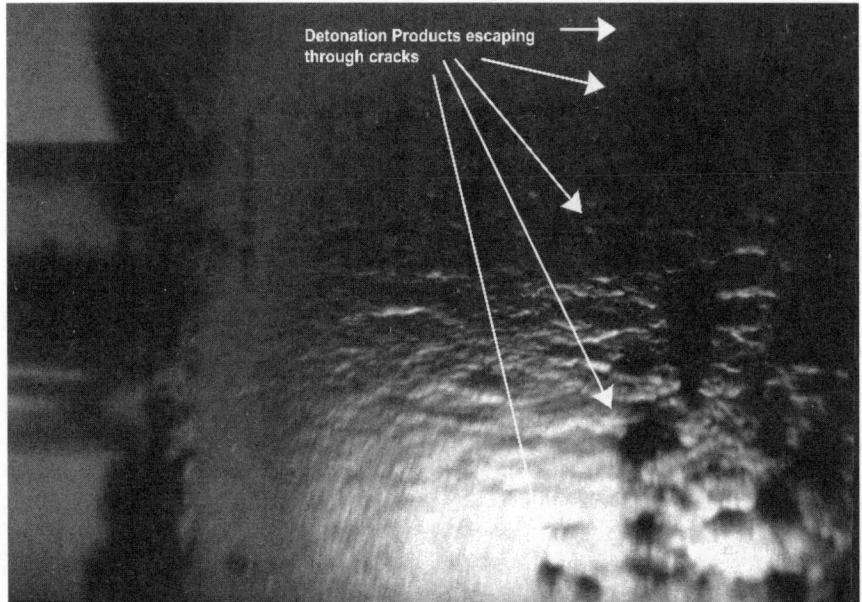

Fig. 4.7. Products of C-4 detonation escaping through cracks in an OFHC copper armature.

Shock theory and timing can be used to see if there is any correlation with the premature fracturing. If timing shows that shocks are in the metal at the time the cracking took place, geometric optics can be utilized as a first-order estimation of how shocks might propagate through the explosive-armature. In addition, the effect of possible Mach stem formation in the detonation products along the explosive-to-metal interface within the armature can also examined.

Clearly, if the result of an application of geometric optics and Mach stem effects to the situation is promising, the next logical step is to employ a detonation hydrocode to analytically study the cracking problem in detail.

4.1.6.1 Shock Waves and Timing

The timing of shock waves within the armatures is examined first. It is necessary to explain some of the details of the test set-up, and of the devices tested, which bear directly on the timing issue. An explosive-driven flux compression generator armature is a hollow metal tube, packed with C-4 high explosive. We initiated the explosive through the use of a single Reynolds RP-501 or RP-80 exploding bridge wire (EBW) detonator, which is located in the center of one of the ends of the armature. The detonator is positioned using a detonator locator and armature end cap. The particular detonator used in the test affects only the initiation timing; timing of the more expensive RP-80 has a tighter standard deviation, but is the same diameter (7.5 mm) and length (21 mm) as the RP-501.

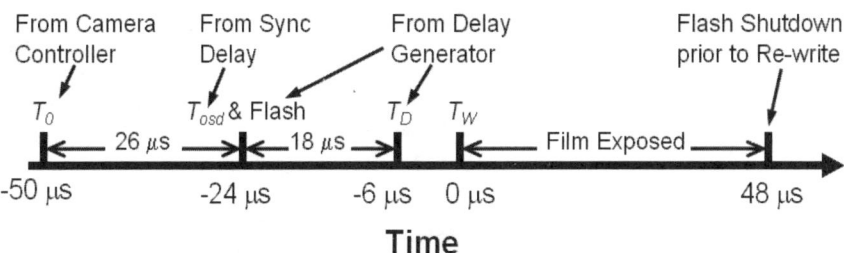

Fig. 4.8. Test photography timeline.

The explosives-loaded armatures are tested inside a detonation tank, using a Cordin rotating-mirror, high-speed camera to photograph the explosive expansion of the armature through a window portal. An array of xenon flash tubes, specially designed and built for this project, is placed in the tank to illuminate the armature with sufficient light for high-speed photography, see Fig. 4.8 for a time-line of the test events. The flash array is fired 24 μsec before the first photographic image is recorded, as it takes at least that long for the light from the tubes to reach its full intensity. In addition, the light from the array has to remain at full intensity for about 100 μsec to allow for full exposure of all of the frames of film. Because the light from the flash system also serves as the camera shutter, it has to be shut-off within 200 μsec of when it first began, in order to prevent double-exposures ("re-write"). The FCG armature photographs were taken at the rate of 1 frame every 2 μsec (in other words, at half a million frames per second).

For these tests, the camera controller sent a master timing pulse, T_0, to the system at 50 μsec prior to the desired camera writing start time. The synchronizer delay was set to 26.0 μsec, so that the synchronizer delay pulse, T_{osd}, was sent to the delay generator at $T_0 - 26$ μsec = 24 μsec. Within the delay generator, the pulse timing for the flash array was set to 0 μsec and the timing for the detonator was set to 18.0 μsec, so that the flash pulse would fire at T_{osd} - 0 μsec = 24 μsec prior to the desired writing start time, T_w. In the same way, the detonator pulse, T_D, would fire at T_{osd} - 18 μsec = 6 μsec prior to T_w. The photography sequence ran in the following manner: when the rotating mirror in the Cordin camera was at full speed, a master timing pulse was sent out from the controller. This pulse caused the flash system to begin firing 24 μsec and the detonator to begin its firing process 6 μsec before the first frame of film was exposed. The film was exposed at the rate of one frame every 2 μsec for 24 to 26 frames, and the flash was shut-down before the mirror rotated back into position to re-write the first frame (150 μsec later).

Given this timing information as a starting point, the next task was to determine when the detonation wave from the detonator initiated the main charge reaction. Fig. 4.9 is a digital photo taken of an RP-501 detonator, 5 μsec after the initiation pulse was applied to the detonator. The 60 nsec exposure, obtained through a gated image intensifier camera, clearly shows the input end of the detonator at the top, and the end with the explosive charge at the bottom.

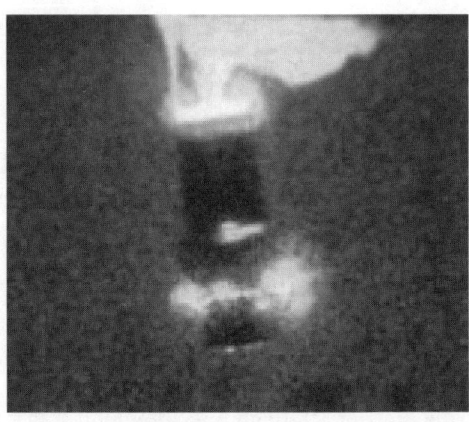

Fig. 4.9. RP-501 detonator, before and 5 µs after an electrical pulse was applied to the exploding bridge wire detonator EBW, which is ~ 7.5 mm in diameter and ~ 21 mm in length (50 ns exposure time).

The initiation process causes the detonator casing to open-up as the hot gases start to expand, and the plastic molded plug in the head end of the detonator (top of photo) is fluorescing due to shock energy from the detonation.

The manufacturer's specification for the RP-501 states a typical function time of 2.8 µsec with a standard deviation of 0.5 µsec, and the timing of the event depicted in Fig. 4.9 demonstrates that the detonator will begin to produce a detonation wave in the main charge explosive at about 4 µsec after T_D. Therefore, it is reasonable to assume that in the armature tests the detonation process in the main charge began 1 to 2 µsec before the Cordin camera started writing its first frame (at T_w).

Figures 4.10 through 4.13 are taken from the Cordin camera photographs of a test firing of an explosives-loaded 6061-T6 aluminum armature with a smooth, polished surface finish. The explosives were only loaded in about half of the length of the armature (to about 9 cm of a 15 cm-long armature) in each of the tests studied, to avoid frame re-write due to detonation product afterburning. Fig. 4.10 shows the armature as it appeared in frame 3, exposed over the time period 4 to 6 µsec after T_w. The short, black circumferential marks are 1 cm divisions that begin from the detonator end of the armature. These reference marks and the writing on the armature were made with a permanent ink marker, and their presence did not affect the performance or the expansion of the armature. Note the luminous material being ejected from the right-hand (detonator) side of the armature. Unfortunately, the film developer cut-off a portion of frame four, see Fig. 4.11, but there is enough to see that longitudinal cracking had started while this frame was being exposed during the time period from 6 to 8 µsec after T_w. Therefore, the cracks begin at about 8 µsec after detonation starts in the main charge. The photograph in Fig. 4.12 is frame 5 of the series, an exposure beginning 8 µsec and ending 10 µsec after T_w. The longitudinal cracks are more evident, and extend between about 2 cm and 3.5 cm from the detonator end of the armature. Note also that the armature is expanding with a bell shape, rather than the conical shape desired for a flux-compression generator, and that the apex of the bell is at about the 3 cm mark. Fig. 4.13 clearly shows the longitudinal cracking, on an exposure which spanned the 10 to 12 μsec time period after T_w. This series of events and their timing are typical of each of our FCG armature shots, irrespective of the surface finish (as received, polished, or purposely roughened), temper, and armature material (aluminum or copper). Figures 4.14 through 4.16 depict part of the armature expansion sequence for an OFHC copper armature; again, the bright line along the armature is flash lamp reflection. The photograph in Fig. 4.14 was the fourth frame of the series, again spanning 6 to 8 μsec after T_w, and the cracking is just beginning to appear. The cracking is more evident in Fig. 4.15 (frame 5), and is very evident in Fig. 4.16 (frame 7, exposed during the 14 to 16 μsec time period after T_w).

The question is, where should the detonation wave first appear at the armature surface, and when? Assuming that the detonation front moves radially outward from the end of the EBW detonator, the wave should emerge about 2 to 3 cm from the detonator end of the armature. Determining when the wave front should appear requires knowing its velocity. According to the Cheetah run for Accurate Arms C-4, the CJ velocity at TMD is 7.818 mm/μsec, as compared to the detonation velocity at TMD of 8.37 mm/μsec [Dob85].

Fig. 4.10. Expanding aluminum armature, 4 to 6 μs into camera write sequence. Detonator initiation point is located approximately 2 cm from the right armature end. Armature diameter is 38.1 mm and wall thickness is 3 mm.

Fig. 4.11. Expanding aluminum armature, 6 to 8 μs into camera write sequence.

Fig. 4.12. Expanding aluminum armature, 8 to 10 μs into camera write sequence.

Fig. 4.13. Expanding aluminum armature, 10 to 12 μs into camera write sequence.

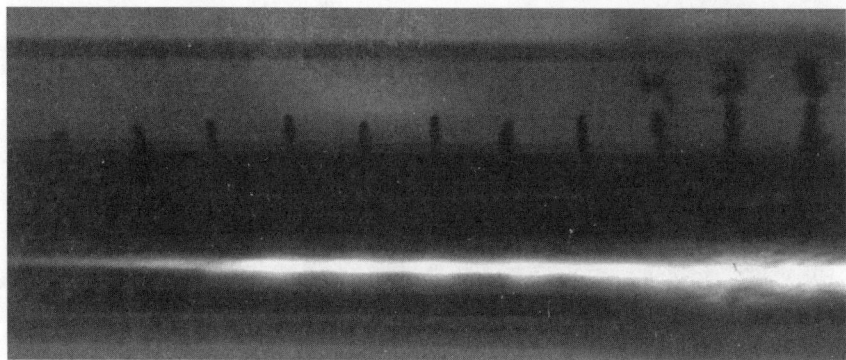

Fig. 4.14. Expanding OFHC copper armature, 6 to 8 μs into camera write sequence. Detonator initiation point is located approximately 2 cm from the right armature end. Armature diameter is 38.1 mm and wall thickness is 3 mm.

4.1 Armature Dynamics 77

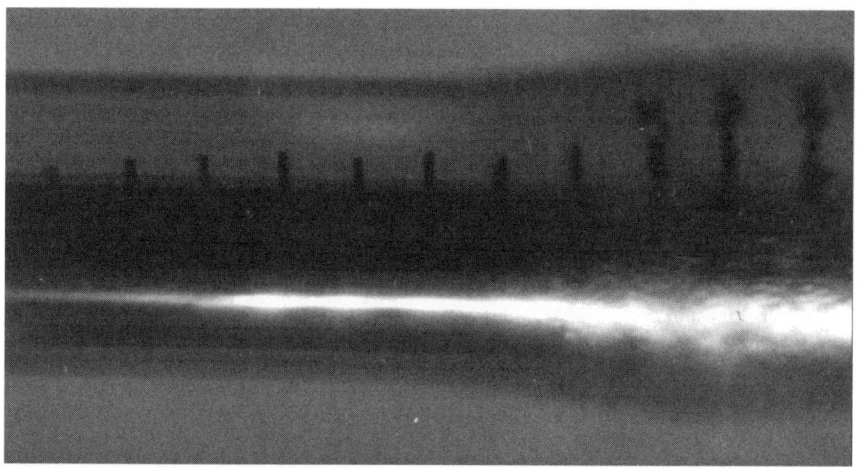

Fig. 4.15. Expanding OFHC copper armature, 8 to 10 microsec into camera write sequence.

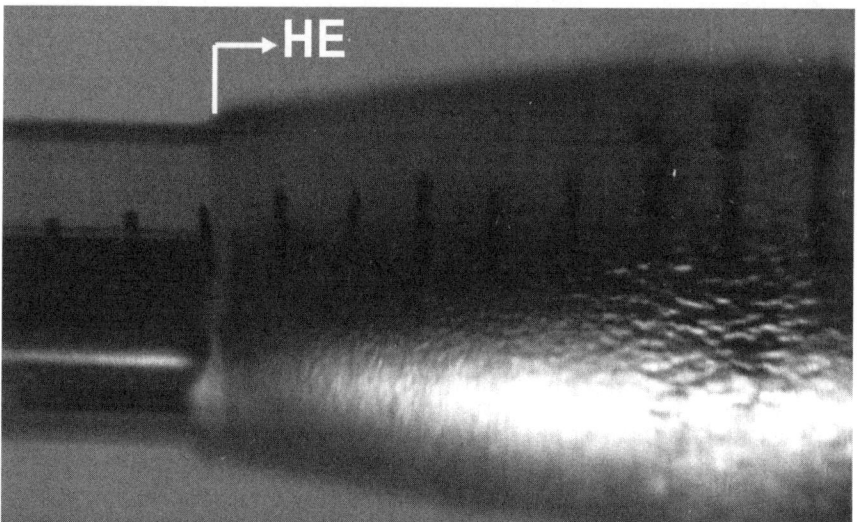

Fig. 4.16. Expanding OFHC copper armature, 14 to 16 µs into camera write sequence. As indicated, only the right half of the armature was filled with HE.

Assuming that the ratio of these two velocities (0.934) holds for densities different than TMD, then the detonation velocity at a C-4 density of 1.58 g/cm^3 would be

$$\frac{7.565 \text{ mm}/\mu s}{0.934} = 8.10 \text{ mm}/\mu s. \tag{4.29}$$

This compares favorably with the detonation velocity computed by referring to the Cordin photographs, which is about 8 mm/µsec. The acoustic velocity of 6061-T6 aluminum is ~ 5.24 mm/μsec and the distance that the detonation wave has to travel, see Fig. 4.17, is between 12.25 and 13.23 mm in the explosive, and 3 mm in the aluminum. Therefore, the total wave travel time can be estimated as

$$\left(\frac{12.25\,\text{mm}}{8.1\,\text{mm}/\mu s}\right) + \frac{3\,\text{mm}}{5.24\,\text{mm}/\mu s} =$$

$$\text{or} \quad \left(\frac{13.23\,\text{mm}}{8.1\,\text{mm}/\mu s}\right) + \frac{3\,\text{mm}}{5.24\,\text{mm}/\mu s} =$$

(4.30)

$$2.08\ldots2.21\,\mu s \ ,$$

and the shock wave appears on the surface of the armature at about the middle of the first frame in the series of photographs, or two frames before Fig. 4.10.

Up to this point, the wave being examined is the detonation wave front from the detonation of main charge explosive, and during its transit of the explosive it is a compressive wave. As it passes into the armature tube, it is moving into a higher density medium, so some of it is reflected back as a compressive wave into the explosive charge/detonation products, and some of it is transmitted as a compressive wave into the metal. This initial shock wave results in a pressure rise on the order of 25 GPa (the CJ pressure) in the metal. It has been shown that for such pressures the entropy change across the shock is negligible [Hos65]. Therefore, even if the shock strength were to decay along its path, the hydrodynamic "flow" in the metal behind the shock is nearly isentropic, so the velocity of subsequent waves is unaffected by the leading shock passage.

When the wave reaches the outside of the armature tube it experiences a tremendous drop in density because it is at the interface between the metal of the armature and the surrounding atmosphere. Accordingly, a little of the energy is transmitted to the surrounding gas, but most of it is reflected as a rarefaction (a Taylor expansion fan, or wave). The passage of this reflected rarefaction back through the metal of the armature puts the metal into tension. This rarefaction travels through the tube until it meets the original interface between the explosive charge and the tube, which is now an interface between metal and gas (detonation products). So, the rarefaction is next reflected as a compression shock from the inner surface of the armature tube, 3 mm / 5.24 mm/μs, or about 0.5 μsec after the first shock reflection. Each reflection of a shock from within the metal results in additional shocks and rarefactions within the metal, at intervals of about 0.5 μsec, until the metal begins to expand and break-up.

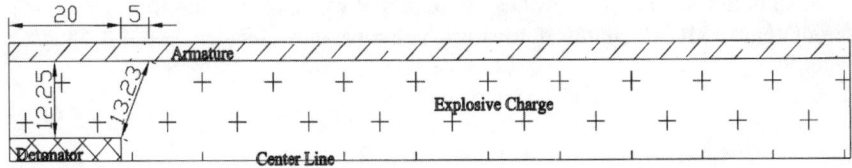

Fig. 4.17. FCG armature, layout, dimensions in mm

4.1 Armature Dynamics   79

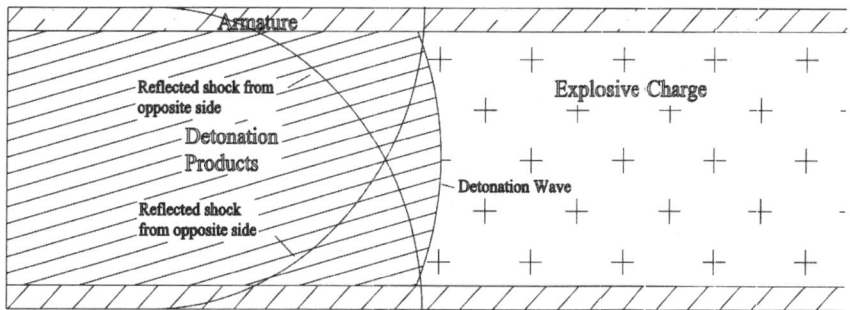

Fig. 4.18. Shock waves within an armature during firing

As can be seen in Fig. 4.18, the next compressive wave from the explosive detonation to appear at this point on the surface of the armature is the detonation wave reflection from the explosive-metal interface on the opposite side of the charge.

If the armature has not started to expand, the wave travels 12.25 mm through unreacted explosive, 32.38 mm through the explosive reaction products, and 3 mm through the metal of the armature tube. Its total travel time is

$$\frac{12.25\,\text{mm}}{8.1\,\text{mm/}\mu\text{m}} + \frac{32.38\,\text{mm}}{7.565\,\text{mm/}\mu\text{m}} + \frac{3\,\text{mm}}{5.24\,\text{mm/}\mu\text{m}} = 6.37\,\mu\text{s} \quad (4.31)$$

This corresponds to a shock wave arrival after exposure of the third frame had begun. The reflected rarefaction from the outer surface of the armature caused by this shock wave has an additive effect when considered with rarefactions already in the metal. According to this time analysis, the hydrostatic stresses within the metal fluctuate between high compressive and high tensile magnitudes at a frequency greater than 10^6 cycles per second for several microseconds.

So, both shock theory and timing show a correlation between the time longitudinal cracking appeared, and the time at which the hydrostatic stresses due to shock effects were present in the armature metal. Shock wave shape and direction are the next pieces of the puzzle to be found. The concept of geometric optics, combined with low-cycle fracture mechanics and other insights already developed as to the dynamic stress field within the armature tube holds clues necessary for understanding shock effects within the armature.

4.1.6.2 Longitudinal Cracking

As the explosive charge detonation progressed along its length, the armature tube began expanding and longitudinal cracks appeared on the expanding surface of the tube. This is to be expected if the cracking is caused by strains due to gas pressurization from the detonation. Of importance, though, is that the cracking began as the tube expanded to only about 1.15 times its original diameter; much less than the factor of two predicted in magnetic flux-compression generator theory [Kno70].

Logically, if pressurization of the tube by explosive detonation products and the tube's subsequent expansion is the cause of these longitudinal surface cracks, the cracks should grow and extend along the length of the tube as the tube continues its expansion. Examination of the photograph in Fig. 4.16, however, reveals that the armature tube continued to expand past the 1.5 expansion ratio without the cracks extending farther down the tube. In fact, the cracking is evident up to midway between the 4 and 5 cm marks on the armature in Fig. 4.15, but in Fig. 4.16 the equivalent expansion location (at about 9 cm on the armature) still appears smooth. The photograph in Fig. 4.16 clearly illustrates that whatever process is behind the armature tube cracking does not affect the remainder of the armature tube. Why was the cracking restricted to the first two diameters of the armature tube length, and why did the tube not begin to crack at other places farther along on the tube surface as the detonation progressed along the explosive charge?

4.1.6.3 Optical Properties of Detonation Waves

In two-dimensional charges of limited radius, there is a region where acoustic waves exhibit behavior similar to that of light waves moving in transparent media. In an explosive, a detonation wave with a single initiation point is spherical within this "optical" behavior region, and when the wave meets an inert homogeneous material within the region it refracts according to the laws of geometric optics established by Descartes, Fermat, and Huygens [Bus70].

For example, in a given cylindrical charge of radius r_i with a single detonation initiation point, I, the detonation wave front will theoretically be spherical within a region surrounding I, according to the application of geometric optics to explosive events, see Fig. 4.19.

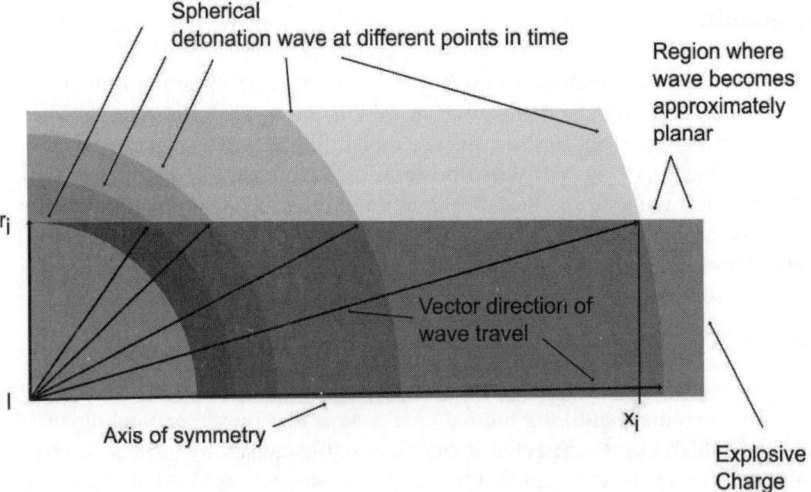

Fig. 4.19. Geometric optics applied to the detonation wave in a cylindrical charge of radius r

The radius of curvature of a spherical detonation wave segment generated at I increases linearly. When the wave reaches a point with the distance x_i from I, the radius of curvature of the wave front no longer increases; it assumes a steady-state value. The wave then advances along the length of the charge by translation. According to theory, the value of x_i is generally confined to $4r_i \leq x_i \leq 7r_i$ and depends on the explosive type [Bus70].

According to the optical behavior concept, in our armature tests, the detonation wave front within the explosive would have expanded spherically at the detonation velocity, until it reached a point somewhere between 6.4 and 11.2 cm from the initiation point. The real region of spherical expansion would have been slightly less, but there definitely was a short region beyond the detonator where the detonation wave front would have impinged directly on the inner surface of the armature tube. As the wave traveled down the tube, the angle of impingement would have then decreased to about 10° within about two charge diameters of the detonator. A closer look at the detonation wave and its associated pressure field as it traveled along the explosive charge helps explaining how this happened.

In Fig. 4.20, the detonation wave is represented by a spherical wave of increasing radius of curvature, traveling along a boundary between relatively lower-density explosive and higher density inert material (such as copper in our flux-compression generator). Each point on the detonation wave travels along a vector perpendicular to the tangent to the wave at that point, as long as the wave is traveling through a homogeneous medium. The explosive is considered homogeneous at the macro level in this example; in reality, density variations in real explosives will alter the wave travel, depending on the amount of variation and the distribution of the variations. Where the detonation wave meets the boundary, part of it is reflected as a compressive wave, and the portion of the wave transmitted into the higher density material is refracted as a compressive wave. The reflection and refraction angles depend on the detonation wave angle of incidence, as well as on the ratio of acoustic velocities between the inert and explosive media (see Fig. 4.20). According to geometric optics, this relationship is embodied in two mathematical formulae [Whi68]

$$\begin{aligned} i &= a \\ \sin r &= n \sin i \end{aligned}, \qquad (4.32)$$

where i is the angle of incidence, a is the angle of reflection, r is the angle of refraction, and n is the ratio of the velocity in the incident medium to the velocity in the refraction medium.

As the incidence angle of the detonation wave decreases, the angle of its reflection also decreases. It approaches zero as the incidence angle approaches zero. Also, at an incidence angle of 0° the angle of refraction is 0° (N.B. In Fig. 4.21, the angles of incidence and of refraction are angles measured from the normal to the interface of the materials, but the angle of incidence is equal to the angle between the interface and the tangent line to the wave at the intersection). As the angle of incidence increases toward 90°, the angle of refraction increases but does not reach 90°; it is limited by the ratio n, cf. Table 4.1).

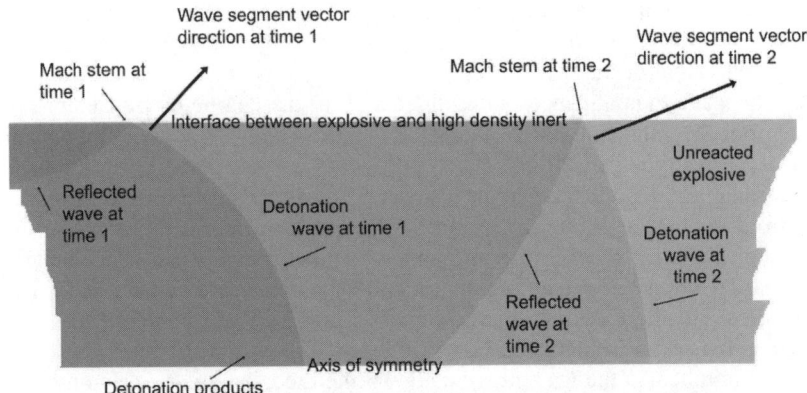

Fig. 4.20. The detonation wave and its reflection from the boundary between explosive and confinement.

In brief, the detonation wave becomes planar as it moves down the armature, and its transmitted shock is refracted into the armature tube at an angle of up to 30°, depending on the type of metal.

4.1.6.4 Mach Stem Effects

Previously, the question was raised as to whether the formation of Mach stems would affect armature cracking. Before that question can be answered, it must be determined whether or not Mach stems are formed in the detonation products within the armature.

Recall that the concept of geometric optics demonstrates the tensile and compressive waves generated within the armature tube tend to travel in directions normal to the tube surfaces, in the volume surrounding the detonator. As the detonation wave and associated rarefactions travel down the length of the explosive charge, their impingement angle on the inner surface of the armature tube continuously decreases from 90° (i.e., the wave travel is normal to the surface) to near 0°. The detonation wave becomes essentially planar in its travel down the length of the armature. According to Cook's detonation head theory, the angle achieves 0° after traveling about two charge diameters (approximately 7.6 cm) [Coo55]; in reality, some wave curvature always remains at the interface between the detonation products and the metal.

As illustrated in Fig. 4.20, the combination of incident and reflected detonation waves within the armature can form Mach stems. By their nature, Mach stems alter the pressure pattern within the region adjacent to the interface between the explosive and the armature tube. Mach stems form when the angle of incidence, i, reaches between 45° and 49°, see Fig. 4.22 [Lam65]. This is due to gas dynamics: the fluid flow along the interface must be parallel to the interface, and below i values of about 45°, the dynamics of wave reflection allow the flow to be parallel. Under this condition, the point of reflection moves with the point of intersection of the detonation wave and the interface, in what is termed a regular reflection. Beyond i values of about 49°, flow behind the reflected wave can no longer be parallel to the interface, and a normal shock forms between the interface and the intersection of the incident and reflected waves. The intersection of the three shocks is the triple point of the Mach stem. Refer to Fig. 4.2 for a sketch of the Mach stem geometry.

During armature expansion, the triple point has a curved trajectory which pulls away from the interface. The angle between the trajectory and the interface between the explosive and the armature tube begins at 0° for i values of approximately 45°, increases to about 5° for i values of approximately 80°, and drops back to 0° as i approaches 90°, see Fig. 4.23 [Mel93, Dre93, and Sha53].

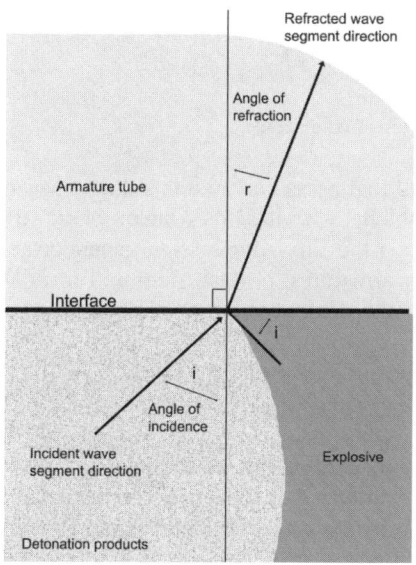

Fig. 4.21. Detonation wave refraction.

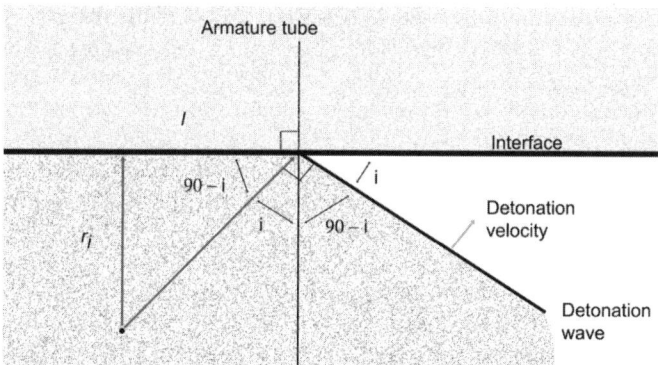

Fig. 4.22. Regular reflection.

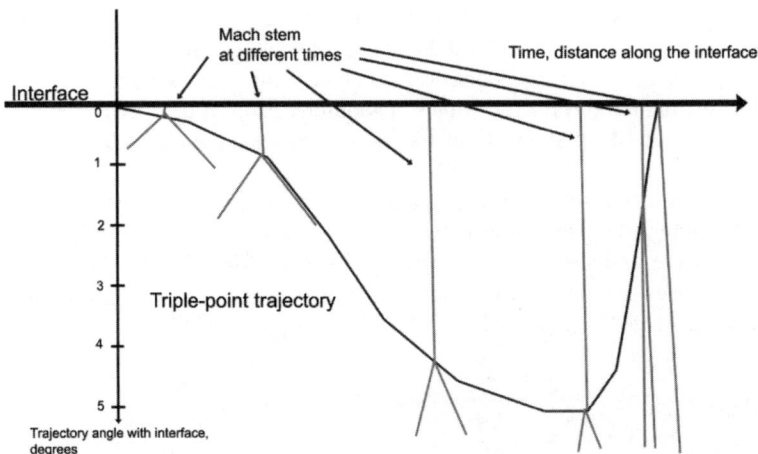

Fig. 4.23. Mach stem triple point trajectory with time and distance.

Recall from the discussion in Chap. 4.1.2 that pressure variations due to Mach stems were believed to be the cause of problems with the Los Alamos Mark 101 armature. Do Mach stems have any effect on the flux-compression generator armatures? According to the geometry of our armatures, a Mach stem will form at about 16 mm beyond the detonation point (the detonation point is about 21 mm into the armature):

$$l = r_i \tan(90° - i), \qquad (4.33)$$

given that $41° \leq 90° - i \leq 45°$ for $r_i = 16$ mm and 13.9 mm $\leq l \leq$ 16 mm.

In accordance with this analysis, the detonation wave reflection is regular until about 3.7 cm (16 mm + 21 mm) down the length of the armature tube, beyond which point the wave reflection becomes a Mach stem. The Mach stem forms and gradually pulls away from the interface as the detonation wave and its reflection progress down the length of the armature. (N. B. - The Mark 101 armature, in contrast, was initiated at several points along its longitudinal axis, and the detonators were not fired simultaneously, so a simple model of its Mach stem structure is not possible.)

In our armature Mach stems are not present along the interface between the explosive and the armature tube until at least 3.7 cm along the tube, so they appear to have no direct role in the formation of longitudinal cracking of interest to this investigation. They may play an indirect role, by removing the incident shock from contact with the interface so that the shock is no longer refracted into the tube. There is one caveat; Cook's analysis did not take the formation of a Mach stem into account. Note that the normal shock part of the Mach stem forms and attaches itself to the interface sooner (at about 3.7 cm, or about one charge diameter) than Cook's estimate of where the detonation wave becomes planar. Therefore, the appearance of a Mach stem in the explosive charge modifies the analysis performed using Cook's theory. As a result, travel of the detonation wave down the armature becomes nearly planar at a point closer to the detonator than would be predicted using Cook's theory. This brings Cook's theory into agreement with our armature photographic evidence.

4.1.6.5 Hydrocode Analysis

At this stage, all that is lacking is a detailed view of the regions of compressive and tensile stress within the armature tube metal. Such information will not only verify the cause of longitudinal cracking, but it also will show whether design changes in the armature could diminish or stop the cracking. To this end, gaining insight into the very dynamic and complex situation occurring within the armature tube as it is being subjected to the passage of shocks and rarefactions is nearly impossible without resorting to the use of a hydrocode model.

In previous sections, structural analysis showed that gas pressurization could not have caused the longitudinal cracking in armatures being tested. In addition, the point was made through shock wave timing, geometric optics, and Mach stem trajectory analysis that shock waves and associated rarefactions are the likely cause of premature longitudinal cracking in those armatures. Because there is no experimental method that directly reveals armature tube shock dynamics, detonation hydrocodes must be employed to study the cracking problem.

The 2DL (two-dimensional Lagrangian) hydrocode has been used in one form or another at Los Alamos for over 40 years. 2DL was ported from machine language to FORTRAN in 1968. TDL is a special version of 2DL written for personal computers by researchers at Los Alamos. It uses finite difference techniques to calculate two-dimensional solutions of Lagrangian reactive flow, and TDL includes solutions for elastic-plastic flow, real viscosity, heat conduction, and gravity effects. Four different methods of burning an explosive can be used to model detonation phenomena in TDL. The burn method used depends on whether the burning is thermally initiated or a constant-volume detonation. It also depends on burn geometry, on whether the explosive is homogeneous or heterogeneous, and on how the user wishes to model shock effects [Mad98].

Numerical methods that incorporate finite differences follow fundamental rules of calculus, whereby differential equations are translated into algebraic equations through approximations to small portions (finite differences) of the derivative. In a finite difference code, the solution of a differential equation is reduced to the simultaneous solution of linear, algebraic equations that are written for each nodal point within a set of boundaries. A node point with its associated boundaries is referred to as a cell, and the group of cells that define a problem is referred to as the mesh.

In this study, TDL was used to model shock dynamics within the armature tube for the first 4 centimeters past the end of the detonator, which is a total of 6 centimeters of the armature. The model was limited in this way to increase the number of cells available within the armature tube, since TDL has a limited number of cells allowed (3000). Both aluminum and copper armatures were modeled. Fig. 4.24 is a drawing of the armature model. The model represents the axisymmetric explosives-loaded armature as a two-dimensional figure, with the left edge of the model being the longitudinal axis of the armature. The right edge of the model is the outer surface of the armature tube, and the detonation progresses radially outward from the end of the detonator charge (Block 2) to consume the main explosive charge.

The output from TDL was compared to what could be seen on the surface of the armatures in photographs of the testing, such as Fig. 4.14 through Fig. 4.16. It was also compared to the results of the timing analysis, to the results of geometric analysis of detonation waves, and to the Mach stem predictions. In each case, applicable parts of the model matched what was seen on the armature surface and what was predicted by the analysis.

Figures 4.25 through 4.28 are graphs of selected portions of the TDL output from the FCG armature model with a copper armature. Comparison of the graphs to Fig. 4.24 shows the correspondence between the model and the graphed data. Coordinates on the graphs correspond to ordered pairs that represent individual cells in the model. Most of the graphs indicate the location of the outer surface of the armature, and Figures 4.25 through 4.28 show the location of the explosive/metal interface on the inside of the armature tube.

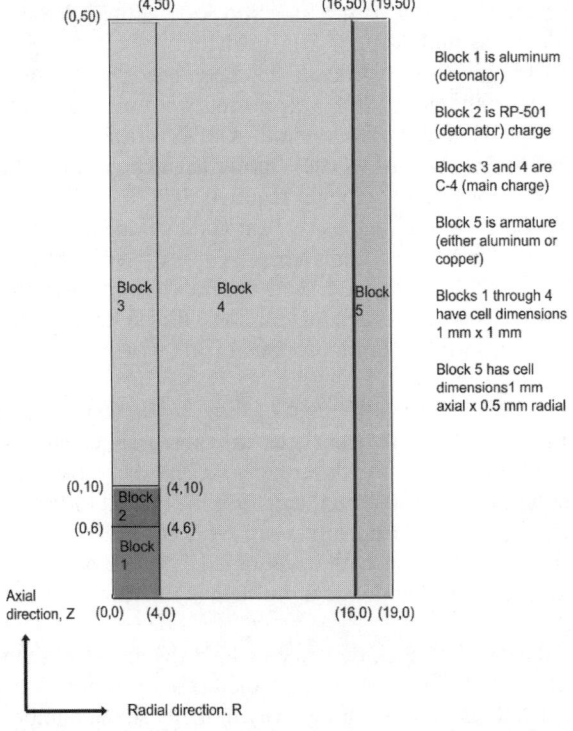

Fig. 4.24. Input model of the FCG armature for TDL hydrocode. Numbers in parenthesis are the coordinates, in millimeters, of the block corners. Note that armature is axis-symmetric, so that a 2d-model is used.

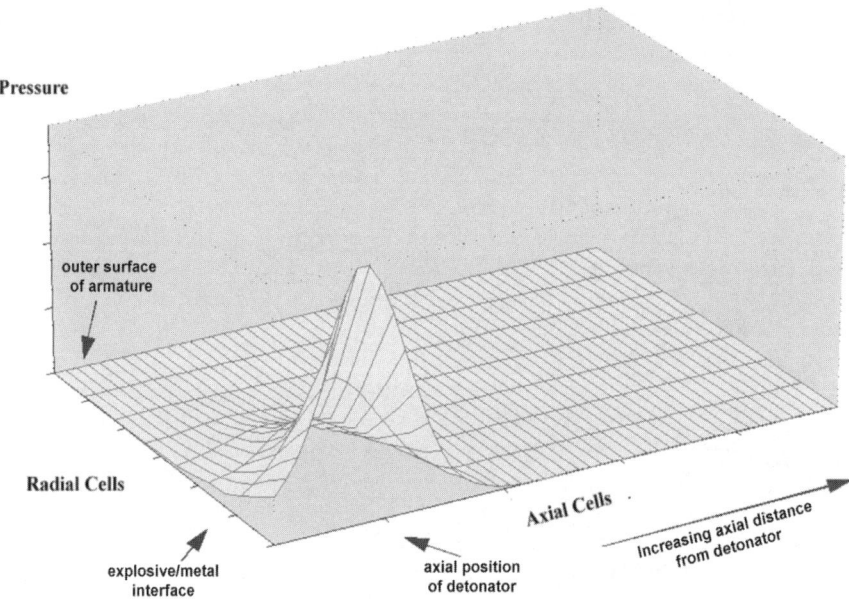

Fig. 4.25.a. Block of cells which includes the explosive/metal interface between I=16 and I=17. Shock has entered the armature tube but has not reached the armature outer surface (from the FCG TDL armature model); $t = 1.7$ μs after detonation.

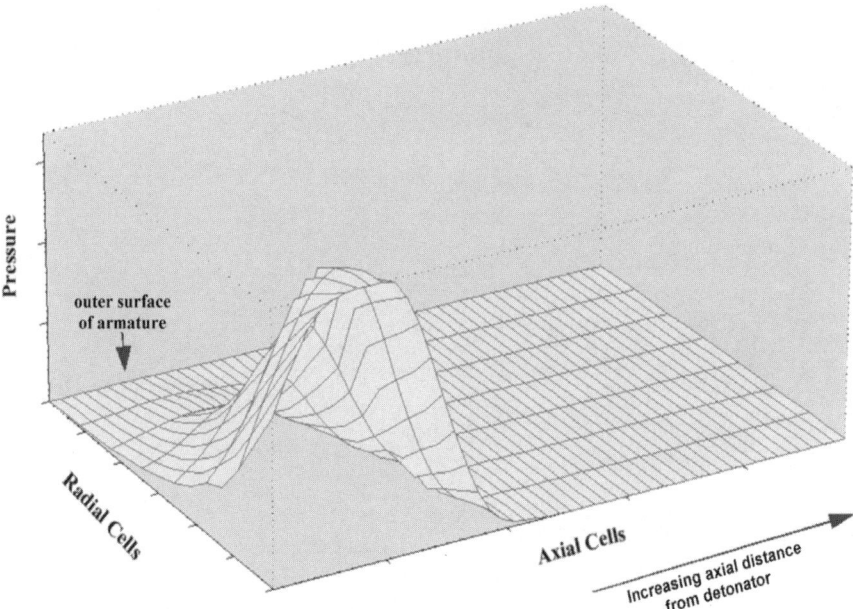

Fig. 4.25.b. Same block of cells; shock has almost reached the armature outer surface; $t = 2.0$ μs after detonation.

88   4 Mechanical Aspects

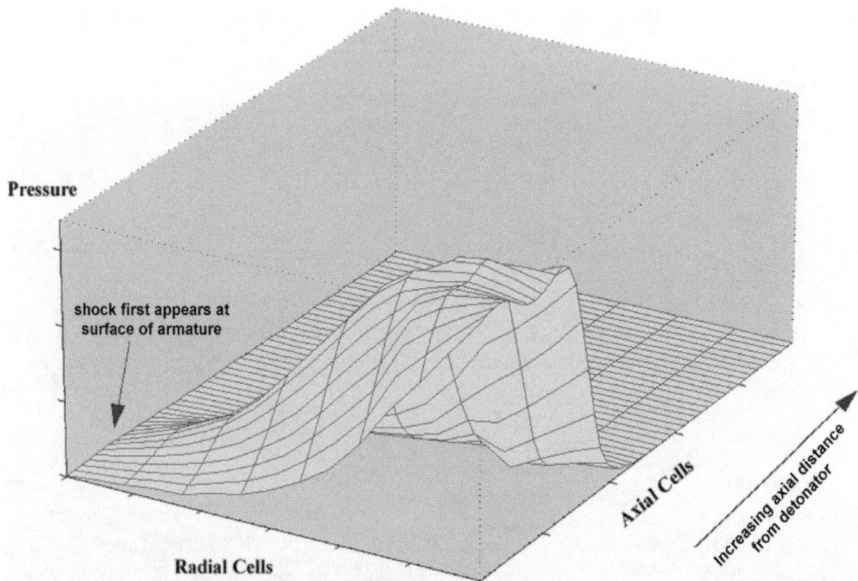

Fig. 4.26.a. Same block of cells at $t = 2.11$ μs after detonation. Shock/detonation wave has reached the outer surface of the armature, directly above the detonator; arrival time corresponds with prediction from Chap. 4.1.6.1 (2.08 to 2.21 microseconds after detonation).

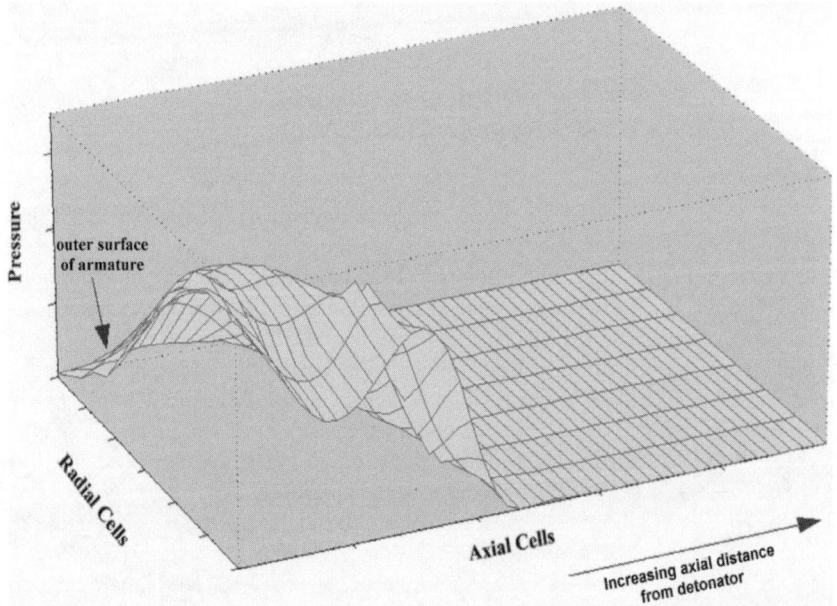

Fig. 4.26.b. Same block of cells. Shock reflecting into metal from armature outer surface, but pressure still compressive; t = 2.3 μs after detonation.

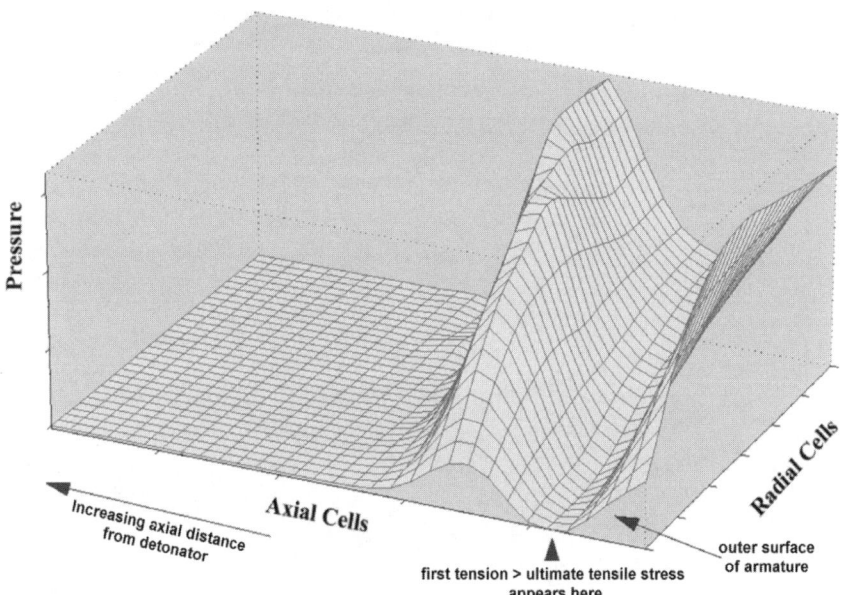

Fig. 4.27. a. Same block of cells, but viewed from outside the armature (outer surface is in the foreground). Tensile stress immediately under the armature surface has exceeded the tensile ultimate strength of the metal; $t = 2.7$ μs after detonation.

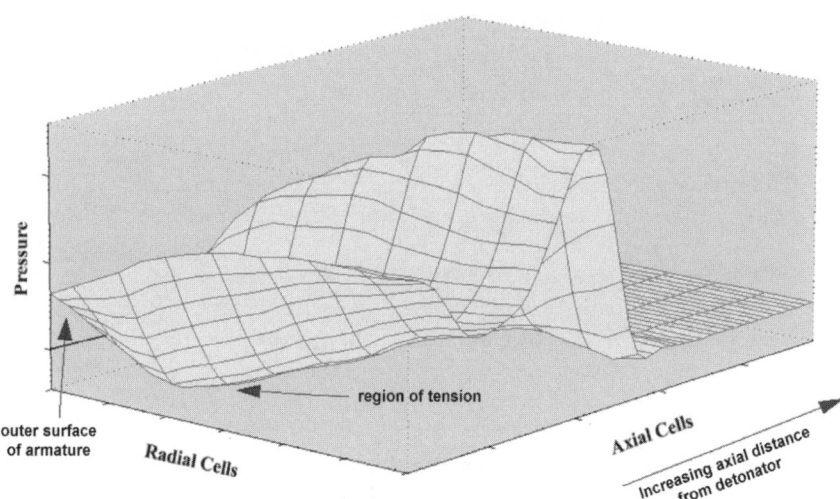

Fig. 4.27. b. Same block of cells, but from original viewpoint. Tensile region expanding, and includes armature outer surface above detonator. Expansion wave reflecting from armature inner surface; $t = 2.8$ μs after detonation.

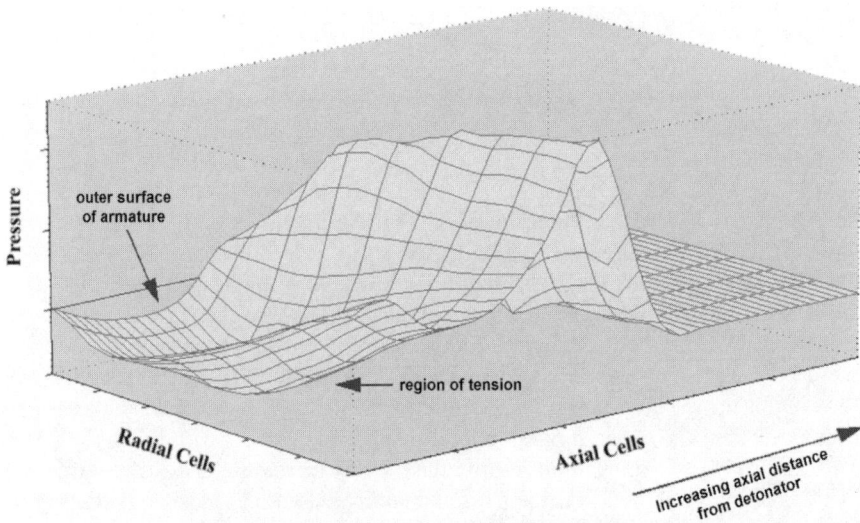

Fig. 4.28.a. Same block of cells. Region of tensile stress is expanding and moving down the armature; $t = 3.0$ μs after detonation.

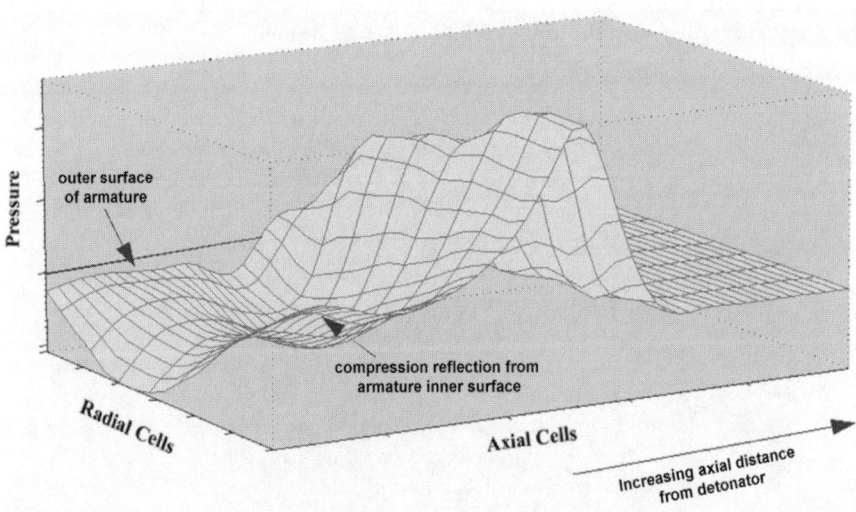

Fig. 4.28.b. Same block of cells. Tensile stress region no longer expanding, but still moving down the armature. Its compressive reflection from the inner surface of the armature is moving into the metal; $t = 3.3$ μs after detonation.

4.1.6.6 Simulation Results

The hydrocode analysis showed the first arrival of the shock wave at about 2.10 μsec after detonation, see Fig. 4.26a, which corresponds to the arrival time of between 2.08 and 2.21 μsec found by the timing analysis in Chap. 4.1.6.1. The analysis also revealed tensile regions caused by shock reflections within the metal tube of the armature. These regions formed near the armature outer surface, and then retreated towards the inner surface of the armature tubing as they traveled down its axis. The outer part of the region stays at the surface and moves down the tube until it travels axially about 1.5 cm beyond the end of the detonator, at about 3.5 μsec after detonation. At that point, it moves off the surface into the armature wall. Both the timing and trajectory of the tensile region in the simulation correspond well with the armature experiments. Photography of the armature tests shows that the armature tube begins to expand under gas pressurization about 5 μsec after the main charge starts its detonation, cf. Fig. 4.10 through Fig. 4.13. This expansion also appears in the simulation. At that point, shock pressurization effects within the tubing become marginal compared with explosive expansion effects. Beyond 5 μsec, the hydrocode analysis shows the expansion process, but not to the extent that is evident in the photography. For example, the simulation for an OFHC copper armature tube shows a maximum of 0.84 mm expansion at 10 μsec after detonation, but the photographs show 4.7 mm expansion at the same time. The difference is likely to be spalling; spalling was not modeled in this armature simulation, but the combination of high tensile stress and its position within the armature tubing certainly leads to spalling of the outer layers of the armature very soon after imposition of the stress.

Therefore, the TDL simulation verifies the results of the structural and timing analyses, and it allows visualization of the armature fracturing process. Tensile stresses within the first 3.6 cm along the outer surfaces of the armature proceed from shock dynamics in the first 5 μsec after detonation. These stresses damage the metallic crystal structure of the armature tube, and then are replaced by intense hydrostatic compressive stresses. This compression holds the tube structure together until gas pressurization expands it and causes immediate fracturing of the damaged metal. These fractures only occur in the damaged area, as the undamaged remainder of the armature tube is ductile enough to expand to about twice its original diameter before cracking.

If a more detailed simulation is desired, a Method of Characteristics hydrocode can be used to model the shocks. According to published research on this type of hydrocode, "This method of solution eliminates potential confusion of material dissipation with artificial dissipative effects [*i.e., the artificial viscosity that must be utilized in TDL*] inherent in finite-difference codes, and thus lends itself to accurate calculation of elastic-plastic deformation, shock-to-detonation transition in solid explosives, and shock-induced structural phase transformation." [Joh91] For the presented research, TDL was sufficient; for follow-on investigations that require greater accuracy in the location of shock waves, a Method of Characteristics code such as CHARADE can be used.

4.1.6.7 Metal Fatigue

When explosive expansion begins, the passage of high tensile, wave-induced stresses through the outer layers of the armature tubing has already damaged the metallic crystal lattice structure. This damage creates fracture initiation points; as the armature expands, the longitudinal fractures seen in the photographs are in all probability due to the coalescence of many of those initiation points. According to Broek, "Under the action of cyclic loads cracks can be initiated as a result of cyclic plastic deformation. Even if the nominal stresses are well below the elastic limit, locally the stresses may be above yield due to stress concentrations at inclusions or mechanical notches. Consequently, plastic deformation occurs locally on a microscale, but it is insufficient to show in engineering terms." [Bro82]

The limit of this initiation point coalescence can be seen in Fig. 4.29, a composite of photographs from two different armature tests. In the photograph of the copper

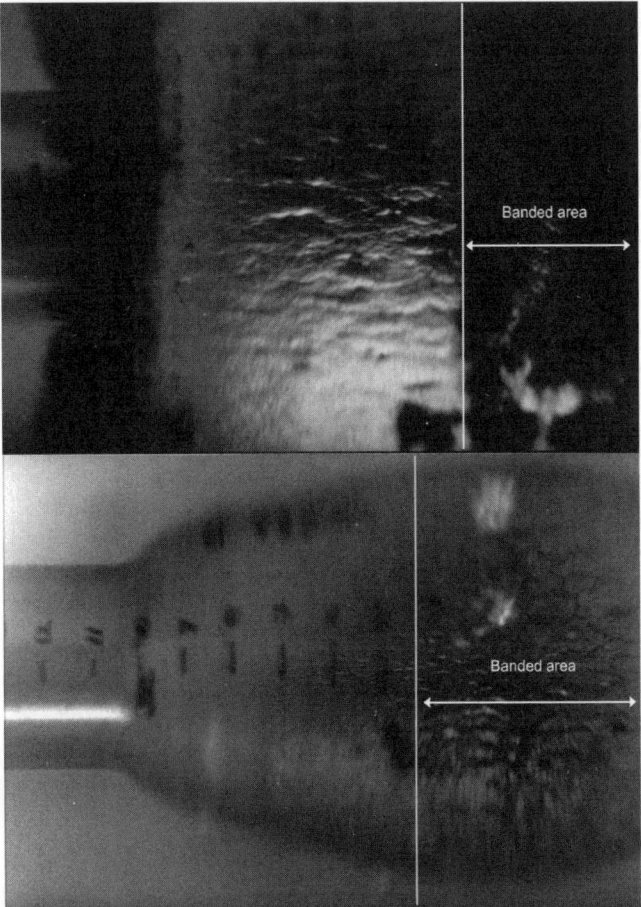

Fig. 4.29. Longitudinal "crack bands" on armatures from two separate tests.

armature at the top of the figure, detonation products are escaping from cracks that pass completely through the metal. In the bottom photograph of an aluminum armature, some of the escaping gas is hot enough to glow. The left edge of the longitudinal cracking "bands" marks the limit of longitudinal cracking due to wave-induced stresses. What causes the coalescence of fracture initiation points?

Evidently, the passage of shock waves and rarefactions through the metal within the first 4 cm of our armature tube results in a greatly varying stress field. This causes a type of metal failure called fatigue; metal fatigue is the sudden separation of the structure into pieces due to the application of fluctuating loads over a period of time. When the loads are large enough to cause failure of the structure in fewer than 1×10^5 cycles, the phenomenon is called low-cycle fatigue [Col81]. In the case of the armature tests, hydrostatic stresses vary between high compressive and high tensile magnitudes at a frequency greater than 10^6 cycles per second for several microseconds. As long as the levels of compressive and tensile hydrostatic stress are below the failure limit for the metal (its yield point, in this case), the stress cycles cause low-cycle metal fatigue.

Given the alternating compressive and tensile stresses in the armature as predicted by TDL, it is reasonable to expect that longitudinal cracks form where tensile detonation wave reflections cause crack initiation points. These points form a pre-conditioned, damaged area through low-cycle fatigue. The points coalesce into cracks, which rapidly extend through the damaged area as soon as explosive expansion begins. The undamaged areas of the armature do not crack until much greater expansion takes place, later on in the test. This is what causes the "banded" distribution seen in Fig. 4.29, with a regular network of cracks in the pre-damaged area of the armature and a more random distribution of structural cracks formed later in the remainder of the armature.

4.1.7 Simulations and Tests of Multi-Layer Armatures

4.1.7.1 Impedance Matching

When a shock wave moving through a medium arrives at a boundary, it may partially or totally reflect back into the incident medium as a compressive wave or as a rarefaction, or it may be entirely transmitted into the adjoining medium. What happens to the wave depends on the degree of shock impedance mismatch between the media at the boundary. The shock impedance of a material is equal to the product of its pre-shocked density and the velocity of the shock in the material. When it impinges on a boundary, the shock wave will reflect as a compressive wave if the adjoining medium has higher shock impedance than the incident medium or as a rarefaction if the adjoining medium has lower shock impedance than the incident medium. No reflection will take place if the impedances are equal. The greater the impedance mismatch, the greater the portion of the shock energy reflected, and the smaller the portion transmitted. This has been verified experimentally [Coo74].

We postulated that including low-density layers between high-density materials through the thickness of the armature would reduce the tensile stress generated within the outer layer by reflection of the shock as it leaves the armature. To verify this, we performed simulations of a multi-layer armature, composed of inner and outer 1mm layers of OFHC copper with a 1mm inner layer of acrylic polymer (Plexiglas®), using TDL.

The simulation investigated tensile stress reduction in the outer, metal layer through impedance mismatch within the armature thickness. The simulation predicts lower tensile stresses at the armature surface than in metallic armatures, but the stresses still exceed ultimate strength for the copper outer layer. Additionally, the region of tensile stress spans most of the outer layer, and it moves along the length of the tube as the simulation progresses. This is in contrast to the single layer armature, wherein the tensile stress region follows a trajectory away from the armature surface.

As another check of this postulate, shock impedances were calculated and the expected outcome of testing based on those results was compared to actual test results generated by explosive testing of multi-layer armatures.

The shock impedances of the materials involved in the testing were determined in the following manner [Coo96, Coo74]. Determination of the pressure and particle velocity at a material interface allows calculation of shock impedance:

$$Z = \rho_0 U$$
$$P = \rho_{0u} U$$
$$\therefore Z = \rho_0 \left(\frac{P}{\rho_{0u}} \right) = \frac{P}{u} \quad (4.34)$$

where Z is shock impedance, ρ_0 is initial density, u is interface particle velocity, U is shock velocity, and P is interface pressure.

Note that in this set of calculations:
- o Shock pressure at the Accurate Arms C-4 to cylinder interface was determined to be 38.65 GPa [Shk03].
- o The detonation/shock wave travels from left to right across the material interface.
- o The governing Hugoniot equations for waves traveling from left to right or right to left are, respectively:

$$P = \rho_0 C_0 u + \rho_0 S u^2 \quad (4.35)$$

$$P = \rho_0 C_0 (u_0 - u) + \rho_0 S (u_0 - u)^2 \quad (4.36)$$

C_0 and S are unreacted Hugoniot coefficients, and along with the initial density, they were determined to be:
- o acrylic (Lucite®/Plexiglas®) [Coo96]:
    - $\rho_0 = 1.181$ g/cm^3, $C_0 = 2.260$ km/sec, $S = 1.816$
- o 6061 Aluminum (Mar80):
    - $\rho_0 = 2.703$ g/cm^3, $C_0 = 5.350$ km/sec, $S = 1.340$

- OFHC Copper (Coo96):
  - $\rho_0 = 8.930$ g/cm^3, $C_0 = 3.940$ km/sec, $S = 1.489$
- Atmosphere (air, helium, etc.) [Coo96]
  - $\rho_0 = 0.001$ g/cm^3, $C_0 = 0.899$ km/sec, $S = 0.939$

The technique used to generate the impedances is to work from the known shock pressure at the inner surface of the armature, outward through the armature layers and their interfaces, calculating unknowns and using them in Eqs. (4.35) and (4.36) to find interface pressures and particle velocities. Those pressures and velocities are then used in Eq. (4.34) to find the shock impedance between the two layers. First, the particle velocity in the innermost layer of the armature (the material next to the explosive charge) is found using the explosive shock pressure in Eq. (4.35), the right-traveling wave equation. Substituting the derived particle velocity and the explosive shock pressure into Eq. (4.34) produces the impedance at the explosive-innermost layer interface. Next, since the pressures within the innermost layer and the next (second) layer are equal at their interface, the equation of the wave reflected from the interface, Eq. (4.36), is set equal to the equation of the wave transmitted through the interface, Eq. (4.35), and the combined equation solved for the particle velocity at the interface. Using the resulting value for particle velocity in the second layer, Eq. (4.35) is solved for the shock pressure within the second layer. As before, use of the particle velocity and the shock pressure in Eq. (4.34) produces the shock impedance at the second interface.

This procedure, when utilized in a step-wise fashion through the thickness of the armature, gives the shock impedances for each layer of the armature, and for the atmosphere surrounding the armature. Comparison of the final two impedances determines the impedance mismatch seen by the shock as it leaves the armature. As a figure of merit, the smaller the absolute value of the impedance mismatch, the less shock energy reflected back into the armature material, and the lower the tensile stress generated within the outer layer of the armature.

Examination of the results indicates that, contrary to what might be thought, an armature with copper inside and aluminum outside has the lowest impedance mismatch, followed closely by a simple, single-layer aluminum armature. Including an acrylic layer or an atmosphere layer within the thickness of the armature produces no apparent advantage.

We conducted another simulation, designed to investigate multi-layer armature expansion angles and radial velocities and utilizing AUTODYN-2D software. Table 4.1 contains data from that simulation set.

We concluded that aluminum armatures have the best expansion behaviors among the different armature constructions simulated. More quantitative data regarding the end-effect is presented in Chap. 4.2.5 Given that a smaller end effects region and a larger expansion angle define "good" expansion behavior, the simulated three-layer armatures exhibited good expansion behavior, approaching that of single-layer aluminum armatures. The simulation also revealed that the expansion angle is strongly a function of armature material density. This agrees with the Taylor approximation,

Table 4.1. Data from simulation of armature expansion behavior

Armature	End effect (mm)	Expansion angle	Radial velocity (km/s)
3.175 mm CU	26.3	6.35	1.05
2 mm Cu, 1.175 mm Al	18.75	10.28	1.65
1 mm Cu, 2.175 mm Al	17.5	11.95	1.87
1 mm Cu, 1 mm (Teflon®), 1.175 mm Al	17.5	12.13	1.88
1 mm Cu, 1 mm (Lexan®), 1.175 mm Al	17.5	12.49	1.94
3.175 mm Al	10.7	13.55	2.27

$$V_A = D \tan \theta, \qquad (4.37)$$

with V_A the apparent velocity of the metal (velocity radially outward from the major axis of the cylinder), D the detonation velocity, and θ the expansion angle [Wal89].

(N.B. The expansion angle is commonly termed the "Gurney angle," which not quite correct; the Gurney equations assume that the detonation wave encounters the metal at normal incidence. Grazing incidence of detonation waves is modeled by the Taylor angle approximation, which uses the Gurney equations to provide estimates of expansion angle and velocity components.)

V_A is related to V, the metal velocity, through geometry. Via the Gurney equations, V is a function of explosive characteristics, the explosive mass, and the mass of the metal being expanded, cf. Chap. 2.1.4. Therefore, the expansion angle would be a function of the metal density if the simulation is correct.

In an attempt to validate these simulations, a series of explosive expansion tests on multi-layer armatures was conducted and photographed, utilizing armatures of three different constructions.

Longitudinal cracking of the sort photographed during the earlier armature testing was evident during this multi-layer test series. Regrettably, photographs of one of the sets of cylinders were unusable for comparison to the TDL simulation because of problems with the polymeric layer.

Comparison of the test data to Table 4.1 shows good agreement of tests and simulations, and appears to validate this portion of the simulations, cf. Fig. 4.34.

4.1.8 Conclusion Shock Analysis

Late in the research work, we obtained a new form of HMX explosive that can be easily hand-loaded into armature cylinders. Initial testing with this explosive in aluminum cylinders at the T6 annealed state of hardness was promising. These tests proved testing design changes that improved the quality of high-speed photography, and surprisingly, showed that this explosive/armature material combination has promise for reductions in armature cracking. Unfortunately, project time and funding ran out at this point, and no simulations or calculations were run on the combination.

There is much more work to be done, primarily investigating the shock dynamics of armature expansion and their impact on generator performance. Nevertheless, the discussed shock physics work provides a basis for future investigations that should allow increased efficiencies in explosive-driven flux-compression generators, in addition to expanding the state of knowledge of shock wave interactions at explosive/metallic interfaces.

4.2 Analysis of Armature/Stator Contact

4.2.1 Introduction

The efficiency of a helical FCGs, is highly dependent on the expanding characteristics of the exploding armature and the nature of contact between the armature and the surrounding stator coil [Gov79]. In a helical FCG, the energy conversion process includes two steps: primary energy of the high explosives to kinetic energy of the expanding armature, and the kinetic energy of the armature to the final electromagnetic energy [Kno70]. The current design of FCGs is very inefficient in converting the explosive energy into electrical energy, typically less than 10% due to loss of magnetic flux. We also note that trying to utilize close to all of the available explosive energy would lead to armature slow down and thus increased operation time and losses, making it impossible to attain close to 100% efficiency, cf. Chap. 2.1. In an attempt to address the inefficiency issue, a number of studies have been conducted to numerically simulate the electromagnetic behavior of helical FCG [Der98, Fre92, Ric89, Jon79], but few have explored the contact characteristics between the armature and the stator. We have shown that some loss mechanisms are likely attributable to the expanding characteristics of the armature and the contact characteristics between the armature and the stator, cf. Chap. 3.2. Therefore, exploring the contact characteristics between the armature and the stator is a step in the right direction for optimal generator design.

Although high-speed X-ray photography is a viable technique for experimental observation of the contact characteristics between the armature and the stator, such experiments are quite cumbersome and expensive. Numerical simulation techniques, such as the finite element method (FEM), provide a relatively inexpensive and easy alternative to physical experiments. However, analyses relying on FE techniques are quite susceptible to misinterpretation of the results and erroneous conclusions unless such models are verified against experimental observations.

To this end, an experimental/numerical study was designed to study the contact characteristics between the armature and the stator in an FCG. A hydrodynamic Finite Element (FE) model was developed to simulate the expansion characteristics of the armature and its ensuing impact with the stator. A number of template

FCGs were designed and manufactured in our laboratory for obtaining a series of experimental data, which were in turn used to fine tune an FE model for numerical investigation of the contact behavior between the armature and stator. Specifically, the radial displacement of the armature, as well as the axial velocity of the armature/stator contact point, were measured experimentally and compared with numerical results showing excellent agreement between the two. The results indicated that the radial and axial velocity with which the armature impacted the stator did not change throughout the length of the armature. However, the results showed that the velocity with which the contact point between the armature and the stator traveled along the length of the armature decreased as the explosion process went on. As expected, the axial propagation velocity of the contact point was found to be at its highest value (2.25 X detonation velocity) in the region close to the detonation end while approaching the detonation velocity at points away from the detonation end.

4.2.2 Finite Element Model

During the explosion process, the armature and stator materials are subjected to shock loading at high strain rates. The constitutive equations of materials at high strain rates are extensively researched [Har89, Bon94]. In most materials, the yield stress usually increases with increasing strain rate. However, previous research on the constitutive equation of several metals, including 6061-T6 aluminum and copper, has shown that there exists a strain rate threshold beyond which any increase in the strain rate has minimal effect on the yield strength [Ste80, Ste86, Ste87]. A number of shock wave experiments [Wil73, Las90] have verified this phenomenon. Since the pressure produced by the explosive shock loading usually is several orders of magnitude higher than the yield strength of the material in a typical FCG, the materials will behave more like a liquid rather than a solid. Therefore, an "equation of state" for the material along with its constitutive equation is needed to describe the behavior of a material subjected to shock loading.

The FE model used in this study employed the Steinberg material constitutive model and state equation to describe the mechanical behavior of the armature and the stator materials [Rud98, Ste91].

$$Y = \left[1 + \frac{AP}{(1+\mu)^{1/3}} - B(T-300)\right] \times \begin{cases} Y_0[1+\beta(\varepsilon_p + \varepsilon_i)]^n & \text{for} \quad \varepsilon < \varepsilon_c \\ Y_{Max} & \text{for} \quad \varepsilon > \varepsilon_c \end{cases} \quad (4.38)$$

where Y and Y_0 are the yield strength and the yield strength at the Hugoniot limit respectively. Y_{Max} is the work-hardening maximum of the yield strength. ε_p and ε_i are equivalent plastic strain and the initial plastic strain, respectively. T is the temperature, μ ($\mu = \rho/\rho_0 - 1$)) is the relative compression ratio, ρ and ρ_0 are the density and the initial density, respectively. A, B, β, n, Y_0, Y_{Max} and ε_C are available phenomenological parameters for several metal material in reference [Ste91].

The Steinberg equation of state, which relates pressure P to the relative compression ratio μ and internal energy per unit volume E_v to that at standard conditions ($P = 0$ Pa, temperature = 300 K), is given by:

$$P = \begin{cases} \dfrac{\rho_0 C_0^2 \mu [1 + (1 - \dfrac{\gamma_0}{2})\mu - \dfrac{\gamma_0}{2}\mu^2]}{[1 - (S_1 - 1)\mu - S_2 \dfrac{\mu^2}{\mu+1} - S_3 \dfrac{\mu^3}{(\mu+1)}]} + (\gamma_0 + a\mu)\rho_0 E_v & for\ \mu > 0 \\ \rho_0 C_0^2 \mu + (\gamma_0 + a\mu)E_v & for\ \mu \leq 0 \end{cases} \quad (4.39)$$

where S_1, S_2, S_3, γ_0, α and C_o are available phenomenological parameters for several metal materials [Ste91].

We used the Jones-Wilkins-Lee (JWL) equation of state [Dob85, Lee68] to describe the behavior of the high explosives Composition C-4. The LS-DYNA-3D hydrodynamic code [Liv97] was utilized to perform the FE simulations.

$$P = A(1 - \dfrac{\varpi}{R_1 V})e^{-R_1 V} + B(1 - \dfrac{\varpi}{R_2 V})e^{-R_2 V} + \dfrac{\varpi E_D}{V} \quad (4.40)$$

where P is the pressure; A and B are linear coefficients, respectively. R_1, R_2 and ϖ are the nonlinear coefficients. V is the volume ratio between the detonation products and un-detonated high explosive. E_D is the detonation energy per unit volume. The physical properties of C-4 are available, for instance, in reference [Dob85]. Eight-node solid elements were used to model one quarter of the armature and the explosive (due to axial symmetry). The sliding-only contact-impact algorithm was used to treat the interface between the explosive and armature, which was specially designed to treat the interface between the gaseous detonation products and the solid material. The FE model consisted of six different components including Comp C-4 high explosives, solid 6061-T6-aluminum end plug, 6061-T6 armature, Lexan cylinder, copper crowbar ring and 6061-T6-aluminum stator as shown in Fig. 4.30a.

4.2.3 Experimental Verification of FE Model

In order to verify the FE model, explosive experiments utilizing a 6061-T6-aluminum armature were conducted to record the armature's deformation contour at various post-detonation time intervals. The experimentally observed deformation profile of the armature was subsequently compared with the deformation profile obtained by FE simulations of the same process. The outer diameter, the wall thickness, and the length of the armature used in this study were 38.1 mm, 3.175 mm, and 152.4 mm, respectively. For post-mortem material analysis, the armature was half sometimes only half-filled with high explosive Comp C-4, cf. Fig. 4.31b.

An intensified gated camera and a rotating mirror camera were used to capture a single shot (15 ns exposure) or a sequence of 26 images covering a time interval of 50 μs, cf. Fig. 4.31a) and b), respectively.

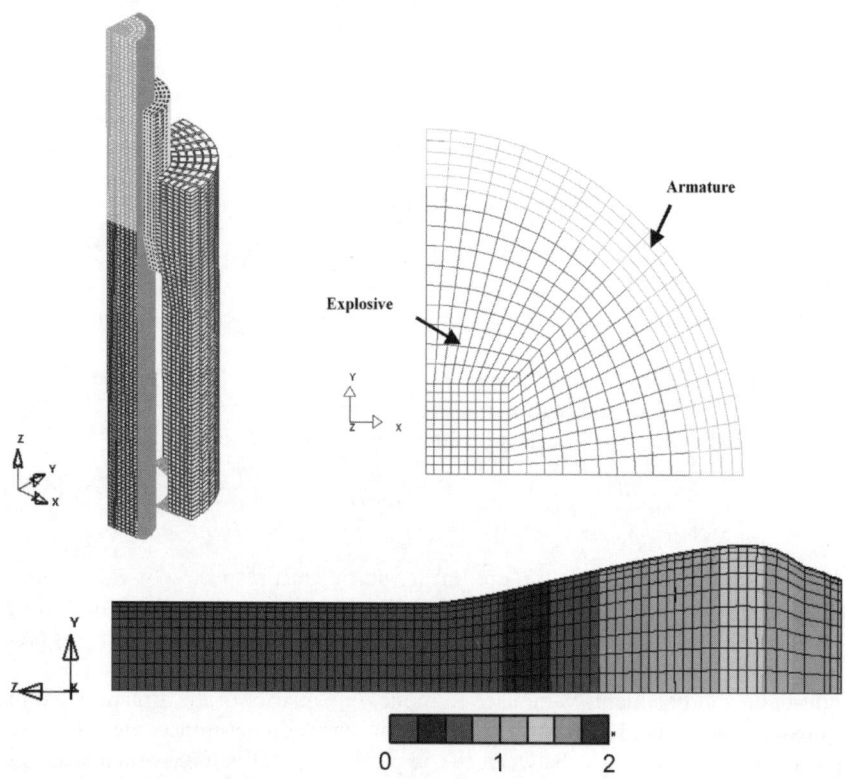

Fig. 4.30. Left to right: **a)** FE model of the template generator, **b)** FE mesh of the armature and explosive, **c)** deforming aluminum armature at t = 10 microseconds showing the total displacement in units of cm.

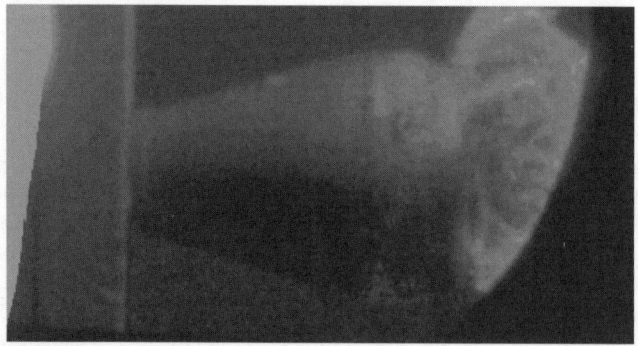

Fig. 4.31.a. Armature expansion during explosion, 15 ns exposure time 6061-T6 aluminum.

Fig. 4.31.b. Armature expansion during explosion, OFHC copper. Only right armature half filled with HE.

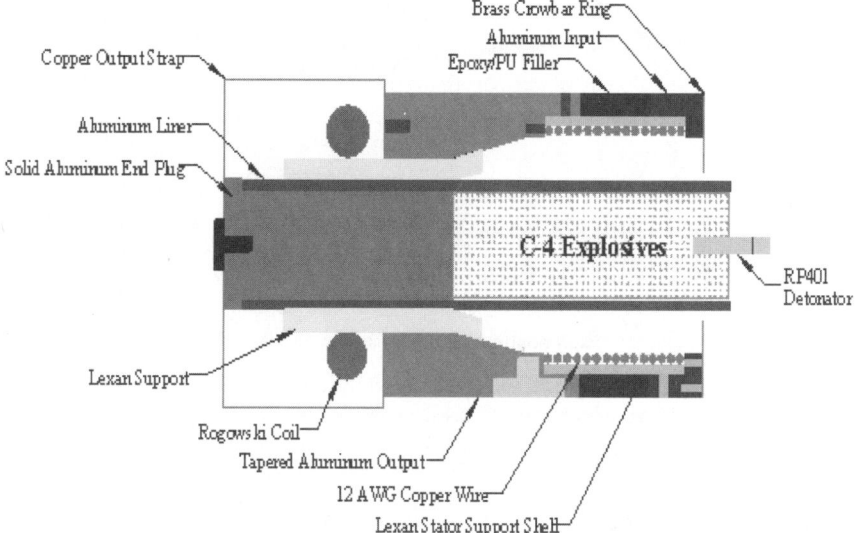

Fig. 4.32. Schematic of the template generator.

The template magnetic flux compression generator is depicted in Fig. 4.32, while the experimental setup is shown in Fig. 4.33. The template generator used in this study consisted of a stator with 16 turns of insulated 12 AWG copper wire initially wound on a 64 mm diameter aluminum 6061-T6 mandrel. The first turn of the copper coil was placed at the stator axial position of 8 mm. The pitch of the copper coil was 3 mm. The input plane (crowbar) was a brass ring made from a 45-mm inner diameter, 0.2 mm thick, slotted brass disc. The load side of the generator had a 14 degree tapered output plane, which reduced the inner diameter from 64 mm to 51 mm.

Fig. 4.33. Experimental setup for the template generator. Flash lamp array is visible in the foreground.

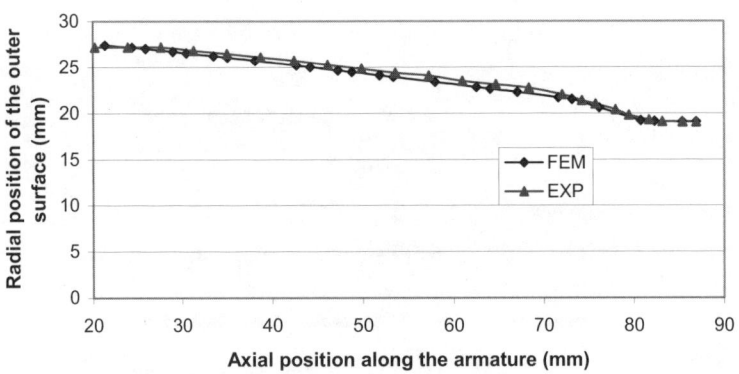

Fig. 4.34.a. Comparison of FEM and experimentally observed deformation of the outer surface of an OFHC copper armature at post-detonation time of 10 μs.

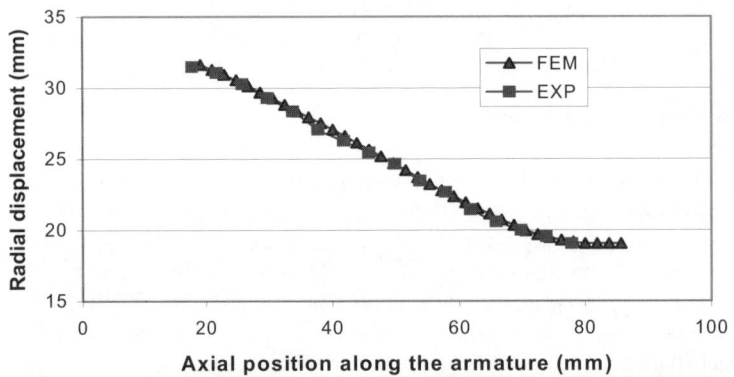

Fig. 4.34.b. Comparison of FEM and experimentally observed deformation of the outer surface of an aluminum armature at post-detonation time of 7 μs.

The 6061-T6 aluminum armature (38.1 mm-OD, 3.175 mm- wall) and the stator were at this point separated by a 6 mm-wall Lexan cylinder. The armature was stepped outward 16.5 mm with respect to the stator. The detonator was inserted at the center of the armature and into the explosive along the armature axis by 12.5 mm. The total length of the armature was 203 mm, but only a partial length of the armature, 125 mm, was filled with the Comp C-4 high explosive. A solid 6061-T6-aluminum end plug measuring 15.875 mm in outer diameter and 78 mm in length was inserted into the armature at the other end to fill the space in the rest of the armature.

The FE results indicated an average armature expansion angle of 13.1 degrees, compared to a value of 13.4 degree as measured from actual experiments (difference of 2.2%). As shown in Fig. 4.34a/b, the excellent agreement between the FE and experimental results on the armature deformation contour was used to verify the accuracy of the FE model.

4.2.4 Deformation and Kinematics of Armature/Stator Contact

The radial and axial displacement histories of a point on the armature located 37.7 mm from the detonation end is shown in Fig. 4.35. The radial and axial expansion velocity histories of the same point along the armature is depicted in Fig. 4.36. The black dots in both figures depict the point in time when the armature first contacts the stator. Fig. 4.37 displays the radial and axial impact velocities between the armature and the stator for various points along the length of the armature.

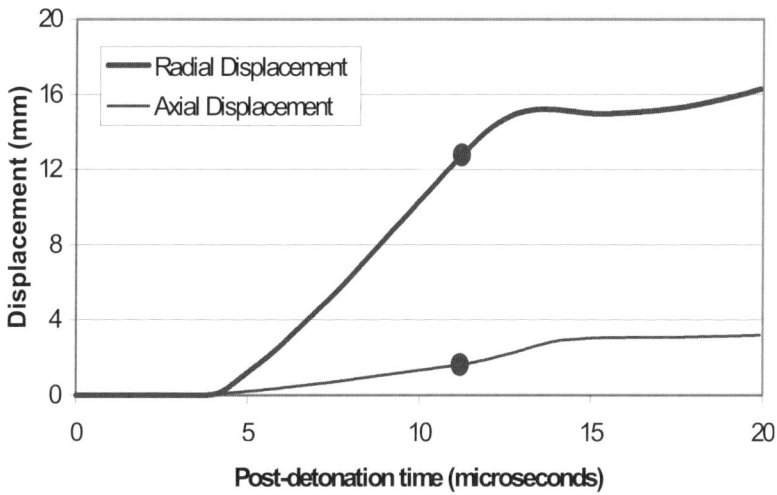

Fig. 4.35. Radial and axial displacements of a point at armature axial position 37.7 mm.

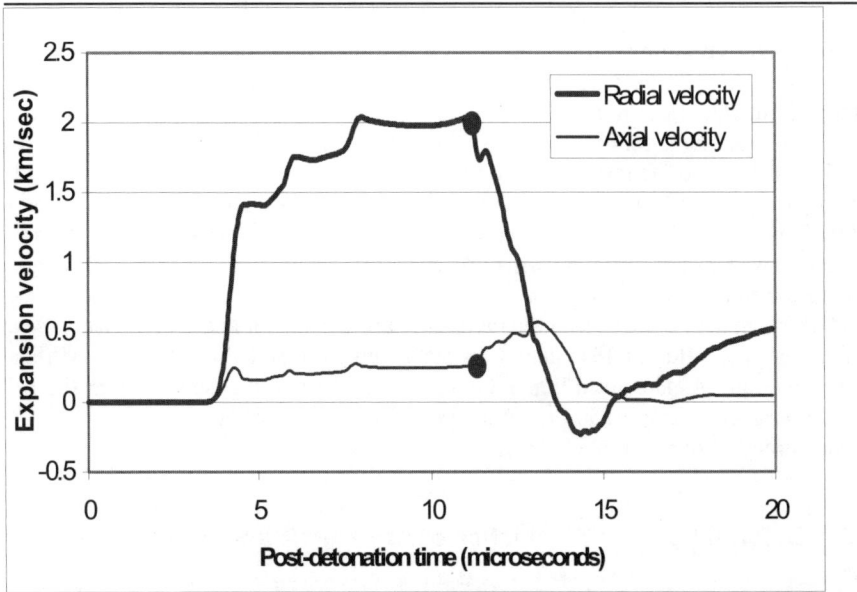

Fig. 4.36. Radial and axial expansion velocities of a point at armature axial position 37.7 mm.

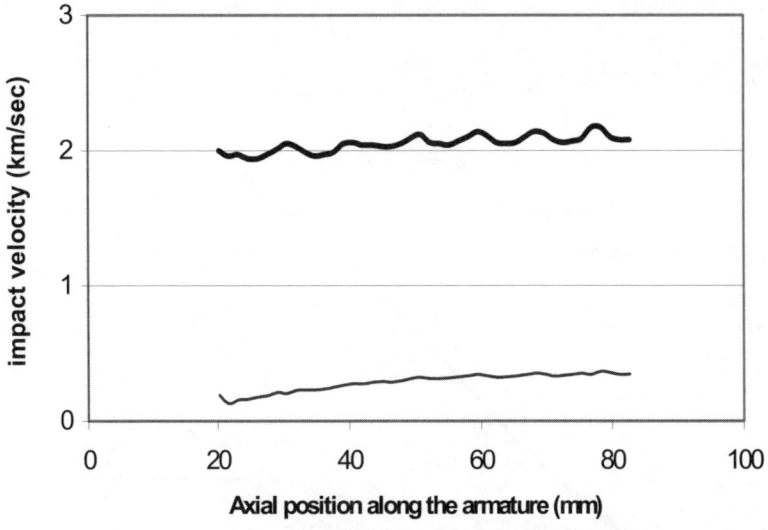

Fig. 4.37. Radial (top curve) and axial (bottom curve) impact velocities between the armature and the stator.

As expected, the radial component of the impact velocity is much larger than the axial component with both velocity components staying relatively constant throughout the length of the armature. Fig. 4.38 depicts the velocity with which the armature/stator contact point moves as a function of axial stator position. This figure shows that the velocity of the contact point decreases as it moves away from the detonation end. It can be seen from Fig. 4.38 that the contact velocity is initially much larger than the detonation velocity of the explosive (8.2 km/sec), approaching the explosive detonation velocity of 8.2 km/s as it moves away from the detonation end. Table 4.2 shows the comparison of results obtained from FE simulations and experimental measurements conducted on the velocity of the contact point between the armature and the stator. As shown in Table 4.2, the relatively close agreement between experimental and FE results provides further verification of the accuracy of the FE model employed in this study for describing the behavior of the armature and stator under conditions of shock loading.

Numerical and experimental results presented in Fig. 4.38 and Table 4.2 indicate that the velocity of the armature/stator contact point initially exceeds the detonation velocity (~ 8.2 km/s).

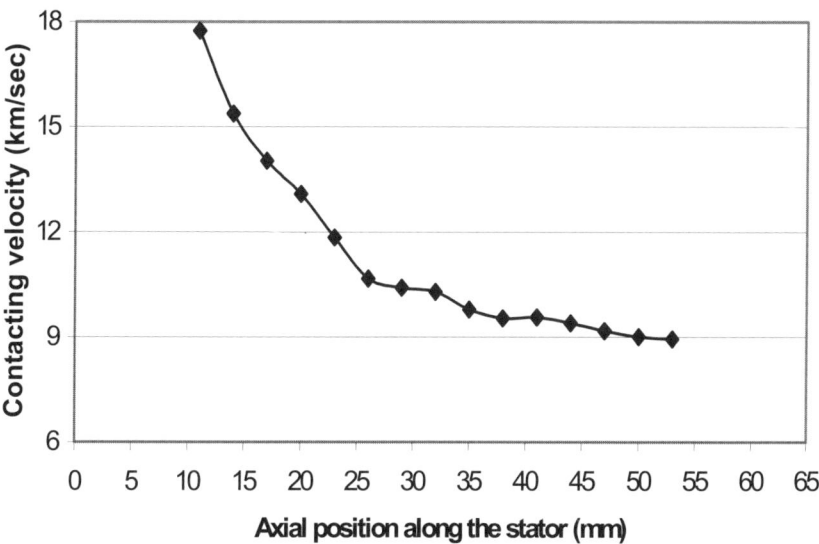

Fig. 4.38. Axial velocity of the armature/stator contact point as a function of stator axial position.

Table 4.2. Velocity of contact point at different axial locations

Axial locations used for axial velocity measurements	Armature/Stator Contact Velocity		
	Experiment Results	Numerical Results	%Relative error
22 mm and 32 mm	11.11 (km/sec)	10.75 (km/sec)	3.24%
32 mm and 44 mm	10.27 (km/sec)	9.84 (km/sec)	4.19%
44 mm and 65 mm	9.19 (km/sec)	8.75 (km/sec)	4.78%

4.2.5 Armature "End-Effect"

As indicated above, the armature/stator impact velocity is higher near the detonating end of the armature. The main reason for this phenomenon is the severe "end-effect" observed near the detonation end as shown by the experimental results in Fig. 4.31, and by the schematic and numerical simulation results shown in Fig. 4.39. The armature impacting the crowbar plate caused the downward spike shown in Fig. 4.39b. Because of this serious end effect, the expansion angle of a point on the armature would be smaller, the closer that point is to the detonation end which in turn causes the velocity of the contact point at points near the detonation end to be higher than the detonation velocity (8.2 km/sec), see Fig. 4.38.

Experimental observations, as well as FE simulations of the armature expansion, clearly show the existence of an "end-effect", resulting in the bell shape contour of the armature near the detonation side, as shown in Fig. 4.31 and Fig. 4.39. This phenomenon can be attributed to the fact that in helical FCGs, the detonation end is normally open to outside air. When the explosive is detonated, some of the detonation pressure escapes through the open end, which in turn reduces the outward expansion of the portion of the armature near this area. Therefore, the length and severity of the armature's "end-effect" is very important in minimizing magnetic flux losses resulting from lack of contact between the armature and the stator within this area. To avoid losses due "end-effect", the first turn of the stator's coil

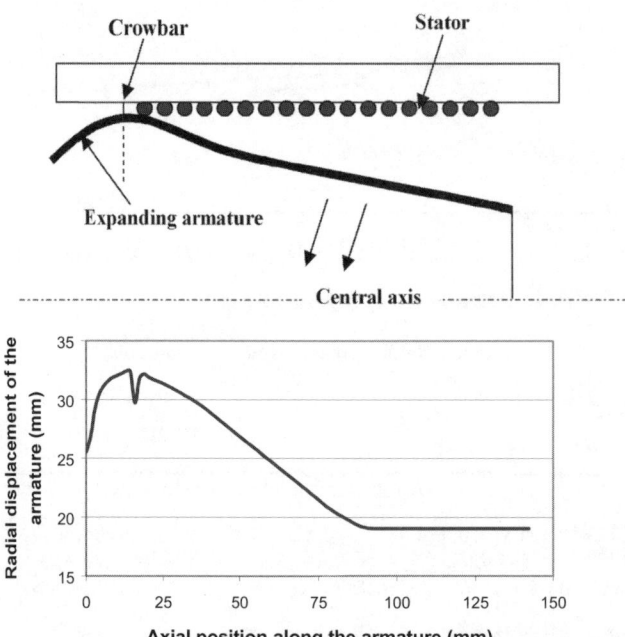

Fig. 4.39. Top to bottom: **a)** Schematic of an expanding armature showing the "End-Effect", **b)** Deformed contour of an Al armature at post-detonation time of 10.039 μs. The dip at ~ 15 mm is due to the crowbar disk.

must begin at a point that is away from the detonation end by a distance equal to the length of the "end-effect" for the specific armature under consideration.

Our studies have shown that under similar conditions, the geometry of the armature plays a significant role in the length of the corresponding "end-effect", as well as the armature's expansion angle. Theoretical analysis and parametric numerical simulations revealed that for the same length of detonation end-effect, as well as the armature's expansion angle, is only a function of the armature's wall-thickness ratio (wall thickness to outer radius ratio), assuming no change in armature and explosive materials.

Defining the length of the "end-effect', L_D, as the distance along the armature's axial direction between the detonation point and the top of the "bell shape" section in Fig. 4.39. The armature wall-thickness ratio, a, is defined as the ratio of its wall thickness, W, to its outer radius, r_A, as given by:

$$a = \frac{W}{r_A} \quad (4.40)$$

The end-effect ratio, υ, defined as the ratio of the length of the end-effect to the outer radius of the armature is given by

$$\upsilon = \frac{L_D}{r_A} \quad (4.41)$$

A numerical parametric study, using wall thickness ratios, a, in the common range of 0.04 to 0.2, was conducted on 152.4 mm-long armatures to study the effect of wall thickness ratio on the length of the resulting armature end-effect and expansion angle. See Table 4.3 for the results of this parametric study.

Fig. 4.40 is a graphical representation of above data showing the effect of armature wall thickness ratio on the length of the end-effect. Similarly, the effect of wall thickness ratio on the armature expansion angle is depicted in Fig. 4.41. It is interesting to note the near perfect linear relationships between the wall thickness ratio, and the ensuing end-effect ratio or expansion angle of the armature. Therefore, for copper and aluminum armatures using C-4 explosive, the following relationships define the severity of the armature's end-effect and its corresponding expansion angle.

Table 4.3. Detonation end effect ratios at different wall thickness ratios.

Wall thickness ratio a	End-Effect Ratio, υ / Expansion Angle, θ	
	OFHC Copper	Aluminum 6061-T6
0.04	0.560 / 14.88	0.320 / 22.58
0.08	0.715 / 10.47	0.401 / 18.90
0.12	0.960 / 8.75	0.480 / 16.62
0.16	1.36 / 6.65	0.561 / 14.08
1/6	1.38 / 6.35	0.562 / 13.55
0.20	1.6 / 6.08	0.577 / 13.17

For OFHC copper, the length of the armature's end-effect and its expansion angle are given by Eqs. (4.43) and (4.44), respectively.

$$L_D = 6.863W + 0.219r_A \text{ (mm)} \tag{4.43}$$

$$\theta = 15.85 - 54.68\frac{W}{r_A} \text{ (degrees)} \tag{4.44}$$

For 6061-T6 aluminum, the length of the armature's end-effect and its expansion angle are given by Eqs. (4.45) and (4.46), respectively.

$$L_D = 1.72W + 0.226r_A \text{ (mm)} \tag{4.45}$$

$$\theta = 24.25 - 60.76\frac{W}{r_A} \text{ (degrees)} \tag{4.46}$$

The above results indicate that the severity of the armature's end-effect is directly proportional to its wall thickness ratio, while its expansion angle is inversely proportional to the wall thickness ratio.

Therefore, for optimal performance (small end-effect and large expansion angle) the wall thickness ratio must be minimized. As expected, Fig. 4.41 shows that aluminum armatures expand far more easily than copper armatures of the same geometry. Conversely, in order to create the same expansion angle (for example approximately 15 degrees), the wall thickness ratio of the aluminum armature can be much higher (up to 0.16) than its copper counterpart (only up to 0.04). This means that for FCG designs where the size of the armature (in terms of its outer radius) must be limited, aluminum armatures are preferable over copper armatures.

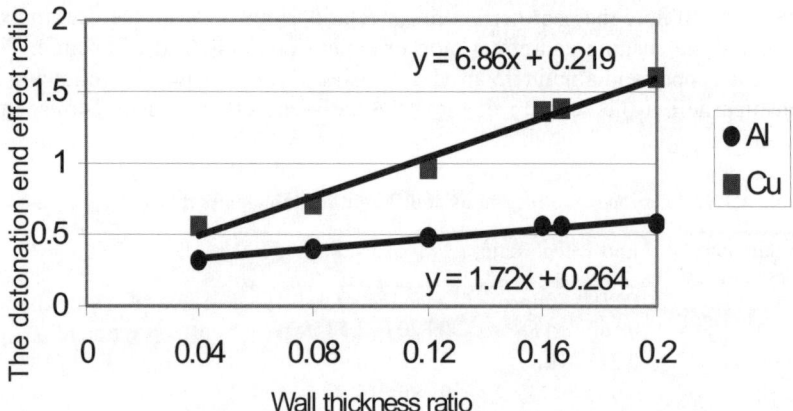

Fig. 4.40. Effect of wall thickness ratio on the severity of armature's end-effect.

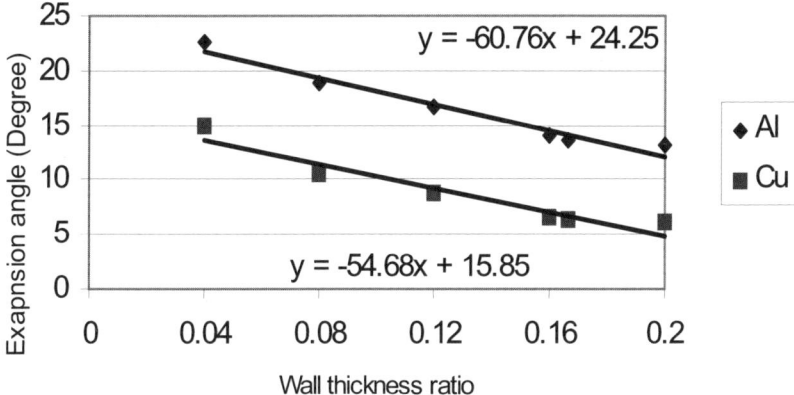

Fig. 4.41. Effect of wall thickness ratio on the severity of armature expansion angle.

4.3 Criteria for Prevention of Armature "Turn-Skipping"

4.3.6 Introduction

As discussed earlier, the efficiency of small helical FCGs can be quite low, largely due to magnetic flux and resistive losses, cf. Chaps. 2 and 5.3. Additionally, the performance of helical FCGs can be significantly limited by non-uniformities in the armature, as well as armature-stator misalignment resulting in the leading edge of the contact (or just simply contact point) between the armature and the helical coil to jump from one turn to another. This phenomenon is commonly referred to as "turn-skipping" [Kno70] and governed by manufacturing (armature wall thickness tolerance), assembly (eccentricity tolerance of the armature with respect to the stator), pitch of the helical turns, and the expansion angle of the armature. The following sections present an overview of conditions resulting in "turn-skipping", and design guidelines for optimum design and prevention of "turn skipping".

4.3.7 Eccentricity Tolerance

In a perfectly aligned and concentric armature/stator assembly, and assuming that the expansion of the armature is uniform, there is only one contact between the armature and the helical coil at any instant. During the operation of helical FCGs, the leading edge of the armature/stator contact moves along the helical coils to compress the magnetic flux into the load coil resulting in current and energy amplification. Any distinct manufacturing or assembly-induced eccentricity of the armature, with respect to the helical coils, would result in the leading edge of the contact jumping either partially within one turn or to skip one, possibly several, turns of the stator altogether.

Previous work reported in the literature is based on the assumption that the armature and stator are perfect in geometry (manufacturing), and that the eccentricity of the armature, with respect to the stator, is mainly caused by assembly imperfections. A study of the literature revealed three criteria, or equations, for formulating acceptable eccentricity tolerance between the armature and the stator. In 1970, Heinz Knoepfel [Kno70] presented the first of such equations, see Eq. (4.47), for determining the maximum allowable eccentricity tolerance of armatures with respect to the helical coil

$$\Delta a = \frac{P}{2} \tan \theta , \qquad (4.47)$$

where Δa is the maximum allowable eccentricity tolerance, P is the pitch of the helical coils, and θ is the expansion or Gurney angle of the armature.

Other researchers [Che86, Pin89] have developed similar equations for quantifying the maximum allowable eccentricity tolerance, as given by Eqs. (4.48) and (4.49).

$$\Delta a = \frac{P}{2\pi} \tan \theta \qquad (4.48)$$

$$\Delta a = \frac{P}{4} \tan \theta \qquad (4.49)$$

Although the magnitude of the maximum allowable eccentricity tolerances obtained from Eqs. (4.47) through (4.49) are quite different, all three equations suggest that the maximum allowable eccentricity tolerance is independent of the armature and the helical coils' radius, and only governed by the expansion angle and the pitch of the helical coils. Based on above equations, in order to prevent "turn skipping", one must incorporate the largest stator pitch and the largest expansion angle possible.

According to the schematic of armature and helical coils shown in Fig. 4.42, the maximum allowable eccentricity tolerances can be derived as Eq. (4.49). In Fig. 4.42, r_H is the inner radius of the helical coils (stator). When the maximum eccentricity is larger than the allowable value determined in Eq. (4.49), the leading edge of the contact (or simply contact point) will jump at least a half-turn ahead and the magnetic flux trapped in between will be lost.

In order to determine which of the Eqs. (4.47) through (4.49) is most suitable for determining the maximum allowable eccentricity tolerance, the following geometrical approach can be used. In Fig. 4.42, C is the contact point between the armature and the helical coils. With respect to the coordinate system X-Z, oriented as shown in Fig. 4.42, the radial position (X-coordinate) of any point located on the helical coil can be described as:

$$X = r_H \cos\left(\frac{2\pi}{P} Z\right) . \qquad (4.50)$$

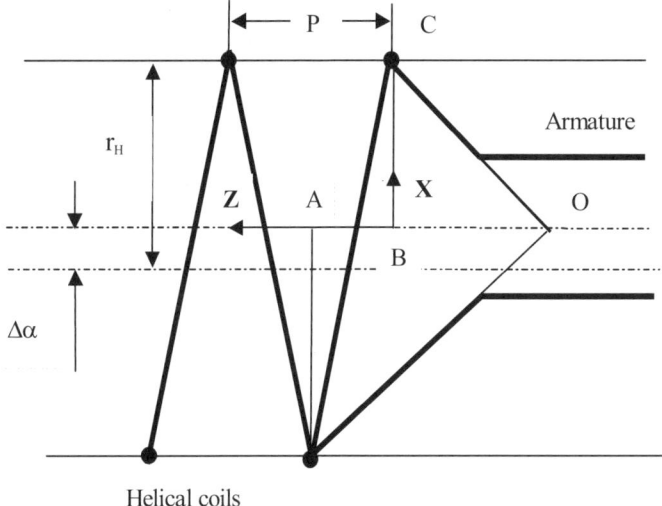

Fig. 4.42. Geometry of armature eccentricity with respect to stator.

Assuming a constant radius r_H for the helical coil, the coordinates X and Z of (4.50) correspond to a unique point on the helical coil. The radius of expansion of the armature can be written as:

$$r_A = (r_H - \Delta a) + X \tan\theta \ . \tag{4.51}$$

The expanding armature intersects the inner surface of the helical coil cylinder along a line. A point along this interaction line satisfies an equation given by:

$$r_H^2 - X^2 = r_A^2 - (X - \Delta a)^2 \ . \tag{4.52}$$

Substituting for r_A from Eq. (4.51) in Eq. (4.52) yields the equation describing the armature/stator interaction line:

$$X = r_H - \left(\frac{r_H - \Delta a}{\Delta a}\right) Z \tan\theta - \frac{(Z \tan\theta)^2}{2\Delta a} \ . \tag{4.53}$$

The intersection point between the two curves prescribed by Eq. (4.50) and Eq. (4.53) defines a second contact point between the expanding armature and the helical coils. In this research, FCGs with pitch and expansion angle of 3.046 mm and 13.55 degrees, respectively, were manufactured and tested. The maximum allowable eccentricity tolerances for the exemplar FCGs, as obtained from Eqs. (4.47) through (4.49), are equal to 0.367, 0.117, and 0.184 mm, respectively. Fig. 4.43 shows the behavior of Eqs. (4.50) or (4.53) with three different eccentricity values obtained from Eqs. (4.47) through (4.49).

112   4 Mechanical Aspects

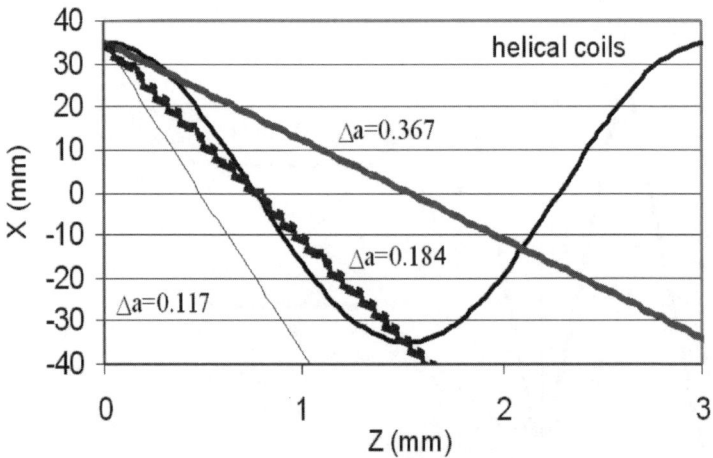

Fig. 4.43. Comparison of three different criteria for determining the maximum allowable eccentricity tolerance.

Referring to Fig. 4.43, the eccentricity value obtained from Eq. (4.47), 0.367, will cause three simultaneous contact points between the armature and the helical coils resulting in the contact points to jump more than half of a helical turn. The eccentricity value obtained from Eq. (4.48), 0.117, will assure that there is no "turn-skipping". The eccentricity from Eq. (4.49), 0.184, will also cause three simultaneous contact points between the armature and the helical coils, resulting in the contact point to jump only half a turn as expected.

Based on the above analysis, the criterion for maximum allowable eccentricity from Eq. (4.47) is too loose, while the allowable eccentricity obtained from Eq. (4.48) is too strict. The criterion for eccentricity obtained from Eq. (4.49) is reasonable, and recommended for use as the design criteria for prevention of "turn-skipping" as a result of eccentricity values exceeding an "allowable" tolerance.

4.3.8 Armature Wall Thickness Tolerance

Another parameter influencing the "turn-skipping" phenomenon is the armature wall thickness tolerance. In order to focus on this parameter, we assume that the eccentricity of the armature with respect to the helical coils is zero, and the armature wall thickness tolerance is δW. It is further assumed that the armature is a perfect cylinder, thus its expansion is uniform with a constant angle. In this case, the worst condition is shown in Fig. 4.44.

Compared with Fig. 4.42, the armature wall thickness tolerance, δW, is equivalent to an eccentricity tolerance of $\delta W/\cos(\theta)$. Therefore, the criterion for maximum allowable wall thickness tolerance can be written as:

$$\frac{\delta W}{\cos(\theta)} = \frac{P}{4}\tan(\theta) \cdot \qquad (4.54)$$

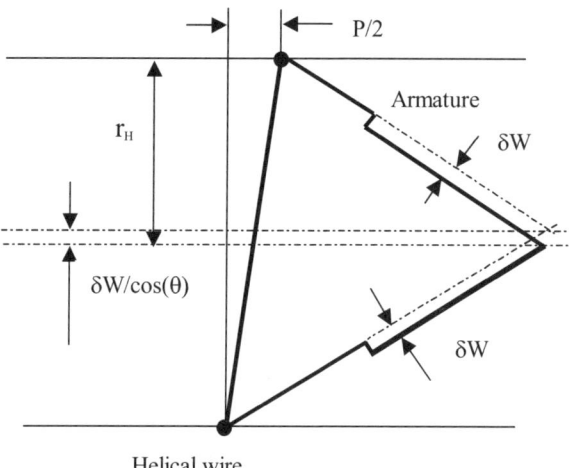

Fig. 4.44. Geometry of variation in armature wall-thickness.

Rewriting the above equation yields:

$$\delta W = \frac{P}{4}\sin(\theta)\cdot \quad (4.55)$$

Eq. (4.54) indicates that the armature expansion angle and the pitch of the helical coils govern the maximum allowable variation in armature wall thickness. A larger pitch and a larger expansion angle are helpful in preventing or minimizing "turn skipping."

4.3.9 Combined Eccentricity and Wall Thickness Tolerances

The worst situation occurs when both armature eccentricity and wall thickness variations are present in a given helical FCG, as shown in Fig. 4.45. In this case, the maximum combined allowable eccentricity and wall thickness tolerance can be written as

$$\Delta a + \frac{\delta W}{\cos(\theta)} = \frac{P}{4}\tan(\theta)\cdot \quad (4.56)$$

In summary, the magnetic flux loss due to "turn skipping" phenomenon can dramatically decrease the efficiency of helical FCGs. The most suitable criterion for maximum allowable eccentricity tolerance is given by Eq. (4.49). Variations in the armature's wall thickness also affect "turn skipping".

Eq. (4.54) can be used to determine the maximum allowable wall thickness tolerance. The criterion for combined eccentricity and wall thickness variations is described by Eq. (4.55).

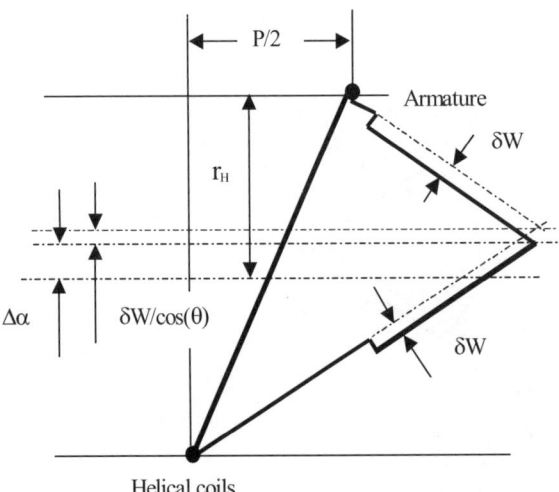

Fig. 4.45. Geometry of combined armature eccentricity and wall-thickness variation.

In general, the phenomenon of "turn-skipping" is directly influenced by: 1) the magnitude of the armature eccentricity with respect to the helical coils, 2) the armature's wall thickness tolerance, 3) pitch of the helical coils, and 4) the armature's expansion angle. The maximum allowable eccentricity and wall thickness tolerances are achieved when the pitch of the helical coils and the armature expansion angle are maximized. The above tolerances are independent of the helical coils' radius, due to the assumption of constant armature expansion angle.

4.4 Scaling of the Armature Expansion Angle

Most of the research reported in the literature has been conducted utilizing FCGs of a fixed size without considerably varying the overall size in a single investigation. It is therefore important to determine the applicability of such results when the physical dimensions of the FCG are scaled up or down. We address in the following the scaling of the armature's expansion angle with the armature's physical dimensions.

In any FCG, the geometrical parameters of the cylindrical armature can be defined as: the outer diameter, D_0, the inner diameter, D_i, and the length, L. The armature's wall thickness is denoted as W. A characteristic non-dimensional parameter, namely, "the wall-thickness-to-radius-ratio", a, can be defined as

$$a = \frac{W}{D_0/2} = \frac{2W}{D_0} \quad . \tag{4.57}$$

The inner diameter of the armature can then be expressed in terms of the wall thickness-to-radius ratio, a, as follows

4.4 Scaling of the Armature Expansion Angle

$$D_i = D_0(1-a). \tag{4.58}$$

The theoretical analysis of the "scaling" problem can be derived from the theory used to predict the velocity of an explosive-driven metal. [Gur43] originally proposed this simple technique in predicting the expansion rate of cylindrical bomb casings during detonation, cf. Chap. 2.1. [Ken70] extended the technique by applying it to other geometries. For the case of a cylinder, the Gurney equation [Gur43] is given by

$$\frac{V}{\sqrt{2E}} = \left(\frac{M}{C} + \frac{1}{2}\right)^{-\frac{1}{2}} \quad \text{for } 0.2 < \frac{M}{C} < 10. \tag{4.59}$$

where V is the final velocity of the metal in a direction normal to the armature's surface. $\sqrt{2E}$ is called "the Gurney velocity" of the explosive, while E is the specific energy of the explosive. The Gurney velocity can be approximated using the following equation:

$$\sqrt{2E} = \frac{V_D}{3}. \tag{4.60}$$

where V_D is the detonation velocity of the explosive. This estimation is usually within 7% of experimentally observed values.

In Eq. (4.59), M is the mass of the cylinder, while C represents the mass of the explosive. For the analysis of a cylindrical armature, the mass of the cylinder, M, and that of the high explosives, C, can be represented in terms of their geometrical parameters and densities as follows:

$$M = \rho_A \frac{\pi}{4} D_0^2 (2a - a^2) L \tag{4.61}$$

and

$$C = \rho_E \frac{\pi}{4} D_0^2 (1-a)^2 L. \tag{4.62}$$

Therefore,

$$\frac{M}{C} = \frac{\rho_A}{\rho_E} \frac{(2a - a^2)}{(1-a)^2}. \tag{4.63}$$

Substituting Eqs. (4.60) and (4.63) into Eq. (4.59) yields

$$V = \frac{V_D}{3} \left(\frac{\rho_A}{\rho_E} \frac{(2a - a^2)}{(1-a)^2} + \frac{1}{2} \right)^{-\frac{1}{2}}. \tag{4.64}$$

The non-dimensional equation for the expansion angle of the armature is therefore be given by

$$\theta \approx 2\sin^{-1}\left(\frac{V}{2V_D}\right) = 2\sin^{-1}\left[\frac{1}{6}\left(\frac{\rho_A}{\rho_E}\frac{(2a-a^2)}{(1-a)^2}+\frac{1}{2}\right)^{-\frac{1}{2}}\right]. \qquad (4.65)$$

4.4.1 Finite Element Analysis of the Armature Expansion Angle

The materials in an FCG are subjected to explosive shock loading. Our previous research [Le02] has shown that the Steinberg material constitutive equation and equation of state [Ste91] can be effectively used to describe the behavior of armature material in shock loading situations.

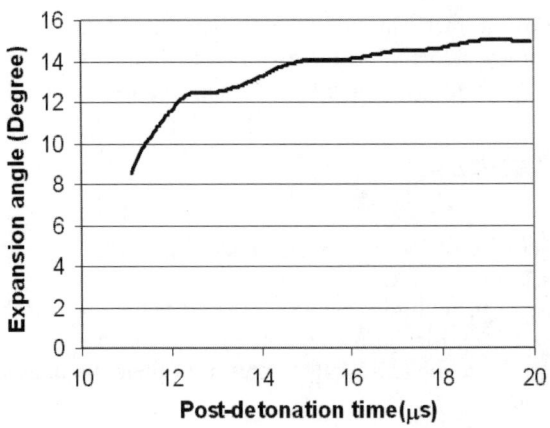

Fig. 4.46. Armature expansion angle vs. post-detonation time (Aluminum armature, $D_0 =$ 38.1 mm, $a = 1/6$).

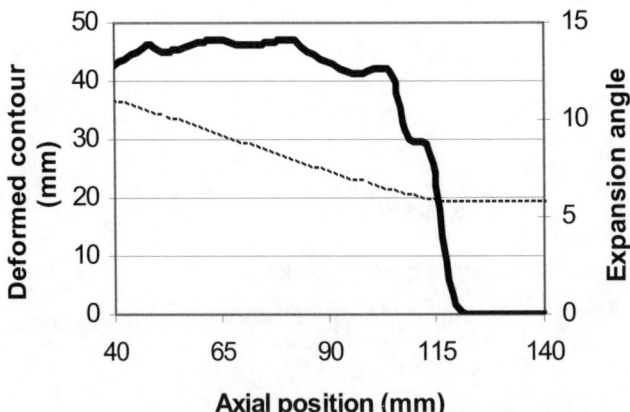

Fig. 4.47. Armature's expansion angle (thick line) with axial position and armature's deformed contour (thin line) (Aluminum armature, $D_0 = 38.1$ mm, $a = 1/6$).

4.4 Scaling of the Armature Expansion Angle

Our previous works [Le01, Ras01] also indicate that the Jones-Wilkins-Lee (JWL) equation of state [Dob85] can be used to effectively and accurately describe the behavior of the explosive material. For our numerical calculations, we used the experimentally verified Finite Element models to investigate the effect of "scaling" on the armature expansion angle in an FCG. A commercial finite element code, LS-DYNA [Liv97], was used to conduct the finite element simulations.

The expansion angle of the armature throughout the explosion process has generally been treated as a constant by some investigators [Jon79, Her79, Che96], but our finite element results show that the expansion angle changes with post-detonation time and at different axial positions of the armature. Fig. 4.46 shows the variation of the expansion angle of a segment of the armature vs. post-detonation time. The results indicate that the expansion angle increases with post-detonation time, but it eventually reaches a steady value. In Fig. 4.47, the dotted line shows the deformed armature contour while the solid line shows the expansion angle at corresponding points along the deformed armature contour. Fig. 4.46 shows the variation in the armature's expansion angle along the armature. The results indicate that the deformed contour of the armature is not exactly a straight line since the expansion angles changes from one axial position to the next and with post-detonation time. However, the assumption that the deformed contour has a constant expansion angle has the advantage of simplifying many analyses.

Since most of the data utilized in this research was obtained through numerical simulations of the explosion process, we had the luxury of knowing how the expansion angle changes with time and at various armature locations. For the purpose of this research and the material presented in the following sections, we used the average of all expansion angles along the armature's length to represent the expansion angle of the armature. The impact of scaling on the armature's expansion angle is discussed in the following.

The following properties were assumed for copper, aluminum and high explosive C-4:

$$\rho_{Explosive} = 1.601 \frac{g}{cm^3} \quad V_D = 8.193 \frac{cm}{\mu s} \quad (4.66)$$

$$\rho_{Copper} = 7.8 \frac{g}{cm^3} \quad \rho_{Aluminum} = 2.703 \frac{g}{cm^3}$$

The dimensions of the armature are set to

$$D_0 = 38.1 \text{ mm}, \ D_i = 31.75 \text{ mm}, \ L = 152.4 \text{ mm}, \ a = 1/6. \quad (4.67)$$

Then, the expansion angles for copper and aluminum armatures from the Gurney equation, cf. Eq. (4.65), are:

$$\theta_{Copper} = 11.77^0 \quad \theta_{Aluminum} = 17.20^0. \quad (4.68)$$

The above Gurney estimations of armature expansion angles are somewhat different than those obtained in this research experimentally or by FE simulations. Explosive experiments, coupled with high-speed photographic techniques, yielded an expansion angle of 6.57° for copper armatures, while numerical simulations resulted in an armature expansion angle of 6.35°. Similarly, the experimentally obtained expansion angle for aluminum armatures was 13.40°, while numerical simulation results yielded an armature expansion angle of 13.55°. As one can see, there is an excellent agreement between experimental and FE simulation results, both of which point out to an over-estimation of the expansion angle by the Gurney equation. The discrepancy between the results obtained through the Gurney equation, Eq. (4.65), and those obtained in this research could be attributed to Gurney equation's assumption of a completely confined explosive, which is not necessarily the case in an open-ended, single end initiated, FCG.

4.4.2 Analysis of Scaling

Eq. (4.65) indicates that the expansion angle is only a function of the densities of the armature and the explosive, as well as the armature's wall-thickness-to-radius ratio. Therefore, it can be concluded that the expansion angle is independent of the scaling factor. In other words, if we use the same armature material and the same explosive, the expansion angle of different size armatures having the same wall-thickness-to-radius ratio, would be the same. The finite element analysis was used to verify this conclusion.

Utilizing our experimentally verified finite element models, two groups of numerical calculations were conducted. In the first group, different size armatures having the same wall-thickness-to-radius ratio were considered.

The pertinent information regarding the FE model as well as the results of numerical calculations are shown in Table 4.4 where $\theta_{Numerical}$ denotes the expansion angle obtained from numerical calculations, while θ_{Gurney} represents the expansion angle obtained from Eq. (4.65). The data indicate that the expansion angle is independent of the scaling factor, however, with some difference in the numerical result for the two calculation models.

The second group of numerical calculations was conducted on armatures having different wall-thickness-to-radius ratios. Based on the data available in the literature, an FCG armature's wall thickness-to-radius ratio ranges from 0.04 to 0.2.

Table 4.4. Effect of scaling on the expansion angle of aluminum armatures, first group.

D_0 (mm)	D_i (mm)	W (mm)	L (mm)	$a = 2W/D_0$	Scaling factor	$\theta_{Numerical}$	θ_{Gurney}
38.1	31.75	3.175	152.4		1	13.55	
45	37.5	3.75	180	1/6	1.18	13.76	17.20
63	52.5	5.25	252		1.653	13.53	
76.2	63.5	6.35	304.8		2	13.63	

Table 4.5. Geometry and expansion angles for aluminum armatures with varying wall thickness-to-radius ratios, second group.

D_0 (mm)	D_i (mm)	W (mm)	L (mm)	$a = 2W/D_0$	$\theta_{Numerical}$	θ_{Gurney}
33.073		0.661		0.04	22.58	23.98
34.511		1.380		0.08	18.90	21.39
36.080	31.75	2.165	152.4	0.12	16.62	19.27
37.800		3.024		0.16	14.08	17.47
38.1		3.175		0.167	13.55	17.20
39.688		3.969		0.20	13.17	15.91

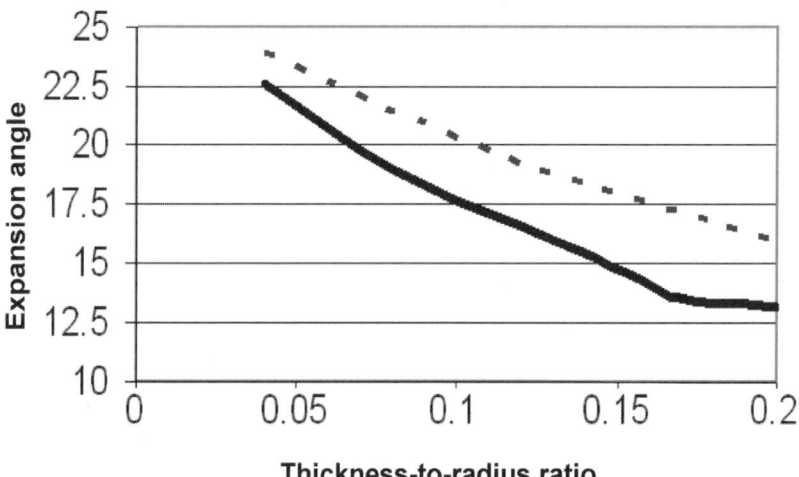

Fig. 4.48. Effect of wall thickness-to-radius ratio on the expansion angle of aluminum armatures – Numerical results (solid line) vs. Gurney estimation (dashed line).

The pertinent information regarding the armature geometries for the second group of numerical calculations and the results is shown in Table 4.5 and depicted in Fig. 4.48 by the solid line.

4.4.3 Modification of Gurney Equation

Results presented in Tables 4.4 and 4.5 indicate that the Gurney equation does not provide an accurate prediction of the armature expansion angle in a helical FCG. However, as expected, both the Gurney equation and numerical results agree in the increasing trend of the armature's expansion angle as the wall-thickness-to-radius ratio of the armature decreases.

Table 4.6. Comparison of $\theta_{Numerical}$, θ_M and θ_{Gurney}.

$a = 2W/D_0$	$\theta_{Numerical}$	θ_M, Eq. (4.69)	θ_{Gurney}, Eq. (4.65)
0.04	22.58	22.22	23.98
0.08	18.90	19.06	21.39
0.12	16.62	16.51	19.27
0.16	14.08	14.39	17.47
0.167	13.55	14.07	17.20
0.20	13.17	12.60	15.91

We modify the Gurney equation, as presented in Eq. (4.65), in order to provide a closer approximation of the armature expansion angle in an FCG. The following equation provides a better agreement with our experimental and numerical results.

$$\theta = Exp(A + Ba)\theta_{Gurney} = 2\exp(A + Ba)\sin^{-1}\left[\frac{1}{6}\left(\frac{\rho_A}{\rho_E}\frac{(2a - a^2)}{(1-a)^2} + \frac{1}{2}\right)^{-\frac{1}{2}}\right]. \quad (4.69)$$

where the coefficients A and B were determined by the least square method to closely match the numerical calculation results.

It was determined that for aluminum armatures charged with C-4 high explosive, the following values for the coefficients A and B in the modified Gurney Eq. (4.69), provide a better estimation of the armature's expansion angle:

$$A = -0.0368, B = -0.984. \quad (4.70)$$

Table 4.6 shows the comparison of armature expansion angles as obtained by numerical calculations, $\theta_{Numerical}$, the modified Gurney equation, Eq. (4.69) with A, B values from Eq. (4.70), θ_M, and the Gurney equation Eq. (4.65), θ_{Gurney}.

4.4.4 Conclusion Expansion Angle

In conclusion, the expansion angle of the armature during the operation of FCGs is not constant. The average expansion angle can be used when a constant expansion angle must be assumed. The expansion angle of the armature in an FCG is independent of the scaling factor. The expansion angle is a function of the densities of the armature and explosive materials, as well as the armature's wall-thickness-to-radius ratio. The armature expansion angle in an FCG, as obtained from the Gurney equation, is much larger than the expansion angles observed experimentally or obtained via numerical simulations. This discrepancy is attributed to the Gurney equation's assumption that the explosive in an FCG is completely confined. A "modified" Gurney equation, Eq. (4.69), is presented for aluminum armatures charged with C-4 high explosive, which more closely matches the experimentally and numerically obtained results.

4.5 Armature treatment

For the armature to expand to its largest diameter without breaking up, a compromise between material softness and strength needed to be found. Full annealing of wrought aluminum alloys results in the greatest amount of ductility in the material and is given the temper designation "O". Full annealing of 6061 aluminum alloys can be achieved by heating the material to a temperature of 415 C (775 F), and maintaining the annealing temperature for 2-3 hours. The cooling should be conducted at a rate of 30 C/hr. (50 F/hr.) from the annealing temperature to 260 C (500 F). Rate of cooling below 260 C is unimportant.

In general, larger expansion ratios lead to better FCG performance since the initial inductance is larger for a given stator diameter. We found that partial annealing of 6061 aluminum and OFHC copper armatures worked best in producing intermediate mechanical properties needed for expansion ratios of up to ~ 3 before breakup. Partial annealing of aluminum armatures (temper designation H-2) is achieved at temperatures below those needed to produce extensive recrystallization, thereby resulting in incomplete softening due to sub-structural changes in dislocation density. For partial annealing of aluminum, we heated the armatures to 345 C (650 F), followed by subsequent cooling to room temperature. No holding time at annealing temperature, or controlled cooling rate is required for this purpose. In the case of OFHC copper armatures, partial annealing was achieved by heating the armatures to 240 C (465 F) for 20 minutes under an inert, nitrogen, atmosphere, followed by uncontrolled cooling to room temperature.

Besides the annealed state of the armature material, the surface finish of the armature might have some effect on the FCG operation. We primarily investigated the effect of armature surface finish on the expansion characteristic of the armature. To this end, as-received aluminum armatures having a nominal inner and outer surface finish of 22 - 24 µm were either highly polished to 3 – 8 µm, or roughened to 48 - 86 µm, representing five different inner and outer surface finish combinations as shown in Table 4.7.

For each surface finish combination, four specimens were prepared with identical surface finish. The armatures were subsequently fired and their expansion angles were recorded.

The experimental results, see Table 4.7, reveal that the difference between the worst-case scenario (surface finish of 86 µm), and best-case scenario (surface finish of 3 µm) is only about ½ of a degree and well within the measurement accuracy. Therefore, we conclude that any effort in improving the armature's surface finish will not produce any significant improvements in the armature's expansion characteristic as measured by the extent of its expansion capability before breakup. Further, the helical FCG performance, current/energy gain, was largely unaffected by polishing the armature. Hence, the majority of helical FCGs had as-received armatures that where carefully inspected, and machined if needed, to be within mechanical tolerances given in Chap. 4.4. If machined, the armature was in a final step machine polished before assembly of the FCG.

Table 4.7. Armatures tested with varying inner and outer surface finish.

Specimen No.	Inner surface finish (μm)	Outer surface finish (μm)	Average expansion angle (degree)
1	24 (as-received)	22 (as-received)	13.54 ± 0.6
2	24 (as-received)	48 (mid rough)	13.68 ± 0.2
3	24 (as-received)	86 (rough)	13.26 ± 0.8
4	24 (as-received)	3 (polished)	13.64 ± 0.5
5	8 (polished)	3 (polished)	13.74 ± 0.8

References

[Bon94] Bonder SR, Rubin MB (1994) Modeling of hardening at very high strain rates. J. Appl. Phys. 76: 2742-2747

[Bro82] Broek D (1982) Elementary engineering fracture mechanics. Martinus Nijhoff Publishers, The Hague

[Bus70] Busco M (1970) Optical properties of detonation waves (optics of explosives) In: Fifth Symposium (International) on Detonation, pp. 513-22

[Cer66] Cernica JN (1966) Strength of materials. Holt, Rinehart and Winston, Inc., New York

[Che86] Chernyshev VK, Zharinov EI, Kazakov SA, Busin VN, Vaneev VE, Korotkov MI (1987) Magnetic flux cutoffs in helical explosive magnetic generators. In: Fowler CM, Caird RS, Erickson DJ (eds) Megagauss technology and pulsed power applications. Plenum Press, New York, pp 455-469

[Che96] Chernykh Y, Nersterov Y, Shurupov A, Karpushin Y, Zolotykh I (1997) Two cascade MCG for generation of rapidly increasing current pulses. In: Chernyshev VK, Selemir VD, Plyashkevich LN (eds) Megagauss and megaampere pulse technology and applications, part 1. RFNC-VNIIEF, Sarov, Russia, pp 327-332

[Col81] Collins JA (1981) Failure of Materials in Mechanical Design. John Wiley & Sons, New York

[Coo55] Cook MA (1951-1955) Detonation wave fronts in ideal and non-ideal detonation. In: First and Second Symposia on Detonation, pp 500-518

[Coo58] Cook MA (1958) The science of high explosives. Robert E. Krieger Publishing Co. Inc., Huntington, N. Y.

[Coo74] Cook MA (1974) The science of industrial explosives. IRECO Chemicals, Salt Lake City, Utah

[Coo96] Cooper PW (1996) Explosives engineering. Wiley-VCH, New York

[Der98] Deryugin YN, Korolev PV, Kargin VI, Pikar AS, Popkov NF, Ryaslov EA, (2004) Numerical simulation of magnetic flux compression in helical cone explosion magnetic generators. In: Schneider-Muntau, HJ (Edtr.) Megagauss magnetic field generation, its application to science and ultra-high pulsed power technology: proceedings of the VIIIth international conference on megagauss magnetic field generation and related topics, Tallahassee, FL, USA, October 18-23, 1998, pp 536-539

[Dob85] Dobratz BM, Crawford PC (1985) LLNL Explosives Handbook - Properties of Chemical Explosives and Explosive Simulants, UCRL-52997 Change 2. Lawrence Livermore National Laboratory, Livermore, CA

[Fic79] Fickett W, Davis WC (1979) Detonation - Theory and Experiment. Dover Publications, Inc., Mineola, NY

[Fre92] Freeman BL, Sheppard MG, Fowler CM (1993) A numerically design, experimentally tested, high-current, coaxial generator. In: Cowan M, Spielman RB (eds) Megagauss magnetic field generatoion and pulsed power applications, part 1. Nova Science Publishers, Commack, NY, pp. 565-572

[Fre00] Freeman BL, E-mail correspondence, 06/28/00

[Fre02] Freeman BL, private communication, 2002

[Fri98] Fried LE, Howard WM, Souers PC (1998) CHEETAH 2.0 User's Manual, UCRL-MA-117541 Revision 5. Lawrence Livermore National Laboratory, Livermore, CA

[Gov79] Gover JE, Stuetzer OM, Johnson JL (1979) Small helical flux compression amplifiers. In: Turchi PJ (ed) Megagauss physics and technology. Plenum Press, New York, pp 163-180

[Gro99] Grove JW (1999) Front tracking in one space dimension. Los Alamos Report, LA-UR 99-3985, also available under http://www.ams.sunysb.edu/~shock/FTnotes/frontier/lecture04/sld001.htm

[Gur43] Gurney RW (1943) The initial velocities of fragments from bombs, shells, and grenades. BRL report 405

[Ham01] Hammond G, E-mail correspondence, 03/16/01.

[Har89] Harding J (1989) The development of constitutive relationships for material behavior at high rates of strain. In: Proceedings of the fourth international conference on the mechanical properties of materials at high rates of strain, Oxford, March 19-22, 1989.

[Her79] Herlach F (1979) Pulsed magnetic filed generators and their practical application. In: Turchi PJ (ed) Megagauss physics and technology. Plenum Press, New York, pp 1-25

[Hoo01] Hood A, Method of characteristics. http://www-solar.mcs.st-and.ac.uk/~alan/MT2003/PDE/node5.html, accessed 1/16/01.

[Hos65] Hoskin NE, Allan JWS, Bailey WA, Lethaby JW, Skidmore IC (1965) The motion of plates and cylinders driven by detonation waves at tangential incidence. In: Fourth Symposium (International) on Detonation, pp 14-26

[Joh01] Johnson JN, E-mail correspondence, 02/18/2001.

[Jon79] Jones M (1979) An equivalent circuit model of a solenoidal compressed magnetic filed generator. In: Turchi PJ (ed) Megagauss physics and technology. Plenum Press, New York, pp 249-264

[Ken70] Kennedy JE (1970) Gurney energy of explosives: estimation of the velocity and impulse imparted to driven metals. Tech. report: SC-RR-70-790, Sandia Laboratories

[Kno70] Knoepfel H (1970) Pulsed High Magnetic Fields. North Holland Publishing Company, Amsterdam, London and American Elsevier Publishing Company Inc., New York

[Lam65] Lambourn BD, Hartley JE (1965) The Calculation of the hydrodynamic behaviour of plane one dimensional explosive/metal systems. In: Fourth Symposium (International) on Detonation, pp 538-552

[Las90] Lassila DH, Lebianc M (1990) High strain rate deformation behavior of shocked copper. UCRL-JC-103469, Lawrence Livermore National Laboratory, Livermore, CA

[Le01] Le X, Rasty J, Neuber A, Dickens J, Kristiansen M (2001) Calculation of air temperature and pressure history during the operation of flux compression generator. In: Reinovsky R, Newton M (eds) Digest of technical papers of the 2001 IEEE Pulsed Power Plasma Science Conference, IEEE Press, Piscataway NJ, pp 939-942

[Le02] Le X (2002) Experimental and finite element study of armature dynamics in helical magnetic flux compression generators. Ph.D. Thesis, Texas Tech University
[Lee68] Lee EL, Horning HC, Kury JW (1968) Adiabatic expansion of high explosive detonation products. UCRL-50422, Lawrence Livermore National Laboratory, Livermore, CA
[Liv97] (1997) LS-DYNA keyword user's manual, version 940. Livermore Software Technology Corporation
[Mad98] Mader CL (1998) Numerical modeling of explosives and propellants, second Edition. CRC Press, New York
[Mar80] Marsh SP edtr (1980) LASL shock Hugoniot data. Los Alamos Series on Dynamic Material Properties, Berkeley: University of California Press, p 182
[Mat01] MatWeb Materials Property Database, "Aluminum 6061-T6; 6061-T651," http://www.matweb.com/SpecificMaterial.asp?bassnum=MA6016&group=General, accessed 2/14/01. MatWeb Materials Property Database, "Oxygen-free high conductivity Copper, Soft, UNS C10200," http://www.matweb.com/SpecificMaterial.asp?bassnum=M102A&group=General, accessed 2/14/01.
[Pin89] Pincosy PA, Abe DK, Chase, J.B. (1990) Design of a high-gain flux-compression generator. In: Titov VM, Shvetsov GA (eds) Megagauss fields and pulse power systems, part 1. Nova Science Publishers, Commack, NY, pp 441-448
[Ras01] Rasty J, Le X, Neuber A, Dickens D, Kristiansen M (2001) Experimental and numerical investigation of the armature/stator contact in magnetic flux compression generators. In: Reinovsky R, Newton M (eds) Digest of technical papers of the 2001 IEEE Pulsed Power Plasma Science Conference, IEEE Press, Piscataway NJ, pp 106-109
[Ras02] Rasty J, X Le X (2002) Dynamic behavior of armature and stator in helical MFCGs. Presented to Explosive-Driven Pulsed Power MURI program review at Texas Tech University, 16 August 2002 (unpublished).
[Ric89] Rickel DG, Freeman BL, Fowler CM, Vorthman JE, Marsh SP (1990) Simultaneous helical generator. In: Titov VM, Shvetsov GA (eds) Megagauss fields and pulse power systems, part 1. Nova Science Publishers, Commack, NY, pp 399-402
[Ric01] Rice G, Telephone conversation, 02/22/01.
[Rud98] Ruden EL, Kiuttu GF, Peterkin RE (2004) Explosive axial magnetic flux compression generator armature material strength and compression effects. In: Schneider-Muntau, HJ (Edtr.) Megagauss magnetic field generation, its application to science and ultra-high pulsed power technology: proceedings of the VIIIth international conference on megagauss magnetic field generation and related topics, Tallahassee, FL, USA, October 18-23, 1998, pp 557-562
[Sha53] Shapiro A (1953) The Dynamics and Thermodynamics of Compressible Fluid Flow, Volume I. John Wiley & Sons, New York
[Shk03] Shkuratov SI, Talantsev EF, Dickens JC, Kristiansen M, Baird J (2003) Longitudinal-shock-wave compression of $Nd_2Fe_{14}B$ high-energy hard ferromagnet: The pressure-induced magnetic phase transition, Appl. Phys. Let. 82: 1248-1250
[Ste80] Steinberg DJ, Cochran SG, Guinan MW (1980) A constitutive model for metals applicable at high-strain rate, J. Appl. Phys. 51: 1498-1504
[Ste86] Steinberg DJ, Breithaupt D, Honodel C (1986) Work hardening and effective viscosity in solid beryllium, Physica B&C 139 & 140: 762-765.
[Ste87] Steinberg DJ (1987) Constitutive model used in computer simulation of time-resolved, shock-wave data, Int. J. Impact Eng. 5: 603-611.

[Ste91] Steinberg DJ (1991) Equation of state and strength properties of selected material. UCRL-MA-106439. Lawrence Livermore National Laboratory, Livermore, CA

[Sun01] (2001) "Multi-Material Flows," http://www.ams.sunysb.edu/~shock/FTnotes/frontier/lecture04/sld001.htm, accessed 1/9/01.

[Ugu79] Ugural AC, Fenster SK (1979) Advanced strength and applied elasticity. Elsevier North Holland Publishing Company, New York

[Von76] Von Holle WG, Trimble JJ (1976) Shaped charge temperature measurement. In: Sixth Symposium (International) on Detonation, pp 691-699

[Wal89] Walters WP, Zukas JA (1989) Fundamentals of Shaped Charges. CMC Press, Baltimore, MD

[Wei00] Weisstein E (2000) "Helmholtz Free Energy," http://treasure-troves.wri.com/physics/HelmholtzFreeEnergy.html, accessed 11/28/00

[Whi68] White HE (1968) Introduction to College Physics. D. Van Nostrand Company, New York

[Wil73] Wilkins ML, Guinan MW (1973) Impact of cylinders on a rigid boundary, J. Appl. Phys. 44: 1200-1206.

[Zuk92] Zukas LA, Nicholas T, Swift HF, Greszczuk LB, Curran DR (1992) Impact dynamics. Krieger Publishing Company, Malabar, FL

5 Basic Physics

Bruce L. Freeman and Andreas A. Neuber

5.1 Shocked Gases within the Flux Compression Generator Volume

The environment within the cavity of a flux compression generator is hostile to insulators and insulating gases. In particular, the shock driven through the armature and into this gas is sufficient to provide significant heating. However, the reflected shocks between the armature and the stator as the armature approaches contact with the stator can severely heat the gas or simply ionize it. With this in mind, a set of experiments was designed to simulate the environment within an FCG and to measure the electrical conductivity of various gases in this environment. Improved electrical conductivity data for gases, particularly sulfur hexafluoride and air, under conditions found within FCG's, may facilitate better FCG modeling and design. These tests provide data toward quantifying the considerations of using SF_6 versus air as the fill gas within an FCG.

5.1.1 Generator Model and Shock Tube Design

A cylindrical FCG with the dimensions listed in Fig. 5.1 was used as the model for the shock tube design. The Gurney method was used to determine the armature expansion velocity and angle. This result was translated to a relative velocity at which the armature approached the stator. The armature angle was calculated to approach the stator at a relative velocity of 3.1 km/s at an angle of 22.6° using the Gurney equations for a cylinder [Ken97]. We note that this value is of the same magnitude as a typical armature radial expansion velocity of ~ 2 km/s or 15° angle for small FCG [Gov79].

These parameters were used to design the shock tube with Gurney equations for a finite asymmetric sandwich with cylindrical tamping [Ken97]. For the experiments, a 2.5-cm ID acrylic tube was used as the shock tube and a 1-mm-thick aluminum disc was used as the flyer plate. About 50 grams of Composition C-4 were used per shot to create a calculated flyer plate velocity of 3.1 km/s.

Table 5.1. Explosive generator dimensions

Component	Material	Inner Radius	Outer Radius
Explosive	Comp C-4	-	3 cm
Armature	Aluminum	3 cm	3.2 cm

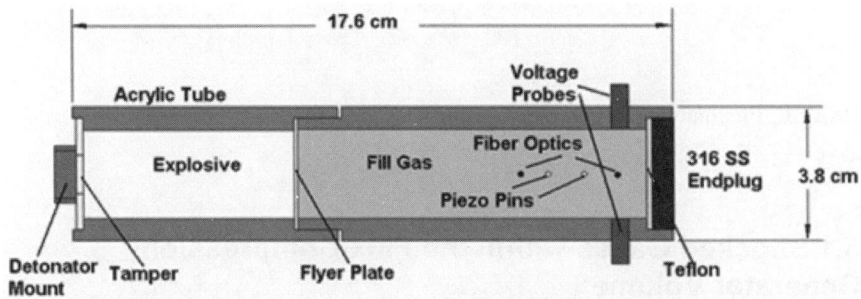

Fig. 5.1. The geometry of the shock tube experiment is shown.

Although the effective explosive mass acting on the flyer plate was only 6 grams, the additional mass, i.e. charge length, allowed the detonation wave to become quasi planar by the time it reached the flyer plate. See Fig. 5.1 for a schematic of the shock tube. RP-501 detonators, manufactured by RISI, were used to initiate the explosive charge.

5.1.2 Simulations

Two dimensional simulations with the Sandia National Laboratory hydrodynamics code, CTH, were performed to verify shock tube operation. Simulations with SF_6 were approximated because of the lack of a good equation of state for this gas. However, several simulations that match well with experimental data were performed with helium.

A plot from the CTH helium simulation is shown in Fig. 5.2. The frame shows the respective shock tube state 28 microseconds after the explosive burn begins. The left hand side of the plot shows material position. The right hand side shows pressure in bars and indicates the position of the shocked regions. The simulations show that the flyer plate becomes slightly curved while being driven, but the initial and reflected shock fronts remain flat.

5.1.3 Diagnostics

The primary measured quantities were voltage and current. A doubly terminated 100:1 capacitive divider circuit was used to measure voltage, as shown in Fig. 5.3. Voltage probes were constructed from 5 mm diameter copper wire. Current was measured with a Pearson transformer at the capacitor bank. The current and voltage measurements were used to calculate the bulk resistance of the region between the voltage probes. The initial voltage applied to the 84 μF capacitor bank was for all experiments set to 2 kV, a typical voltage level observed in the small sized single-staged FCGs. The centerline of the voltage probes was 7.5 mm from the SS end plug.

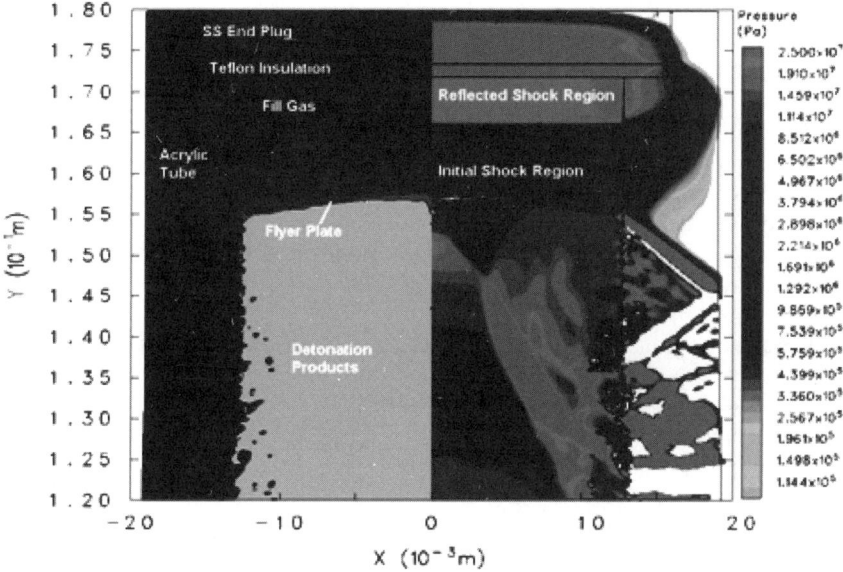

Fig. 5.2. A CTH plot at 28 μsec is shown. The left scale is position and the right scale indicate grey level pressure indication.

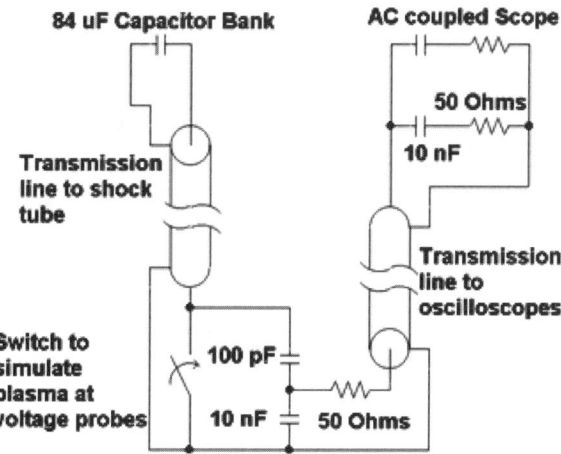

Fig. 5.3. A schematic of the diagnostic circuit used for the resistance measurement is shown.

The use of a capacitive divider is a departure from other conductivity experiments [Ham88], [Mit82], [Wei96], [Nel99]. Series shunt resistors that limit the current and voltage across the shock tube probes improperly simulate the FCG environment, where current through an arc is not internally limited. Also, the resistive voltage dividers, though tried, were unsuccessful because the high resistances required to limit the leakage current produced poor signal to noise ratios.

5.1.4 Synthetic Air Data

Synthetic air (20% oxygen/80% nitrogen) was used at atmospheric pressure and ambient temperature. For the data shown, the shock velocity was measured to be 4.67 km/s using piezoelectric pin data with a 10 mm separation distance. The flyer plate and reflected shock velocities were calculated to be 3.82 km/s and 9.19 km/s, respectively, using the Hugoniot equations. Table 5.2 shows the average properties for all synthetic air shots fired.

Table 5.2. Average measured initial shock velocity and calculated pressure for all synthetic air shots

Shock Properties	Shock Velocity [km/s]	Pressure [MPa]
Initial Shock	4.6 ± 0.1	46 ± 3
Reflected Shock	9.0 ± 0.3	180 ± 10

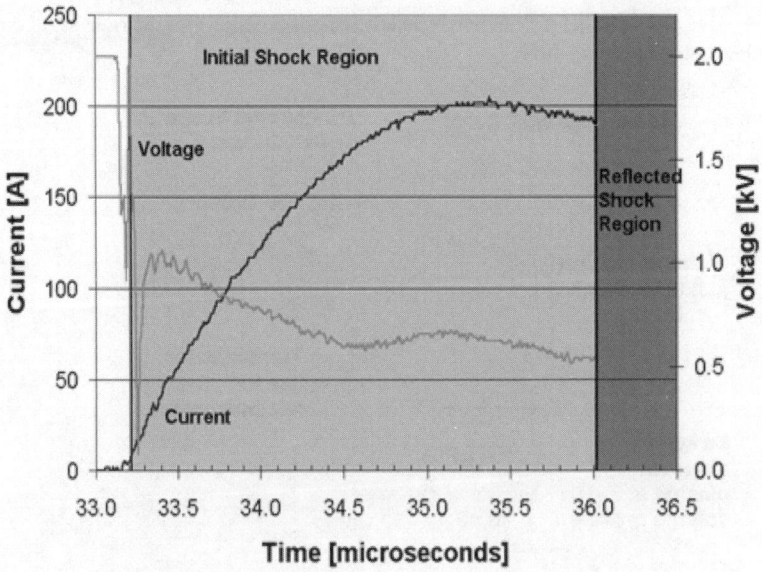

Fig. 5.4. Voltage and current recording for a synthetic air experiment.

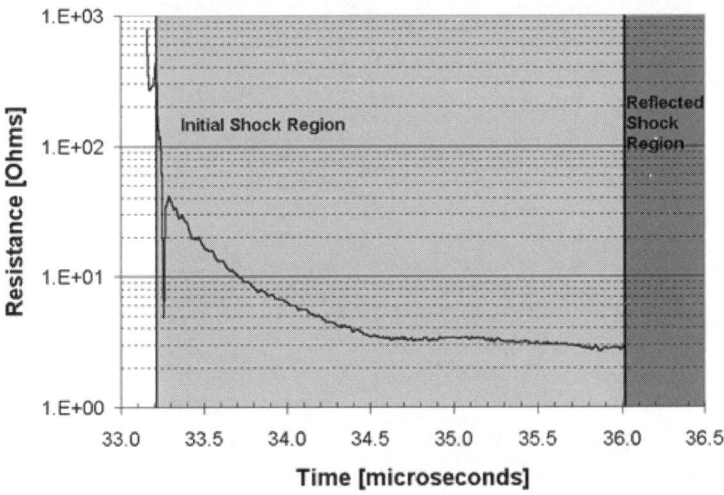

Fig. 5.5. A plot of the computed resistance is shown for synthetic air.

As shown in Fig. 5.4, a voltage and current response occurred as the initial shock front reached the voltage probes, indicating that the initially shocked region is conductive. The shaded areas on the plot indicate the state of the gas between the voltage probes. The useful measurement time ends as the reflected shock wave reaches the voltage probes, heavy shading.

The calculated resistance as a function of time is shown in Fig. 5.5 and indicates that the initially shocked region of synthetic air decreases to a stable resistance of 3 Ω before the reflected shock front approaches. Resistance values above 1 kΩ are lost in experimental noise.

Synthetic air has been shown to ionize to a bulk resistance of ~3 Ω within the initial shock wave generated in the shock tube environment. A preliminary resistivity of 0.03 Ω-cm is calculated, assuming a plasma disk with the diameter of the acrylic tube and the thickness of the voltage probes (0.5 cm).

5.1.5 SF$_6$ Data

The sulfur hexafluoride was initially at ambient temperature and pressure. The initial shock wave compressing the gas within the shock tube was measured at 3.57 km/s with piezoelectric pins 10 mm apart. Other shock properties were calculated with Hugoniot and ideal gas equations. The flyer plate velocity and shock pressure were calculated to be 3.20 km/s and 75 MPa, respectively. The flyer plate begins to pass the voltage probes before the shock wave reflected from the end plug returns. Current and voltage measurements, shown in Fig. 5.6, were used to calculate resistance using Ohm's Law.

Table 5.3. Average measured initial shock velocity and calculated pressure for all SF_6 shots.

Shock Properties	Shock Velocity [km/s]	Pressure [MPa]
Initial Shock	3.8 ± 0.3	84 ± 16
Reflected Shock	8.2 ± 0.8	360 ± 70

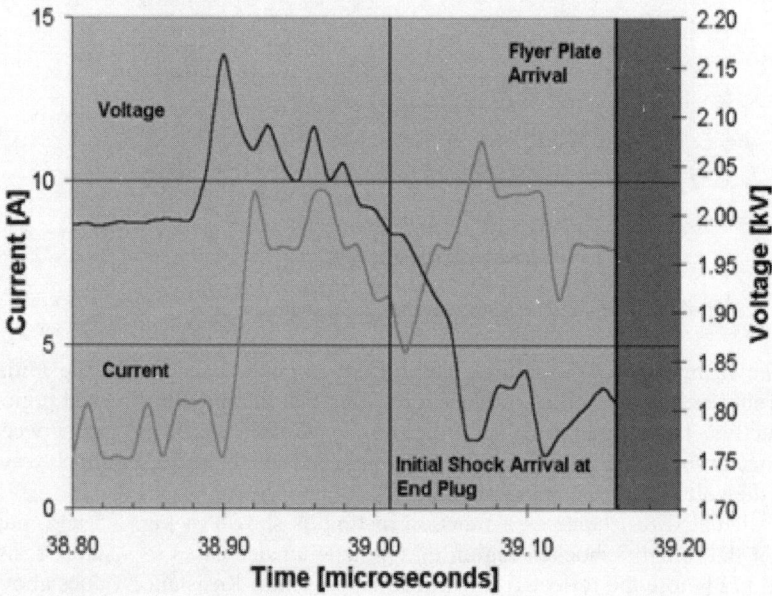

Fig. 5.6. Current and voltage traces for SF_6 experiment.

Based on the piezoelectric pin measurements, the initial shock front is expected to reach the voltage probes at 36.19 µs after the experiment is initiated, although there is no voltage or current response until approximately 38.9 µs. This phenomenon was typical for SF_6, cf. Fig. 5.12, which rarely showed the onset of conductivity in conjunction with a specific timing event such as the arrival of a shock front at the voltage probes. A possible interpretation of this behavior is that under the conditions within the shock tube, SF_6 may be near an ionization threshold. Resistance values, Fig. 5.7, above 1 kΩ are lost in experimental noise. Timing calculations indicate that the initial resistance drop to near 200 Ω after the initial shock front passes the voltage probes corresponds to the arrival of the initial shock wave at the end plug. The decrease in resistance may be attributable to photo-ionization from radiation emitted by the stagnated region. Fiber optic measurements of light intensity with a photodiode at the voltage probes indicate the emission of light well in advance of the initial shock wave.

A minimum resistance of 200 Ω is measured at the time of highest pressure as the reflected shock wave and flyer meet at the voltage probes. A preliminary resistivity of 2 Ω-cm is calculated at this time. Resistivity was calculated in the same manner as for synthetic air

5.1.6 Data Overview

A total of over 30 shots have been fired using SF_6, synthetic air, helium, and argon, and the measured shock velocities and resistances have been different and reasonably reproducible for each gas. While it is conceivable that the induced conductivity is due to the shocked acrylic shock tube wall or the aluminum flyer plate, such a model is considered to be unlikely. If the acrylic dominated the measurements, the bulk resistance for each shot would not have varied according to fill gas. Also, if the aluminum flyer plate had a significant effect, the bulk resistance for each shot could be expected to converge to the same value as the flyer plate passed the voltage probes, which was also not supported by the measurements.

5.1.7 Shock Tube Summary

Our initial tests have shown the shock tube environment to be a reasonable approximation to explosive magnetic flux compression generator environments. The shock tubes have generated average flyer plate and initial shock velocities of 3.4 ± 0.3 km/s and 3.8 ± 0.3 km/s, respectively, in SF_6, and 3.7 ± 0.1 km/s and 4.6 ± 0.1 km/s in synthetic air.

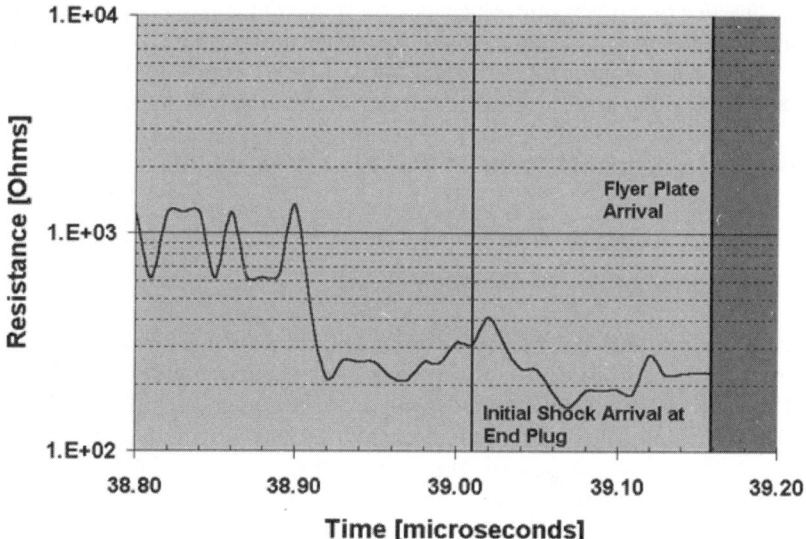

Fig. 5.7. Computed resistance for SF_6 as a function of time.

SF$_6$ has been shown to ionize to a bulk resistance of 200 Ω within the shock tube environment, and a preliminary resistivity of 2 Ω-cm has been calculated. SF$_6$ displayed behavior that may indicate the shock tube conditions are near an ionization threshold. Synthetic air ionized within the initial shock region to a bulk resistance of 3 Ω, and a preliminary resistivity of 0.03 Ω-cm.

For these shock conditions, SF$_6$ maintains better insulation properties than synthetic air, but the flyer plate is driven through the air at a slightly higher velocity. These differences suggest the optimum FCG fill gas may not always be SF$_6$, depending on the design constraints for the generator and its particular application.

5.2 Electrical Breakdown of the Generator Volume

To further enhance our understanding of losses due to electrical breakdown of the volume between armature and stator, we measured the resistance across two probes inside an FCG (armature 38 mm OD, expansion ratio 1.7). For these measurements, we refrained from applying any seed current prior to firing since this would have interfered with the small signal amplitudes.

As described below, the probes were located in the generator end section, where one could argue that the measured quantities are different from positions further upstream towards the initiation end. Specifically, as the armature expands, one might suspect that gas is pushed towards the generator end, resulting in a higher gas density even before the expanding armature cone reaches the end position, an effect that is often referred to as "plowing". However, in all our experiments, we have never observed such plowing in our generators. It is speculated that the absence of plowing for our generators is due to the fact that any axial gas flow ahead of the expanding cone would experience not a smooth, but a structured surface (adjacent circular helix wires, cf. Fig. 5.13). We conclude that this structured surface and the axial and radial velocity of the armature deformation being much larger than the speed of sound in the gas prevents plowing. No further attempts were made to find operation conditions or geometries where plowing would be obvious.

5.2.1 Resistance of the Generator Volume

We tested the generator volume for its susceptibility to voltage breakdown, measured by the volume resistance. Utilizing two 1 mm diameter rigid wire probes inserted radially into the generator end section, we measured the associated resistance between them. The probes were located 90° azimuthally from each other and flush with the generator inside (the end section was machined from non-conducting material in this case), see Fig. 5.8. Based on the simple equivalent circuit shown in Fig. 5.9, we calculate an apparatus limited risetime of about 2 ns.

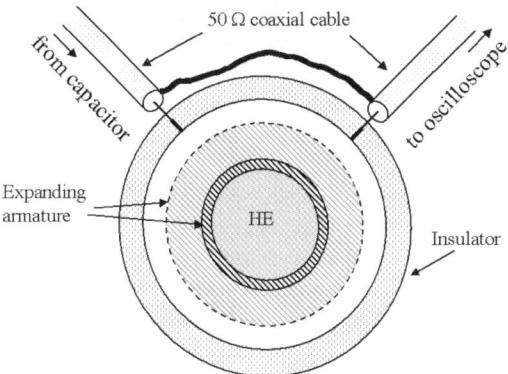

Fig. 5.8. End on view of FCG with rigid wire probes for resistance measurements.

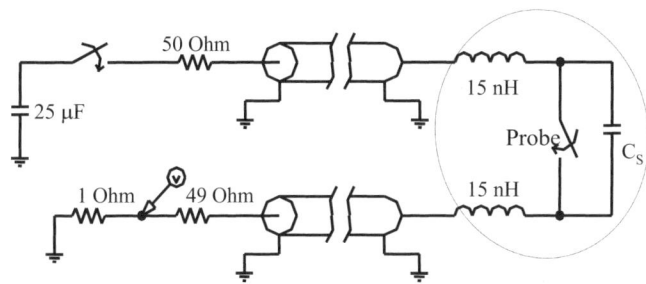

Fig. 5.9. Equivalent circuit accounting for the inductance introduced by the short leads from the ends of the coaxial cables to the wire probes and the current path between the probes. Capacitor charging voltage is 2 kV, the 15 nH inductances are parasitic.

As stated earlier, the applied voltage of 2 kV is typical for the voltage generated in a small single stage generator. The actual resistance was calculated from the voltage signal measured with a 50 Ohm input impedance oscilloscope employing a high quality 30 dB attenuator instead of the resistive divider indicated in Fig. 5.9. We note that the attenuator has a flatter frequency response, DC to ~ 1 GHz, than a simple resistive divider, which would have to be compensated for capacitive parasitics. Based on the matched impedance throughout the circuit, the resistance between the probes is calculated from

$$R = \left(\frac{V_{charge}}{V_{measure}} - 2 \right) \cdot 50\,\Omega, \tag{5.1}$$

with the initial capacitor charging voltage, V_{charge}, (~ 2 kV) and the measured voltage, $V_{measure}$. The switch between capacitor and coaxial cable was typically closed for several 10 μs or longer. This ensured that the upper coaxial cable in Fig. 5.9 is fully charged before the generator is fired. The main capacitor discharged with an RC time constant much longer than the duration of generator operation so long as the changing resistance stayed above a few tens of ohms. It is only during the very last moments before the armature makes physical contact with the probes that the resistance dropped to really small values caused by a rather fast electrical breakdown in the gas.

As far as the temperature/pressure of the gas is concerned, we have observed previously three developmental stages of operation [Neu02]: (1) The initial stage, which can be represented by a freely expanding armature, that shows fairly low gas temperatures, as low as 2,000 K. (2) The intermediate phase up to ~ 4 microseconds before generator burnout that exhibits mainly an atomic copper line transition at about 0.8 eV. (3) The last few microseconds reveal a highly compressed gas with temperatures of about 5,000 K and pressures of about 1,500 bar. As the volume resistance in the first two stages is above ~ 10^6 Ω, our detection limit in this setup, we will focus on resistance changes during the final phase.

As the armature is expanding and approaching the stator, the increasing pressure and temperature of the trapped gas act in different ways on the volume resistance. Since we are with our generators on the high pressure side of the Paschen curve, see Fig. 5.10, a higher pressure means a higher breakdown field. However, a higher temperature means more thermal excitation, eventually all the way up to dissociation and ionization. While temperature and pressure are rising, the electric field between probe and armature is increasing as well due to the shrinking distance from the expanding armature. Hence, we expect eventually that an electric field induced breakdown will occur with a current path from one electrode through gas volume, armature, and gas volume to the other electrode.

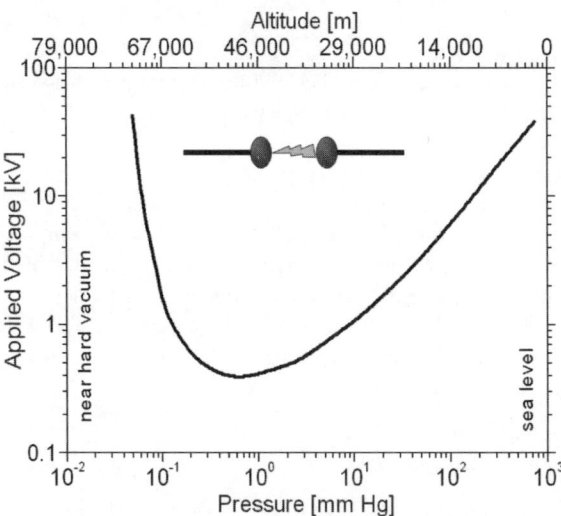

Fig. 5.10. Maximum applied voltage as a function of the pressure at 1 cm electrode separation in air (temperature effects with changing altitude are unaccounted for). Breakdown will occur if the applied voltage is above the depicted line. An electrode separation of 0.5 cm at twice the pressure would have the same breakdown voltage, i.e. the breakdown voltage is proportional to the product of pressure and electrode separation.

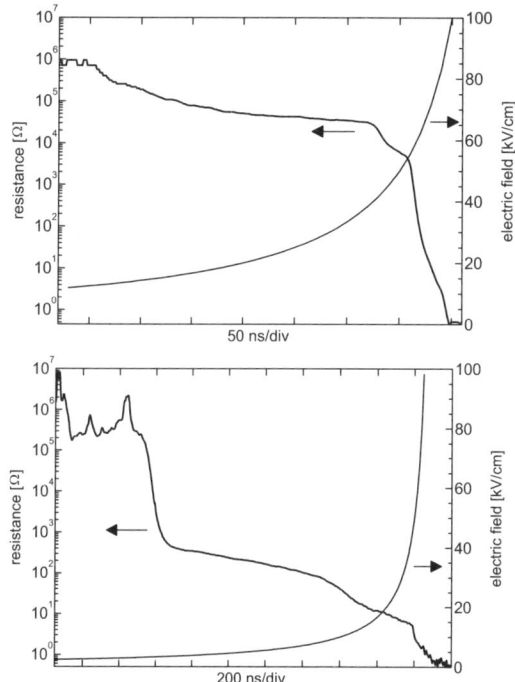

Fig. 5.11. Measured resistance of the compressed generator volume (top: air, bottom: argon) and calculated electrical field amplitude with narrowing armature-stator distance. The armature makes physical contact with the probes in the last portion of the figures.

Both air and argon resistances, depicted in Fig. 5.11, show a plateau at 10^5 or 10^2 Ω, respectively, that lasts about 200 ns or 900 ns, respectively. The associated current during this time ranges from 10 mA to 10 A. All electrons that are generated between the probes due to the rising temperature are moving quickly to the positive electrode. The ions contribute much less to this current since their mobility is typically 3 orders of magnitude lower. The overall current saturates, it only depends on the rate at which new electrons are generated and on the possible formation of positive space charge due to the slowly moving ions. It takes a much higher field to start efficient electron multiplication via collisonal ionization (40 kV/cm or 15 kV/cm for air and argon, respectively), as it is manifested in the external circuit by a sharp drop in resistance to below 1 Ω.

These field amplitudes should be compared with the breakdown field of air and argon at STP, which is 30 kV/cm or 5 kV/cm, respectively. We used Paschen's law to estimate the expected breakdown field, E_b, at the elevated pressure and temperature. The pressure dependence of the breakdown field on the high-pressure side of the Paschen curve (uniform field only) is given by [Nas71]:

$$E_b = \frac{B \cdot p}{\ln\left(A \cdot p \cdot d / \ln\left(1 + \frac{1}{\gamma}\right)\right)}, \tag{5.2}$$

with the temperature dependent constants A, $B \propto 1/T$ and the secondary electron emission coefficient, γ, which is assumed being constant. The expected variation range of the given parameters are $p = 1 \ldots 2{,}000$ atm, $d = 1 \ldots 0.01$ cm, and temperature $T = 300 \ldots 5{,}000$ K. Hence, the breakdown field is expected to be about a factor 30 higher for the highly compressed and shocked gas. However, the observed breakdown field amplitudes increase merely by a factor of less than 3. Clearly, other effects that are not included in Eq. (5.2), such as thermal dissociation, excitation, and ionization of the trapped gases, come into play in the high pressure, high temperature regime.

It is worth mentioning that the behavior of magnetic flux compression generators has been successfully calculated by including electrical breakdown in the numerical model just by using the same breakdown field for the shocked and compressed air as for air at STP [Nov01]. Obviously, our findings provide the experimental justification for such an approach.

The temporal behavior observed for the generator operated in SF_6, cf. Fig. 5.12, is so different from air and argon that it deserves our particular attention. Contrary to air and argon, no drop in resistance is observed before the armature is fully expanded at the axial position of the conductivity probe, indicated by the left arrow in Fig. 5.12, positioned at $t = 25$ μs. Moreover, the resistance remains high up to 2.3 μs after the armature has made "contact" with the probes. Eventually, the resistance drops rather fast (faster than for air or argon) within about 150 ns from a value higher than 10^6 Ω to 1 Ω or less. We conclude that the delay is caused by the layer of shocked and compressed SF_6 being trapped between armature and stator. This behavior is not surprising as we observed for SF_6 in the shock tube and flyer plate arrangement no measurable drop in resistance when the shockwave passed through the regions of the electrodes, cf. Chap. 5.1.5.

The question arises if the SF_6 is still in its gas phase or if it is liquefied by the high pressure of several 1,000 atmospheres (SF_6 is known to liquefy at about 37 atmospheres at room temperature). We have an uncompressed volume/length of about 2,100 mm^3 / mm with an armature diameter of 38.1 mm and a stator inner diameter of 64.3 mm. This translates to an SF_6 mass/length of 0.013 g / mm by substituting 6.16 g / liter SF_6 gas density at 1bar and 20 C. The density of SF_6 in its liquefied state is 1.5×10^3 g / liter, which then results in a layer thickness of 43 μm. The electric field in the gap between armature and probe associated with the 2 kV applied voltage is then about 500 kV/cm, which is approximately the breakdown strength of Teflon or Mica. If we assume that the compressed and shocked SF_6 has about the same electrical properties as SF_6 at STP (a fact that is believed to be true for air), namely 78 kV/cm breakdown field, we calculate a layer thickness of 250 μm at breakdown threshold, which is also quite realistic. The density of the gas would only be a factor 5 different from the liquid in the former case.

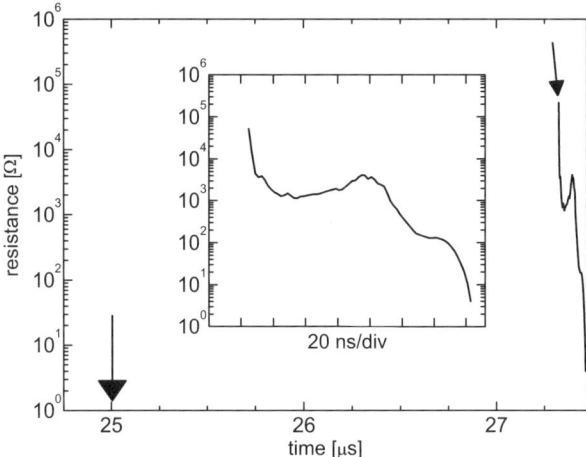

Fig. 5.12. Measured resistance of the compressed generator volume in SF_6. The left arrow indicates the moment when the armature is expected to physically contact the conductivity probes; the insert shows the expanded view of resistance vs. time starting at 27.3 μs. The resistance is larger than 1 Meg-Ohm for times below 27.3 μs.

Obviously, both scenarios have their shortcomings: the temperature will influence the liquid density as well as the breakdown field of gaseous SF_6. However, we can assume for certain that the temperature of the compressed SF_6 is above its critical temperature of 45.6 C. Hence, a distinction between liquid and gas states becomes obsolete and we only keep for the records that the physical density of the maximally compressed SF_6 layer is about three orders of magnitude larger than its gas density at STP.

5.3 Impact of Stator Geometry on Flux Losses

Five different types of end-initiated, constant cross section, capacitively seeded FCGs were tested using low inductance loads. Stator diameters ranged from as small as 19 mm to as large as 89 mm and armature diameters ranged from as small as 10 mm to as large as 38 mm resulting in armature to stator expansion ratios of 1.7 to 2.3. All models tested had constant cross sections, meaning that neither the armature nor the stator was tapered. In an effort to simplify design, all generators were designed with a constant pitch among the stator wire; although it should be noted that a varying pitch is preferable with respect to internal voltage breakdown issues. The FCG armatures were hand packed with up to 260 g of C-4 explosives and the packing density of the explosives varied by less than 2% and was approximately 1.4 g/cm^3. While one might speculate that hand packing explosives introduces voids in the explosives causing jetting and, consequently, flux loss in an FCG system, this was found not to be the case in another study related to the presented work [Bai01], cf. Chap. 4.

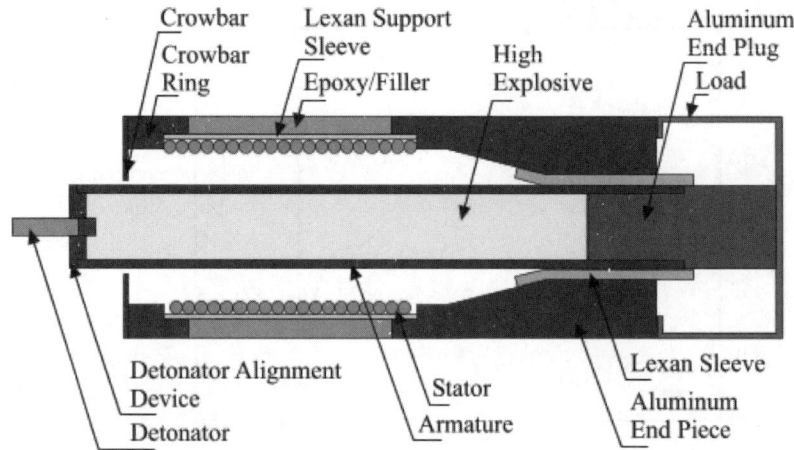

Fig. 5.13. Generic cross section of an FCG representing TTU I, TTU II, and TTU III generators with 38 mm armature outer diameter.

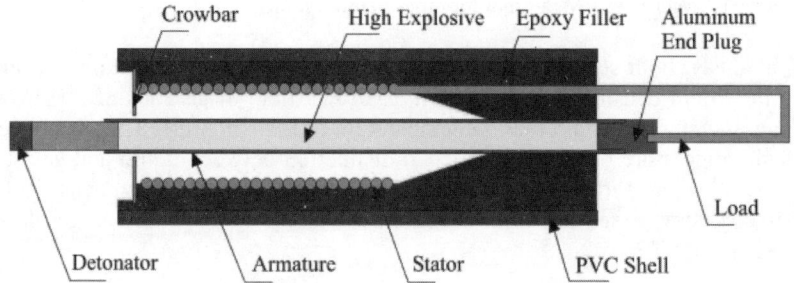

Fig. 5.14. Generic cross section of an FCG representing TTU IV and TTU VIII generators with 16 and 10 mm armature diameter, respectively.

Of the five variations of the generator shown in Fig. 5.13 two (TTU I and TTU II) were tested extensively and were used, along with results found in literature, to establish an initial baseline of performance, cf. Fig 2.10. By using this baseline, modifications to subsequent generator models were made in order to create a more complete picture of the impact of different generator parameters on overall generator performance.

The TTU I model was an FCG constructed with helices that had lengths of 102 mm, inner diameters of 66 mm, and armature to helix expansion ratios of 1.7. The TTU II model FCG's were constructed with larger, 89 mm, inner diameter helices (maintaining the same 102 mm helix length) and armature to helix expansion ratios of 2.3. The TTU III model generators were of a very similar design to the TTU I model generators, with modifications made only to the stator wire pitch. The TTU III model had a stator wire pitch which was half that of the TTU I models, resulting in a shorter stator and faster angular frequency of the armature/stator contact point (or simply "angular frequency"), cf. Eq. (2.27). The TTU I, TTU II,

5.3 Impact of Stator Geometry on Flux Losses

and TTU III models were all manufactured in the same manner and their cross sections are all similar to the generator depicted in Fig. 5.13. The TTU IV and TTU VIII models were smaller generators; constructed to achieve a broad range of angular frequencies (Fig. 5.14 depicts a generic cross section for these model generators). Most generator models were constructed with a single wire helix, but double wire stators for the TTU I and TTU II models and triple wire stators for the TTU I model were also investigated. All generator dimensions and stator wire characteristics are summarized in Table 5.4.

Table 5.4. Physical dimensions of selected helical FCG models TTU I, TTU II, TTU III, TTU IV, and TTU VIII. Dimensions listed are armature outer diameter (AOD), stator inner diameter (SID), stator length (length), stator wire insulation thickness (t_i), and stator wire total diameter (d_w).

#	Model	Wire Type [-]	AOD [mm]	HID [mm]	length [mm]	t_i[mm]	d_w [mm]	pitch [mm]
1	TTU I	Circular Electric Wire PVC Insulated	38.1	66.0	102.0	0.445	3.05	3.05
2		Circular Electric Wire PVC Insulated	38.1	66.0	102.0	0.445	3.05	3.05
3		Circular Electric Wire Non-Insulated	38.1	66.0	102.0	0.000	3.05	3.05
4		Square Magnet Wire Formvar Coating	38.1	66.0	102.0	0.038	2.11	3.05
5		Tin Coated Copper Wire TFE Teflon Insulated *	38.1	66.0	102.0	0.254	1.25	3.05
6		Tin Coated Copper Wire TFE Teflon Insulated *	38.1	66.0	102.0	0.254	1.25	3.05
7	TTU II	Square Magnet Wire Heat Shrink Coating	38.1	89.0	102.0	0.445	3.05	3.05
8		Circular Electric Wire PVC Insulated *	38.1	89.0	102.0	0.445	3.05	6.10
9	TTU III	Tin Coated Copper Wire TFE Teflon Insulated	38.1	66.0	51.0	0.254	1.52	1.52
10	TTU IV	Square Magnet Wire Formvar Coating	15.9	28.7	27.9	0.051	0.64	0.64
11	TTU VIII	Circular Copper Wire PVC Insulated	9.5	19.1	22.9	0.254	0.64	0.64

* denotes FCGs with multi-conductor (2 wires run in parallel) stators.

Table 5.5. Performance summary of FCGs listed in Table 5.4. Performance assessors listed are figure of merit (β), armature/stator contact point angular frequency (ω), compressible volume (V), initial inductance (L_i), final inductance (L_f), seed current (I_i), final current (I_f), and energy gain (G_E). TTU II, with the smallest ω, exhibits the largest energy gain.

#	Model (Shot)	β [-]	ω [rad-MHz]	V [cm³]	L_i [μH]	L_f [nH]	I_i [A]	I_f [kA]	G_E [-]
1	TTU I	0.64	13.2	232	20.2	45.7	97.0	4.7	5.3
2		0.65	13.2	232	20.2	45.7	13.2	0.7	6.5
3		0.70	13.5	232	21.0	45.7	874.0	64.6	11.9
4		0.68	14.2	232	23.1	45.7	884.0	60.5	9.3
5	*	0.71	13.3	232	20.3	45.7	772.0	57.3	12.4
6	*	0.73	13.6	232	21.5	45.7	400.0	35.4	16.6
7	TTU II	0.68	12.7	515	41.6	65.0	9.7	0.8	10.8
8	*	0.82	6.4	515	10.5	65.0	546.0	35.0	25.4
9	TTU III	0.64	21.5	116	26.7	45.7	22.4	1.3	6.0
10	TTU IV	0.55	53.5	13	17.8	90.0	444.0	8.0	1.7
11	TTU VIII	0.50	44.0	5	3.1	32.6	119.0	1.2	1.0

* denotes FCGs with multi-conductor stators.

As documented in Table 5.5, there is little difference in the figure of merit, β, for the TTU I and TTU II model generators, #1 through #7 in Table 5.5. However, the FCG generator TTU II with its stator wound using two parallel running wires (#8 in Table 5.5), which has about twice the compressed volume and half the inductance of the TTU I generators (cf. Table 5.4), produces a figure of merit of $\beta = 0.8$, thus distinctly conserving more flux. We consider this a dramatic change in performance since the β is the exponent of the inductance ratio determining the overall gain, cf. Eqs. (2.20) and (2.21). The intrinsic flux loss parameter, α, as it is defined in Eq. (2.13) changes from $\alpha = 0.80$ to $\alpha = 0.89$. This means that for the larger volume generator with a double helix, only roughly 10 % of flux is instantaneously lost rather than the 20 % for the smaller TTU I (with little variation, no matter if single, double, or triple wound) with the 50 % larger ω.

The TTU IV, TTU V, and TTU VIII models served to illustrate the trend towards lower values of β as ω was increased, as can be seen in Fig. 2.10. It is believed that the performance observed by these smaller volume FCGs was worse than the trend line prediction due to the tighter machining tolerances required with such small dimensions. The maximum displacement, d, the armature can have relative to the stator before partial turn skipping (PTS), sharp periodic spikes in the time derivative of the current, dI/dt, Fig. 5.15, occurs is

$$d \cong \frac{h \cdot \tan \gamma}{2\pi}, \quad (5.3)$$

where h is the stator wire pitch, and γ the expansion angle of the armature [Che86]. Refer to Chap. 4 for more details on armature imperfections or misalignment. The Gurney angle of the armature is roughly 15 degrees for all test groups. For the TTU IV test group, the wire pitch was 0.74 mm, which results in a maximum displacement of the armature relative to the helix of 0.03 mm. For TTU VIII generators, the pitch is 1.14 mm, which yields a maximum allowable displacement of 0.05 mm. Such tight tolerances would require new construction techniques to ensure that PTS does not occur.

It is obvious that, although the fluctuation of the numerical values of β are rather large for a given angular frequency, there is a distinct trend to higher values of β at lower ω as shown in Fig. 2.10. The fluctuations are caused not only by the simplifications made in the derivation of Eq. (2.27) but also by the different share the resistive losses have compared to the total loss. Depending on the specific design, the intrinsic flux loss can play a larger role in how much flux is conserved than the resistive losses do, cf. Chap. 5.4.2.

The reasons for the distinct trend toward higher values of β at lower ω, see TTU II #8 in Table 5.5 and Fig. 2.10 are not fully understood. One speculation is that the rapidly changing magnetic field structure causes strictly azimuthal eddy currents without any axial component in the armature/helix wire near the contact point. Additionally, the smaller generators have higher ohmic losses due to their smaller helix wire cross-sections. For a small cylindrical generator with single pitch (armature 50 mm dia., stator ~ 100 mm dia., load inductance fixed at L_L = 2.3 µH) an optimum wire pitch (note that the generators initial inductance, L_0, inductance increases with decreasing pitch) with the highest energy gain was found to be ~ 1.25 mm [For02]. We will analyze the flux loss situation in the following Chap. 5.4 in more detail.

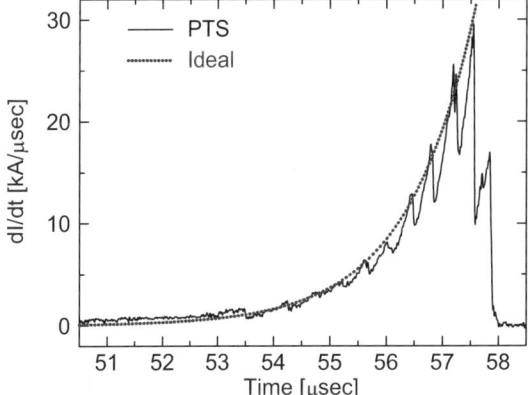

Fig. 5.15. Comparison of partial turn skipping, PTS, with an ideal current waveform free from PTS.

It should be noted that the wire insulation in the TTU I FCG had little influence on the performance of the generator. Excluding electrical breakdown, it was evident that TTU I without any insulation on the helix wire performed for all practical purposes as well as the other generators with insulation, cf. Table 5.5. Therefore, a description of the intrinsic loss with the help of a flux delay parameter as it can be found in the literature, e.g. [Gov79], is questionable for the helical FCGs in this study.

Stators wound with rectangular and square wire were tested as well. It was found that stators wound with square and rectangular wire performed slightly worse than stators wound with circular wire, even while ensuring identical quality of armature/stator alignment, cf. Fig. 2.20. The cause of the diminished performance could be attributed to be breakdown between adjacent windings in the stator or between stator and armature, which might be enhanced at the edges of the adjacent windings [She68]. However, the observed behavior exhibited very little dependence on the current (voltage) amplitude, so that we rather conclude that the square wire geometry displays more intrinsic flux loss.

Since also very little difference in performance was noted for generators with single wire stators when compared with generators with double or triple wire stators with the same effective pitch, an eddy current moving in the wire cross section seems to have negligible impact. Furthermore, we would like to exclude the type and thickness of helix-wire insulation as a dominant factor for flux loss changes since neither parameter, when varied, produced significant changes in results.

5.4 Intrinsic Flux Loss

It is obvious that the resistances of the stator coil, armature, and other parts of the FCG cause some flux loss by reducing the magnitude of the current generated due to ohmic losses. The flux lost as a result of this joule heating is largely accepted as a byproduct of the geometry, and is simply expected to be lost. Some techniques, such as bifurcating wires or making slight modifications to the typical FCG geometry, provide means to reduce the impact joule heating has on the total flux lost in the system. For a given generator size, these methods, while they do produce more efficient FCG operation, are generally accompanied by undesirable consequences such as a operating at a lower voltage thereby producing a smaller output current or having a smaller initial inductance resulting in a smaller overall current gain.

It should also be clear that we will have to find additional loss mechanisms for the helical FCG besides ohmic losses. Quite a number of possible mechanisms have been suggested in the past. We will briefly present the more important ones and list experimental evidence why or why not we consider the suggested mechanism important in our helical FCGs.

Any effect that would cause additional resistance to the flow of current in the FCG (apart from its operating resistance) would result in additional joule heating and flux loss. Axial cracking or other significant axial straining hindering the flow of the circular eddy currents in the armature could be such an effect [Fow75]. We have shown, however, that axial cracking along the armature can be easily avoided if a length of armature that is susceptible to the cracking is allowed to be sheared off by the crowbar during the first stages of generator operation, see Chap. 4.2. It is also interesting to note that FCG resistance increases during the first stages of operation (the time when the armature is expanding but has not yet reached the stator), due to the proximity effect [Cra68]. This increase in resistance cannot be avoided and should be included in the losses that are inherent to the system.

Still, even if the losses associated with joule heating and the development of alternate conducting paths are taken into consideration, it is difficult to make theoretical data (i.e. computer simulations) match data gathered experimentally without the inclusion of some other source of flux loss included into theoretical calculations. This intrinsic flux loss is accounted for in every model known to the authors for simulating helical FCG operation. Each model presents its own unique method of accounting for this intrinsic flux loss, and most of these methods fall into two separate categories. The methods in the first category attribute the intrinsic loss to effects related to electric breakdown inside the FCG active volume (i.e. volume between armature and stator which is collapsed during explosive expansion of the armature). The second category centers on the idea that flux is trapped in conducting layers of the FCG, and therefore lost for compression.

Electric breakdown, member of the first category, is a potential problem common to all pulsed power systems due to the extreme conditions in which they are operated. FCGs are not exempt to this problem, in fact, some regard voltage breakdown as the most important source of flux loss occurring in a FCG. It has been noted in [Nov02] that the breakdown voltage depends not only on the electrical breakdown strength (EBS) of the gas trapped between the stator and armature, but also on that of the stator wire insulation and the quality of the expanded armature surface. The conductivity and pressure of the trapped gas was extensively studied in [Neu02], and it was found that the same breakdown field for air at STP could be used for the shocked and compressed air, but the question of the armature surface and shocked insulator material being conducive to breakdown still remains. It has also been proposed that a finite time is required for insulators inside the FCG to break down electrically (i.e. the insulators around the stator wire). This would imply that the breakdown point remains a certain distance behind the point of contact and would result in several turns of the helix which still conduct current even after full armature expansion. This remaining helical FCG inductance would serve as a limit of magnetic to electric energy transfer [Ste79, Gov79]. When experiments were performed utilizing a simple single pitch generator [Neu01] with different insulation types and thicknesses surrounding the helix wire (insulation thicknesses did not exceed conductor diameter when measured from armature to helix conductor surface), we noted no appreciable difference in performance, see Chap. 5.3. Therefore, it is assumed that another physical process is the cause for the intrinsic flux loss.

It has been widely speculated in the open literature that unrecoverable flux trapped in the conducting layers of the generator components can be lost for compression [Fow75, Kno70, Gov79, Cra68, Pav90]. While the methods for incorporating this loss term vary from computer model to computer model, the basic underlying explanation of this phenomenon seems to be the most plausible cause of the intrinsic flux loss. As elucidated in [Gov79], the magnetic field stored in the skin layers of the stator and armature has a finite decay time, and commensurable time is needed for the current to change directions. The current will gradually turn around over a finite length, and thereby add length to the overall current path [Gov79]. This could be easily incorporated into a computer model as a "loss inductance", as was done in [Cra68], or via other methods, as was done in some of the aforementioned literature. Monitoring the effect this loss term has on experimentally acquired data is extremely important for the detailed understanding of the intrinsic flux loss. We will elucidate in Chap. 5.4.2 that the magnitude of the experimentally observed intrinsic flux loss in single pitch generators can be *a priori* estimated from first principles.

5.4.1 Zipper Contact

As it is generally accepted, the intrinsic flux loss in the generator occurs in the vicinity of the armature-helix contact point. The detailed magnetic field and current distribution are very complicated at this singularity. Although this progressing contact bears some resemblance with a sliding contact, it is in fact more like a motion of mass cells (if we divide the armature into small cells) similar to what we observe in a zipper. However, in the helical FCG, one side of the zipper, the helix, does barely move, while the other side, the approaching armature, undergoes a strong mechanical deformation in the contact point vicinity. It is instructive to project the helix wire and armature of the FCG into a plane and to zoom in on the contact point, such that the curvatures of helix wire and armature do not show. By doing so, we can directly compare the sliding contact and the zipper contact, see Fig. 5.16. We note that real surfaces also exhibit ripples, which are not shown in the figure. Such ripples are believed to indeed affect the quality of the zipper contact effectively causing the contact to jump microscopically from ripple to ripple.

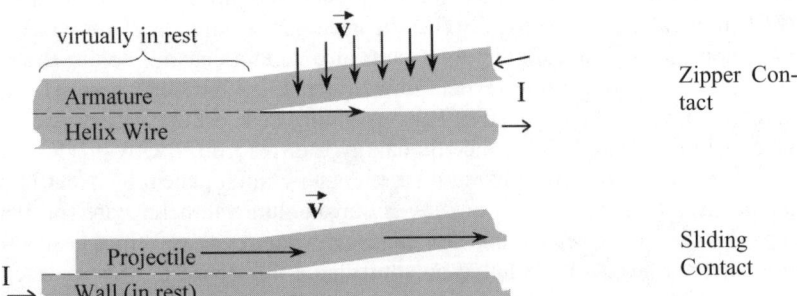

Fig. 5.16. (a.) Zipper contact as found in HFCG indicating the velocity of the mass cells (vertical arrows) and the contact point (horizontal arrow); (b.) Sliding contact, e.g. rail gun with the projectile having a single velocity throughout.

It should be noted in Fig. 5.16 a.) that the section of the armature behind the contact point comes virtually to a stop as its momentum is transferred to the mass of the helix wire and the inertial backing of the helix wire (epoxy, concrete). Hence, on the microsecond timescale, the armature moves only very little beyond its position reached at the moment of contact with the helix wire.

The obvious differences in the motion of the conductor mass cells for the two contacts are amplified with respect to the current flow. The zipper contact has current flow from the right to the left in the armature, through the contact point, and finally from left to right in the helix wire. For the sliding contact, the current is fed in from the left, goes through the contact point, through the projectile, the opposite contact point, and finally returns to the left in a rail gun geometry. In short, the contact point in the zipper contact geometry has major current flowing in the conductors in front of it, whereas the sliding contact exhibits little major current flow in front of the contact.

Although there have been a few computer programs developed that enable calculating the sliding contact, there is to the authors' knowledge presently no code available that would allow calculating the zipper contact geometry as it involves changing boundaries due to the deformation of the armature. To make things worse with regard to resolving the magnetic field structure close to the contact point, the skin depth associated with current and magnetic field diffusion is of the order of less than ~ 0.5 mm in the microsecond timescale. Thus, a generator with an overall dimension of the order of somewhere between 100 mm and 1000 mm would require a grid that has at least $1000^3 = 10^9$ cells. Although the situation can be somewhat improved by utilizing an adaptive grid, solving for the magnetic field of a helical FCG in the time domain remains still beyond the capability of any presently commercially available computer programs. Present computer codes enable at best the calculation of the magnetic field or flux in the frequency domain [Ansoft], see Fig. Fig. 5.17.

However, even the frequency domain calculation has its limits and it is virtually impossible to resolve the detailed magnetic field as it has, for instance, diffused into the helix wire. Nevertheless, the zipper geometry in Fig. 5.16 can be resolved reasonably well in the frequency domain, see Fig. 5.18. We estimate that a flux per unit length of 2.6 µWb/mm at 10kA of current is diffused inside the helix wire and armature as the two conductors approach each other. Even though the frequency domain calculation does not account for any movement or deformation of the conductors, we still observe that the current density and the magnetic field extend somewhat into the region upstream of the contact point. This effect will be much more pronounced in the time domain as the regions upstream of the contact have carried all the current just prior to the approaching contact point shorting the two conductors. Since the angle between approaching armature and helix wire (azimuthally) is extremely shallow, the contact point has a velocity of several 100 mm/µs (about one tenth of a percent of the speed of light). The result will be that the region of non-zero current/magnetic field in the conductors is much extended upstream (towards the initiation end) of the contact point as it takes a finite time for the magnetic field to diffuse out of the conductors.

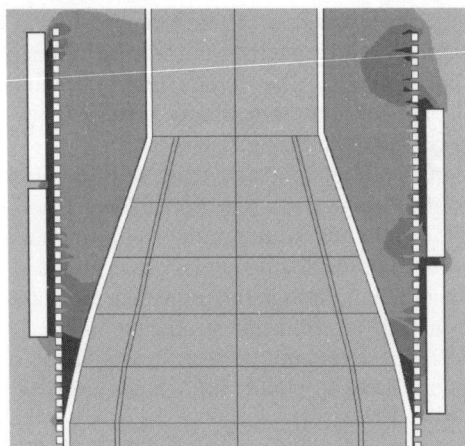

Fig. 5.17. Calculated magnetic field of an FCG in the frequency domain at a frequency of 100 kHz with 10kA current. ▆▆▆ - 4 to 10 Tesla, ▆▆▆ > 2.7 Tesla, ▆▆▆ > 1.3 Tesla, ▆▆▆ ≥ 0 Tesla. The field in the conductors is omitted for clarity of presentation.

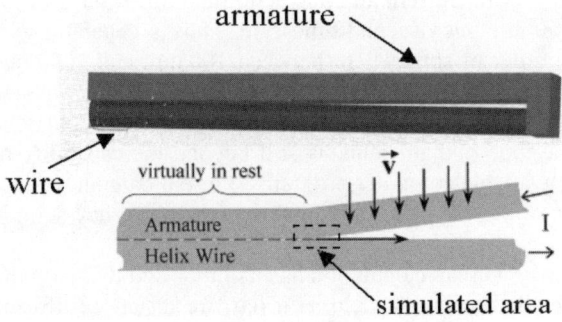

Fig. 5.18. View of the wire-armature structure (top, geometry for field simulation). Contact point details (bottom).

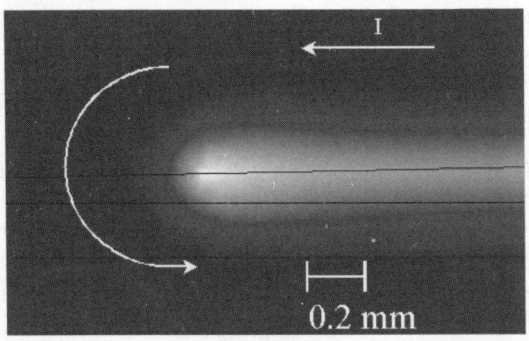

Fig. 5.19. Magnetic flux density zoomed in at the contact point for TTU II at 57 kHz field frequency with 10kA current. ▆▆▆ - 1 Tesla, ▆▆▆ - 3.5 Tesla, ▆▆▆ -10 Tesla.

It should be noted that the magnetic Reynold's number, R_m, the ratio of magnetic diffusion time over the compression time, is large during compression, of the order of 10,000, for our generators. Nevertheless, the amount of flux diffused into the conductors is overall non-negligible as outlined above.

By calculating the flux per unit length close to the contact point, we have determined the maximum amount of flux that will be inside the wire-armature body. It should be noted that the skin depth calculated for the steady state frequency domain differs less than 5 % from the transient solution derived from a first quarter period of a sine wave. The rate of flux loss is calculated by multiplying the flux per unit length with the total length of the wire divided by the time that it takes the armature to make contact from the first to the last turn. To calculate the total flux loss, the integrated current that flows through the wires is needed in order to scale the rate of flux loss inside the wire-armature structure. Clearly, in the extreme case all the flux in the conductors is lost for flux compression. And indeed, we will see in the next section that this extreme case can be used as a worst-case estimate for generator performance.

5.4.2 Magnitude of Ohmic and Intrinsic Losses

In our simple mathematical model for the intrinsic flux loss, which we presented in Chap. 2, the flux left behind in the conductors is expressed as the flux loss parameter α in Eq. (2.13). As we discussed in Chap. 2, α is linked to the rapid azimuthal progression of the armature-helix contact point [Neu02b]. Hence, as suggested elsewhere, it is assumed that $\alpha \sim 1$ for the initial period from the time of crowbar to the moment when the armature makes contact with the first helix turn. After that, α has a smaller numerical value ($\sim 0.7\ldots 0.9$), a value that had to be found by adjusting a so that the calculation matches the experiment. However, the knowledge of α alone fails to provide detailed information as to how the overall loss is distributed between ohmic and intrinsic flux loss. Utilizing a modified version of Eq. (2.13) along with some numerical tools, we have analyzed the behavior of 3 of our FCGs in more detail, see Table 5.6. The larger generators TTU II and X have about the same overall size, with TTU X having a tapered armature. The TTU IV generator is the smallest in the series.

Table 5.6. Physical dimensions of helical FCG models with small overall size, TTU IV, and tapered armature, TTU X.

PARAMETERS	TTU II	TTU IV	TTU X
Helix Length (mm)	70	35	70
Helix Radius (mm)	41	14.4	41
Armature Len. (mm)	120	70	120
Expansion ratio[*]	2	1.8	1.7, 1.4
Number of Turns	23	32	23
Wire Gauge (AWG)	12	22	12
Helix time (µs)	8	4	4

* TTU X has a tapered armature with its largest diameter at the output side, see Fig. 5.20.

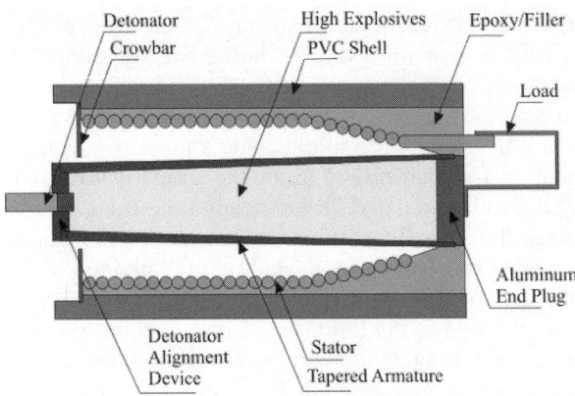

Fig. 5.20. FCG model with tapered armature, TTU X.

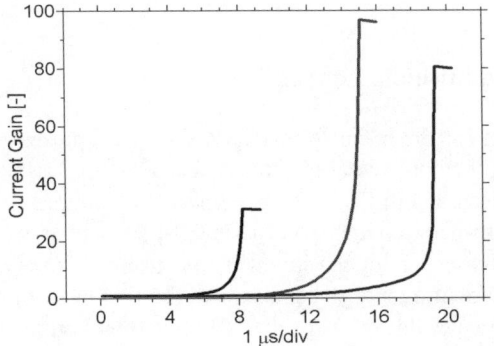

Fig. 5.21. Calculated current gain with α = 0.73, 0.93 and 0.72 respectively. From left to right: TTU IV, X and II. Experimental waveforms (not shown) are within 5 % of the calculated values.

We start our detailed loss analysis by adjusting α for the 3 generators so that their current gain, G_I, matches the experiment, see Fig. 5.21. For this, we have used the incremental version of Eq. (2.13),

$$I(i) = I(i-1) \cdot \left(\frac{L(i-1)}{L(i)}\right)^{\alpha(i)} \cdot \exp\left(-\sum_{i=1}^{N}\left(\frac{R(i)}{L(i)}\right) \cdot \Delta t\right), \quad (5.4)$$

where $I(i-1)$ and $L(i-1)$ are the current and inductance of the previous time step. Besides $\alpha(t)$, $R(i)$, $L(i)$, and the seed current, I_0, are used as input quantities. We have shown previously that this simple model will reasonably well calculate the output of simple single stage, single pitch generators at small current amplitudes, where non-linear diffusion is negligible [Neu01].

The calculated gain (matched to the experiment) for TTU II, IV, and X is depicted in Fig. 5.21. We observed the highest current amplification for TTU X with a current gain of 97, followed by TTU II (with the same stator dimensions as TTU X) with 81, and TTU IV with 31. It should be noted that the three generators have different run times due to their varying axial length and their overall geometry. For instance, the helix of TTU X is wiped out by the armature in only 4 μs rather than in 8 μs for TTU II, which has the same helix diameter, cf. Table 5.6. The helix diameter of TTU IV is only about a third of the two others.

The current in Fig. 5.21 can be used to derive the magnetic flux that is lost during helical FCG operation. Specifically, the relative flux loss at any time step is given by:

$$\Delta\phi'_{CEXP} = \frac{L(i)I(i) - L_0 I_0}{L_0 I_0}, \quad (5.5)$$

which includes both ohmic and intrinsic flux losses.

We also used a modified Eq. (5.4) to determine the individual contributions to the total flux loss by calculating the current for the next time step switching off either ohmic losses or intrinsic losses for the next time step only. This is repeated for each time step, always using $I_{CEXP}(i-1)$ as the base for the previous time step. The resulting incremental loss in flux for R only ($\alpha_{off} = 1$) and α only ($R_{off} = 0\,\Omega$) is then summed over the entire operation time of the generator:

$$\Delta\phi'_y(i) = \sum_{i=1}^{N}\left(\frac{\phi_y(i) - \phi_{CEXP}(i-1)}{L_0 I_0}\right), \quad (5.6)$$

with $y = R$ or α.

As necessary, the sum of the individual losses equals the total flux loss at every time step, see Fig. 5.22, Fig. 5.23, and Fig. 5.24. No flux is lost due to α, $\Delta\phi'_\alpha = 0$, for times prior to the armature's contact with the first helix turn, for instance, up to 4.2 μs in Fig. 5.22. After that, $\Delta\phi'_\alpha$ drops rapidly and even overtakes the ohmic flux loss, $\Delta\phi'_R$, at about 6.2 μs. The parameters and loss balance for all three generators are given in Table 5.7.

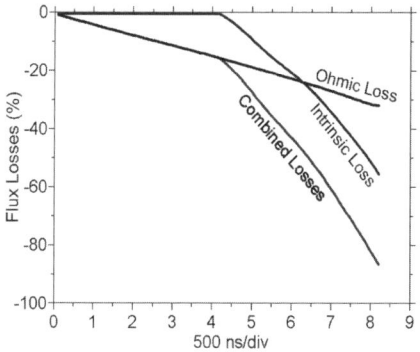

Fig. 5.22. Percentage flux loss, $\Delta\phi'$, in TTU IV. Time to make contact with the first turn is 4.2 μs. Upper curve – $\Delta\phi'_\alpha$, intrinsic loss; middle curve – $\Delta\phi'_R$, ohmic loss; bottom curve – $\Delta\phi'_{CEXP}$, combined losses.

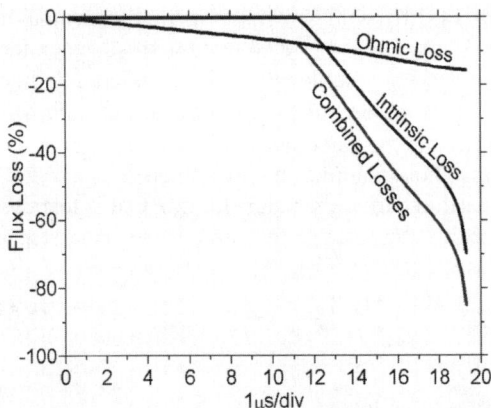

Fig. 5.23. Percentage flux loss, $\Delta\phi'$, in TTU II. Time to make contact with the first turn is ~ 11 μs, helix time is 8 μs.

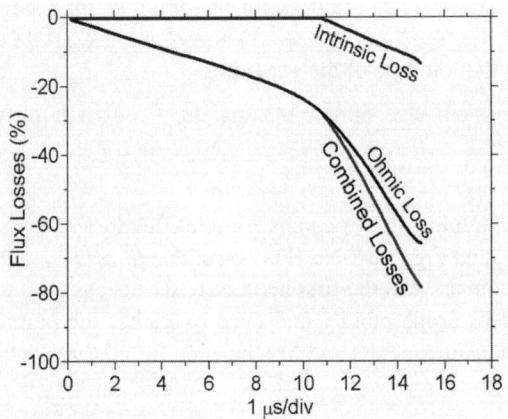

Fig. 5.24. Percentage flux loss, $\Delta\phi'$, in TTU X, tapered armature. Time to make contact with the first turn is ~ 11 μs, helix time is 4 μs.

Table 5.7. Generator parameters and losses.

PARAMETERS	TTU II	TTU IV	TTU X
α, after contact	0.72	0.73	0.93
Mean Freq. (kHz)	57	200	255
Current Gain	81	31	97
Resist. Flux Lost (%)	-16	-32	-66
Intrin. Flux Lost (%)	-69	-55	-13
Total Flux Lost (%)	-85	-87	-79
Flux Conserved (%)	15	13	21[*]

[*] TTU X has a tapered armature with its largest diameter at the output side.

And indeed, even though we made simple assumptions, such as a constant frequency for the calculation of the skin-depth, the flux in the conductors equals the intrinsic flux loss calculated with the aid of the 3D electromagnetic field solver, cf. Fig. 5.19, within ~ 20% for TTU II and TTUIV. Clearly, the resistivity of the conductors can be affected by very large generator current densities due to the accompanying temperature change, changing the skindepth as well. Hence, if such non-linear diffusion becomes predominant, the assumptions will fail, intrinsic flux losses will further increase.

In detail, we believe that the contact point does not progress in a smooth fashion due to microscopic ripples in the armature's / helix surface but rather as a rapid sequence of electrical shorts, cf. Chap. 5.4.1. Thus, the magnetic flux in the wires is left behind and lost for compression, eventually forming eddy currents in the conductors. Imperfections and small ripples in the armature are believed to be the cause for the sequence of shorts. The fact that the armature material is additionally bent and stretched considerably at the contact point does certainly not help the situation. This general idea is not new, however, we were able for the first time to make a quantitative connection between locally stored flux and the intrinsic flux loss.

The tapered generator, TTUX, exhibits a distinctly different behavior, compare Figures 5.23 and 5.24. Comparing the intrinsic flux loss derived from the two methods revealed that the loss derived from the magnetic field diffusion highly overestimates the experimentally observed loss. Hence, magnetic flux is partially recovered from the conductors, we estimate ~ 80 %, for TTU X. Since the tapered generator has a shallower angle with respect to the helix, we speculate that the contact point can run more smoothly as the armature is not deformed as much at the contact point. One could also stress the difference in expansion ratio, which is smallest for the larger end of the tapered generator, cf. Fig. 5.20. The surface of the armature will have smaller ripples at the smaller expansion ratios. This does not mean that constant diameter generators with smaller expansion ratio should be preferred since the initially available inductance, L_0, of the generator and thus the output current will drop considerably with smaller expansion ratio.

It is also known that large FCGs on the meter scale can conserve approx. 90 % of the initial flux. The expansion ratio of these generators is typically also close to 2, and the expansion angle (Gurney angle) is also comparable ~ 15 degree. However, the wall thickness of the armature in these large generators is also larger when the armature contacts the helix. Meaning that the deformation of the armature at the contact point might be considered less critical with respect to having a smoothly progressing contact point. For the typical expansion ratio of 2, the armature's wall thickness at the contact point is only approx. half of its unexpanded value. This also means that a 100 mm diameter armature has ~12 times the wall thickness of our smallest generator TTU IV, since the initial wall thickness has to grow about linearly with the armature diameter if the Gurney angle is to be kept constant.

The distinct difference in behavior between the constant diameter FCGs, see Figures 5.22 and 5.23, and the tapered FCG, see Fig. 5.24, is summarized in Table 5.7. The ohmic flux loss plays in TTU X a much more important role with a 66%

loss as compared to the intrinsic flux loss with only 13%. The total loss of 79% is quite typical for the simple small sized helical FCGs. TTU II with the same stator dimensions exhibits the opposite behavior with 16% ohmic flux loss and 69% intrinsic flux loss, causing 85% total flux lost.

5.5 Shocked Metal Resistivity

An important characteristic of FCGs is the change that occurs in materials of the generator while loaded from the explosive shock wave. It has been shown that materials change their conductivity while being shocked [Bic89]. Extensive hydrostatic testing of the change in conductivity of metals has been conducted as early as the 1940's [Law56]. The changes in conductivity were found to both increase and decrease depending on the type of metal the pressure was applied to.

More recently, experimental studies on the change in resistivity of insulators and semiconductors under hydrodynamic shock loading has also been conducted. This latter work has shown that most insulators and semiconductors become more conductive under shock loading [Bic89].

Theoretical studies of the changes of resistivity of metals under pressure loading have found that the material resistance is dependent upon the electron scattering angle, the peak electron velocity of the Fermi distribution, and the lattice potential energy of the metal [Law56].

To test the shock loading of materials in the FCG, a cylindrical tube was designed to generate an explosive shock wave normal to the surface of the test material. Fig. 5.25 depicts this experimental setup [Hem03]. The tube is 2.5 cm diameter x 20 cm and made out of Plexiglas. The base cradle which holds the test sample is curved to the shape of the shock front profile. This profile has a radius of curvature of 6.125 cm, the distance from the point of initiation for the explosive charge. The tube is filled with 6.125 cm length of hand-packed C-4 explosives. The detonation is initiated using an RP501 exploding bridge wire (EBW) requiring > 200 A to fire.

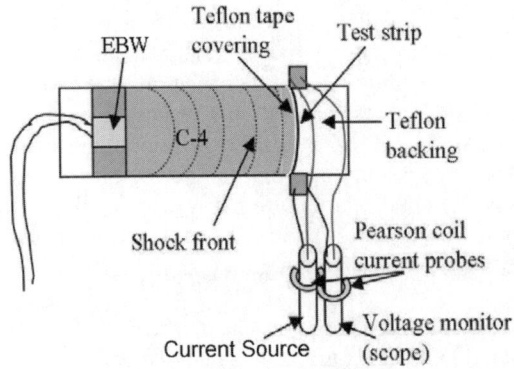

Fig. 5.25. Shocked metal conductivity test setup.

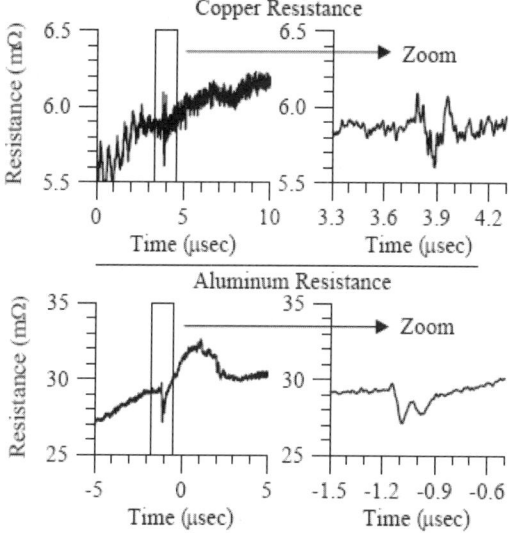

Fig. 5.26. Change in resistance of copper and aluminum samples under shock loading.

Two 50 Ω RG-58 cables are attached to the blocks of the test sample with the inner conductor and insulator extending across the cradle and curved to the profile of the cradle. The pulse current through the sample, ~ 70 A and a few μs long, is supplied by the upper cable while the lower cable is used to monitor the change in voltage of the test strip. The resistance of the sample is determined by the changing ratio of voltage drop across the sample over current through the sample. This setup produces a shock wave of 12 ±1GPa with about 100 ns width.

Fig. 5.26 shows the change in resistance over time for a copper and an aluminum test strip. A long and short (zoom) temporal history of the change in resistance are depicted. The graphs show an initial sharp decrease in resistance for 100 ns, corresponding to the width of the shock wave, followed by an increase in resistance before returning to the original strip resistance value. The slow rise of resistance in the strip prior to and after the shock is associated with resistance change due to joule heating from the current flowing through the strip. Similar results were observed for other samples.

Some of the observed dissimilarities in the behavior of copper and aluminum can be explained by the difference in the thermal conductivity of the two materials with copper having twice the thermal conductivity of aluminum. The rise in temperature of the material may be due to the release of the potential energy stored in the lattice in the form of phonons as the structure returns to its higher state of disorder after the shock wave passed through.

Knowing the geometric shape of the metal samples, 0.2 x 6.5 x 25.5 mm, and accounting for the wire connections, etc., we calculated the drop in resistivity from the measured data to be about 10% for copper and 15% for aluminum. On a longer timescale, one can expect that the resistivity of the shocked metal will increase due to deformation and due a temperature rise of the material caused by a heat transfer from the explosive products into the metal. This temperature rise will depend much on the given shape and geometry of the shocked metal under consideration. For an FCG armature, the resistivity might increase by a factor 2 during the explosively driven expansion before making contact with the stator.

It should be noted that the drop in resistivity in our dynamic loading tests was only about 30% of the observed change in hydrostatic loading at the same pressure level. We suspect that this difference is due to the fact that the shock wave is applying a unidirectional pressure on the metal rather than full volumetric pressure as in the hydrostatic case. If we consider only a third of the vibrational lattice affected with a decrease in the Debye temperature, then the change of resistivity for dynamic loading becomes consistent with hydrostatic loading.

References

[Ansoft] HFSS™, 3D EM Simulation Software by Ansoft.
[Bai01] Baird J, Worsey P, M. Schmidt M (2001) Effects of defects on armatures within helical flux-compression generators. In: Reinovsky R, Newton M (eds) Digest of technical papers of the 13th IEEE International Pulsed Power Conference, Las Vegas, NV, pp 953-956
[Bic89] Bichenkov EI, Gilev SD, Trubachev AM (1989) Shock-induced conduction waves in electrophysical experiments. J. Appl. Mech. Tech. Phys. 30: 291-303
[Che86] Chernyshev VK, Zharinov EI, Kazakov SA, Busin VN, Vaneev VE, Korotkov MI (1987) Magnetic flux cutoffs in helical explosive magnetic generators. In: Fowler CM, Caird RS, Erickson DJ (eds) Megagauss technology and pulsed power applications. Plenum Press, New York, pp 455-469
[Cra68] Crawford JC, Damerow RA (1968) Explosively driven high-energy generators. J. Appl. Phys. 39: 5224-5231
[For02] Fortov, VE (2002) Explosive generators of powerful pulses of electrical current (in Russian). Moscow, Nauka
[Fow75] Fowler CM, Caird RS, Garn WB (1975) An introduction to explosive magnetic flux compression generators. (Los Alamos Report LA-5890-MS)
[Gov79] Gover JE, Stuetzer OM, Johnson JL (1979) Small helical flux compression amplifiers. In: Turchi PJ (ed) Megagauss physics and technology. Plenum Press, New York, pp 163-180
[Ham88] Hamilton DC, Nellis WJ, Mitchell AC, Ree FH, van Tiel M (1988) Electrical conductivity and equation of state of shock compressed liquid oxygen. J. Chem. Phys. 88: 5042-5050
[Hem03] Hemmert D, Mankowski J, Rasty J, Neuber A, Dickens J, Kristiansen M (2003) Conductivity measurements of explosively shocked aluminum and OFHC copper used for armature material in a magnetic flux compression generator. In: Giesselmann M. Neuber, A (eds) Digest of Technical Papers of the 14th IEEE International Conference, IEEE Press, Piscataway NJ, pp. 1073-1076
[Ken97] Kennedy JE (1997) The Gurney model for explosive output for driving metal. In: Zukas JA, Walters WP (eds) Explosives effects and applications, New York: Springer, pp 221-257

[Law56] Lawson AW (1956) The effect of hydrostatic pressure on the electrical resistivity of metals. Chalmers B, King R (eds) Progress in Metal Physics, vol. 6, Pergamon Press, New York, pp 1-44

[Mit82] Mitchell AC, Nellis WJ (1982) Equation of state and electrical conductivity of water and ammonia shocked to the 100 Gpa (1Mbar) pressure range. J. Chem. Phys. 76: 6273-6281

[Nas71] Nasser E (1971) Fundamentals of gaseous electronics. Wiley-Interscience, New York, NY

[Nel99] Nellis WJ, et al. (1999) Minimum metallic conductivity of fluid hydrogen at 140 Gpa (1.4 Mbar). Phys. Rev. B 59: 3434-3449

[Nov01] Novac BM, Smith IR (2001) Explosive-Driven Pulsed-Power Generation (MURI) Program 2D Numerical Simulation of Helical Generators. In: Reinovsky R, Newton M (eds) Digest of technical papers of the 2001 IEEE Pulsed Power Plasma Science Conference vol 1, IEEE Press, Piscataway, NJ, pp 110-113

[Nov02] Novac BM, Smith IR (2002) Fast numerical modeling of helical flux compression generators. In: Selemir VD, Plyashkevich LN (eds) Meggauss-9. RFNC-VNIIEF, Sarov, Russia, pp 564-570

[Neu01] Neuber A, Dickens J, Cornette JB, Jamison K, Parkinson ER, Giesselemann M, Worsey P, Baird J, Schmidt M, Kristiansen M (2001) Electrical behavior of a simple helical flux compression generator for code benchmarking," IEEE Trans. on Plasma Science 29: 573- 581

[Neu02] Neuber AA, Holt TA, Dickens JC, Kristiansen M (2002) Thermodynamic state of the magnetic flux compression generator volume. IEEE Trans. on Plasma Science 30: 1665 – 73

[Neu02b] Neuber AA, Holt TA, Hernandez JC, Dickens JC, Kristiansen M (2002) Geometry impact on flux losses in MCGs. In: Selemir VD, Plyashkevich LN (eds) Meggauss-9. RFNC-VNIIEF, Sarov, Russia, pp 571-577

[Pav90] Pavlovskii AI, Lyudaev RZ, Boyko BA, Mamyshev VI, Mironychev PV, Kol'chatov VM, Yakubov VB (1990) Numerical model for helical magnetic cumulation generators. Titov VM, Shvetsov GA (eds) Megagauss fields and pulse power systems, part 1, Nova Science Publishers, Commack, NY, pp 233-239

[She68] Shearer JW, Abraham FF, Aplin CM, Benham BP, Faulkner JE, Ford FC, Hill MM, McDonald CA, Stephens WH, Steinberg DJ, Wilson JR (1968) Explosive-driven magnetic-field compression generators. J. Appl. Phys. 39: 2102-2116

[Ste79] Stuetzer OM (1979) Theory of small helical magnetic flux compression amplifiers. (Sandia Laboratories Technical Report, SAND79-1075)

[Wei96] Weir ST, Mitchell AC, Nellis WJ (1996) Electrical resistivity of a single-crystal Al_2O_3 shock-compressed in the pressure range 91-220 Gpa (0.91-2.20 Mbar). J. Appl. Phys. 80: 1522-1525

[Wor99] Worsey P, Baird J, Schmidt M (1999) Maximizing resolution of the high-speed photography of explosive-driven generator (EDG) armatures in operation. In: Reinovsky R, Newton M (eds) Digest of technical papers of the 2001 IEEE Pulsed Power Plasma Science Conference vol 2, IEEE Press, Piscataway NJ, pp 1110-1113

6 Generator Modeling

Michael Giesselmann, Ivor R. Smith, Bucur M. Novac, and Andreas A. Neuber

6.1 PSpice Simulation

The performance of Magnetic Flux Compression Generators (MFCG or just FCG) and the performance of the systems of which they are a part of can be modeled using electrical circuit simulators from the SPICE family. SPICE stands for **S**imulation **P**rogram with **I**ntegrated **C**ircuit **E**mphasis [Tui88]. This predecessor of all modern Spice flavors was developed in the 1970's at the University of California at Berkeley to assist in the design of the first analog integrated circuits that were emerging in this era. Since then, SPICE has been developed into an universal electric circuit simulator, refined and commercialized. For the ongoing discussions, we will concentrate on the PSpice® code currently sold by Cadence, Inc. (http://www.cadence.com), using the "Schematics" graphical circuit editor. It shall be noted, that the models described below are easily portable to other current SPICE versions.

The initial discussion will describe a basic model of a magnetic flux compression generator and present results from simulations and theoretical considerations. Later, power conditioning components will be added in the circuits as well as refinements to the FCG model to account for experimentally verified non-ideal behavior.

6.1.1 Theoretical Background

A magnetic flux compression generator is an electro-mechanical energy conversion device. It converts an (explosively driven) mechanical expansion of a cylinder into electrical energy. This form of energy conversion is also the basic principle of the operation of common electrical machines and always happens if the value of an inductance is changed due to some kind of linear, rotational or radial mechanical motion.

The voltage across an inductance for the general case, when both the current and the inductance are changing, is given by:

$$V_L = \frac{d}{dt}(L(t) \cdot I_L(t)) \tag{6.1}$$

If we assume, that the collapsing inductor is connected in parallel to a load resistance and a load inductance be represented by the final value of the inductor, we get circuit Eq. (6.2). For the derivation it is considered that positive current is defined as to flowing into the resistor and inductor.

$$\frac{d}{dt}L(t) \cdot I(t) + L(t) \cdot \frac{d}{dt}I(t) + I(t) \cdot R(t) = 0 \tag{6.2}$$

By separation of variables and integration, we can find the analytical solution shown in Eq. (6.3) for the common current in the circuit as a function of time. $L(0)$ and $I(0)$ represent the initial inductance and current, respectively, cf. Eq. (2.9).

$$I(t) = I_0 \cdot \frac{L_0}{L(t)} \cdot \exp\left(-\int_0^t \frac{R(t')}{L(t')}dt'\right) \tag{6.3}$$

By examination of Eq. (6.3), we can find the maximum theoretical current ratio for a rapid collapse of the inductor with complete preservation of the magnetic flux. In this case all the energy conversion happens before significant energy is dissipated in the load resistor. Again, L_0 and I_0 represent the initial inductance and current whereas L_f and I_f represent the final inductance and current values after the collapse of the inductor is completed at the final time T.

$$\frac{I_f}{I_0} = \frac{L_0}{L_f} \tag{6.4}$$

$$E(t) = \frac{1}{2} \cdot L(t) \cdot I(t)^2 \tag{6.5}$$

Eq. (6.5) is the general expression for the energy in an inductor. If the equation for the current as a function of time (Eq. (6.3)) is inserted into Eq. (6.5), the energy in the collapsing inductor can be derived. The result is given by Eq. (6.6). Under ideal conditions, for a very fast inductor collapse with no magnetic flux losses, the maximum theoretical energy ratio is given by Eq. (6.7).

$$E(t) = E_0 \cdot \frac{L_0}{L(t)} \cdot \exp\left(-2 \cdot \int_0^t \frac{R(t')}{L(t')}dt'\right) \tag{6.6}$$

$$\frac{E_f}{E_0} = \frac{L_0}{L_f} \tag{6.7}$$

L_0 and E_0 represent the initial inductance and energy whereas L_f and E_f represent the final inductance and energy values after the collapse of the inductor is completed at the final time, $t = T$.

The force needed to compress the magnetic field in the inductor and achieve the energy conversion is given by Eq. (6.8). It shows that the force rapidly increases during the flux compression process, since it depends on the square of the current. The actual force needed to collapse the inductor is higher since additional force is needed for the mechanical deformation of the inner cylinder in the typical coaxial FCGs. From the equation of the force Eq. (6.9) can be derived. This equation represents the instantaneous power that is converted from mechanical to electrical energy. The integral of this equation yields the total converted energy that is added to the system during the collapse of the inductor.

$$F(t) = \frac{1}{2} \cdot I(t)^2 \cdot \frac{dL(t)}{dx} \tag{6.8}$$

$$P(t) = F(t)\frac{dx}{dt} = \frac{1}{2} \cdot I(t)^2 \cdot \frac{dL(t)}{dx} \tag{6.9}$$

We note that the equations given above, Eqs. (6.1) to (6.7), are ideal in the sense that they do not account for the intrinsic flux loss as discussed in Chap. 5.4.

6.1.2 PSpice® Implementation of a Basic FCG

The operation of a basic flux compression generator can be illustrated by the PSpice® example shown below. Fig. 6.1 shows the basic circuit containing an explosive flux compression generator ("Expl_Gen1"), a load inductor, a load resistor, and a voltage source ("V_compression") that controls the collapse of the inductor inside of "Expl_Gen1". In this example, a custom symbol, depicting the explosive flux compression generator, was created. This symbol is associated with the sub-circuit shown in Fig. 6.2, which implements the function. Any sub-circuit can be examined by selecting the symbol and descending down in the hierarchy. The interface nodes in Fig. 6.2 connect to the three equally labeled terminals of "Expl_Gen1" in Fig. 6.1. The sub-circuit for the explosive flux compression generator shown in Fig. 6.2 implements the expression shown in Eq. (6.1). It basically represents a voltage controlled inductor, where the control voltage is applied to the "Compression" input terminal with respect to ground. In Fig. 6.1, the control voltage starts at V1 = 1.0 V and decreases linearly to V2 = 0.0 V in 20 μs, providing in this case a linear collapse of the generator inductance. However, we note that the inductance collapse of a realistic generator is generally nonlinear. Nevertheless, the general approach to the simulation is the same, linear or nonlinear.

In the sub-circuit shown in Fig. 6.2, the voltage between the output terminals L+ and L- is the total voltage across an inductor which is changing as a function of time as controlled by the control voltage input. The current in the inductor is entering the L+ node and exiting the L- node. The current controlled voltage source H1 measures the current in L+/L- loop and drives an identical current through L_ref using the voltage controlled current source G1. The voltage drop across the reference inductor is scaled with the control voltage by the multiplication device and fed to one of the inputs of the summing device "ESUM". This portion of the voltage represents the contribution due to the change in current ($L\, dI/dt$).

The ABM1 device "ABM1_Differentiator" is fundamentally a voltage controlled voltage source. This device calculates the derivative of the control voltage through the use of the DDT function and multiplies the result with the value of the reference inductor. The value of the reference inductor is the initial value of the collapsible inductor. The output of the ABM1_Differentiator represents the rate of change of the inductor with respect to time. Note the STP(Time-50ns) function included as an additional factor for the control function for the ABM1_Differentiator device. This function evaluates to zero before 50 ns and has a value of 1 afterwards. It is used to suppress a transient of the differentiator during the startup phase of the simulation. The output voltage of the ABM1_Differentiator device is multiplied with the current to form the $I\, dL/dt$ component and added to the voltage from the $L\, dI/dt$ term by ESUM.

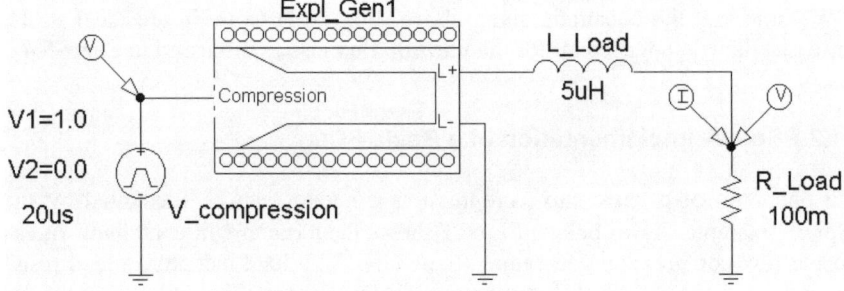

Fig. 6.1. Top level of a basic FCG simulation example.

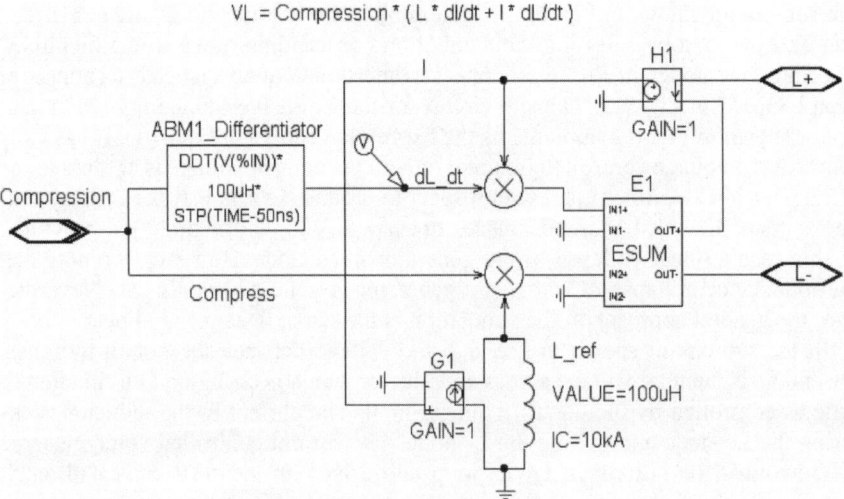

Fig. 6.2. Sub-circuit for the voltage controlled inductor "Expl_Gen1" in Fig. 6.1.

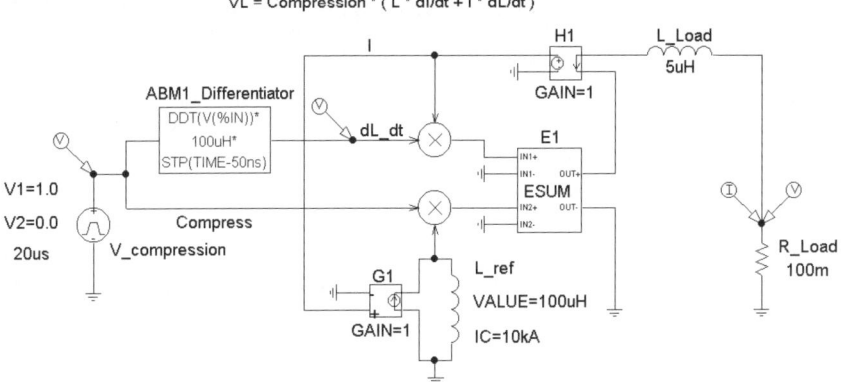

Fig. 6.3. Alternative single page circuit using only basic PSpice® elements.

Although the custom symbol for the FCG generator shown in Fig. 6.1 is graphically appealing and creates an uncluttered system view of the circuit at a hierarchical level above the sub-circuit for the voltage controlled inductor, the creation of such symbols requires a higher proficiency than many first-time users possess. For more details on the creation of custom symbols the reader should be referred to the software manual for the PSpice® program. To avoid the additional complexity of generating custom symbols, we created a simulation model that is contained on a single page, using only standard devices. This schematic is shown in Fig. 6.3. Both models yield identical results with the PSpice® model being more flexible in accepting arbitrary input waveforms.

The results of the simulation of the basic FCG generator are shown in Fig. 6.4. In the upper plot of Fig. 6.4, the current in the inductor is shown. The initial seed current is 10 kA. This value is set by selecting the reference inductor in Fig. 6.2 and Fig. 6.3 and specifying a value for the "IC" (Initial Condition) attribute. During the compression of the magnetic flux, which happens during the first 20μs, the current climbs to approximately 190 kA, just below the maximum theoretical value of 200 kA. Afterwards the current decays to zero while the energy is dissipated in the load resistor.

The lower plot of Fig. 6.4 shows the initial energy in the inductor and the additional energy that is inserted into the system during the flux compression process. Since energy is already being dissipated during the flux compression process, the total energy is slightly below the theoretical maximum of 100 kJ. The S() function is used in the plot to calculate the integral of the converted power as given by Eq. (6.9) and the integral of the power that is dissipated in the load resistor. It can be seen, that the total energy dissipated in the load after 400 μs is equal to the sum of the initial and converted energy.

In order to validate the PSpice® model, we have compared the results obtained from the analytical solutions given above. Both the PSpice® and analytical solutions were obtained using only one linearly collapsing inductor and a resistive load. We slightly modified the PSpice® circuit to match these conditions. The circuit used for verification is shown in Fig. 6.5. This circuit was simulated for 19 μs during which time the inductor collapses to 5% of its original value. The results of this simulation are shown in Fig. 6.6. The printer symbol in the schematic of Fig. 6.5 prints the current values in selectable regular intervals to an ASCII file. To compare the results, we read the contents of the ASCII file into MathCAD® and plotted them along with the results for the analytical solution. The results are shown in Fig. 6.7. It shows that the results from the analytical calculations are a match to the imported results from the PSpice® model. Of course, the PSpice® model can be easily applied to any nonlinear, hence more realistic, inductor collapse, where an analytical solution would fail. All that is required is adjusting the voltage waveform of V_compression in Fig. 6.3.

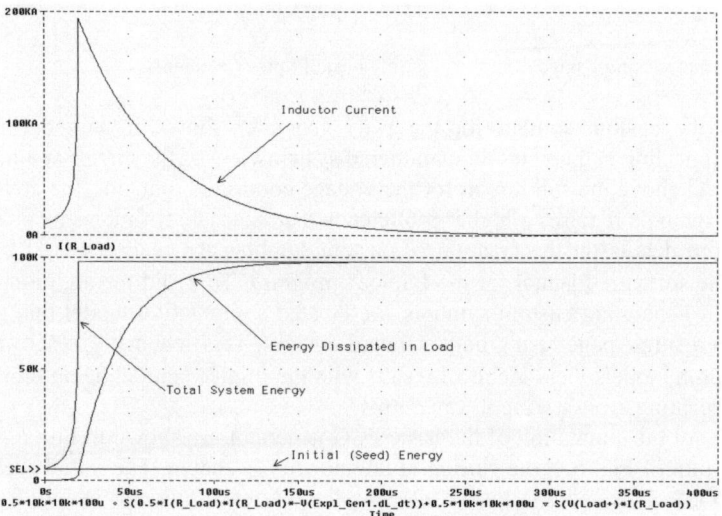

Fig. 6.4. Simulation results for the basic FCG circuit example.

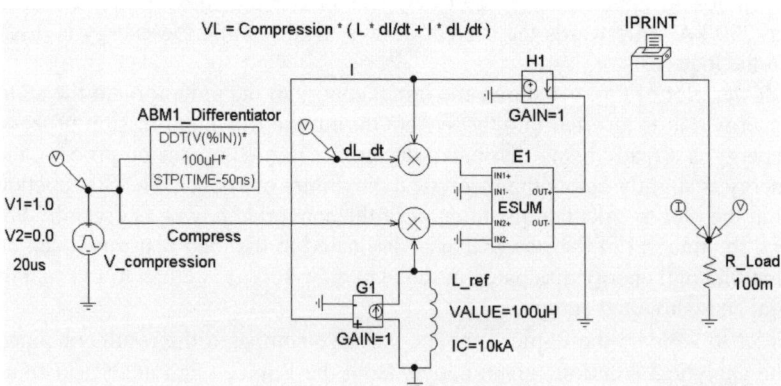

Fig. 6.5. Slightly modified basic FCG Generator circuit for model validation.

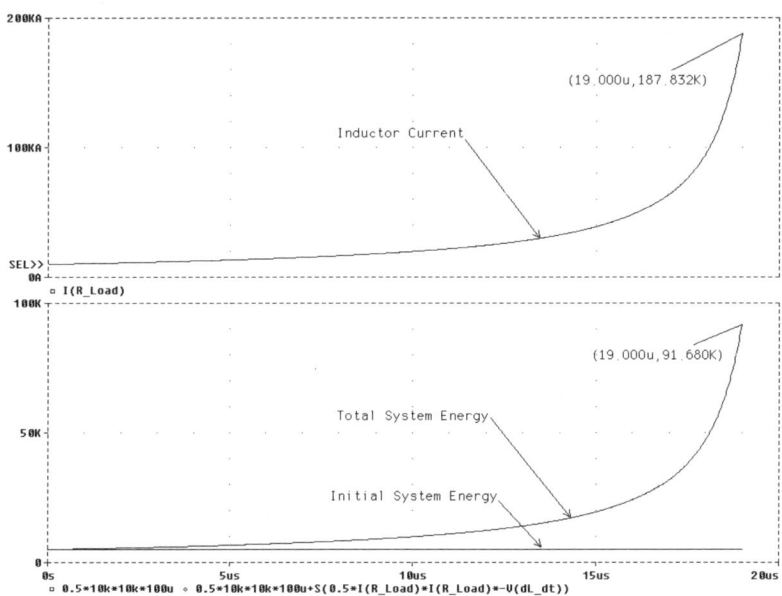

Fig. 6.6. Results from modified basic FCG circuit for model validation.

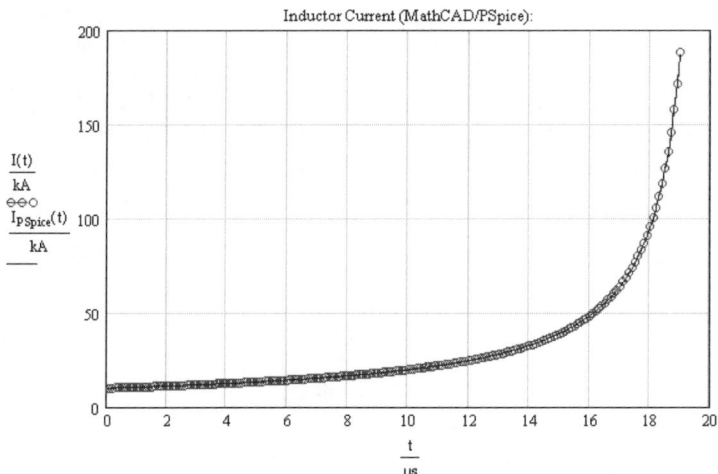

Fig. 6.7. Comparison of results from PSpice® and MathCAD® calculations.

6.1.3 PSpice® Modeling of a Power Conditioning Circuit

Models of magnetic flux compression generators can be used to study the performance of stand-alone generators that are connected to very simple, generic loads. However, the models can also be used to predict the behavior of systems that contain magnetic flux compression generators and other circuitry used for power conditioning. Typically, magnetic flux compression generators are low voltage, high current sources. Power conditioning circuits can be used to convert the output power of FCGs to feed high voltage loads. In this context, high voltage refers to voltages in the range of 100's of kV. A typical power conditioning circuit uses an energy storage inductor in series with a fuse to generate a high voltage pulse from an FCG, cf. Fig. 1.2 in Chap. 1. The load inductor of the FCG is the storage inductor of the power conditioning system. Initially, the current that flows through the storage inductor completes the circuit through an exploding wire fuse. The fuse is initially almost a short circuit and current buildup in the storage inductor is unimpeded. After the fuse blows, that current is commutated into a load that typically has a much higher impedance than the fuse and the voltage across the load will therefore be much higher. The "one-shot" characteristic of the fuse (requires replacement after every test cycle) is not a problem when used with an FCG since it is also inherently a "one-shot" device. The exploding wire fuse is made of a number of thin wires embedded in sand. The sand is used as an arc-quenching medium. Fig. 6.8 shows a drawing of a fuse with an active length of 0.5 meters that we used in some of our experimental work.

The behavior of the fuse can be modeled by PSpice$^©$, which makes it possible to predict the performance of circuits containing an exploding wire fuse and to aid in choosing the correct number of fuse wires and cross section. For the model presented here, it is assumed that the current rise-time in the fuse is in the μs regime. In this time regime, the heating of the fuse wires is almost completely adiabatic; meaning that almost no heat transfer to the environment takes place before the explosion of the fuse. This would correspond to a very high overload factor (current is orders of magnitude higher than the rated current) of a conventional fuse. The fuse model that is described here should therefore **NOT** be used to predict the performance of protective fuses in electric power systems.

The explosion of the fuse wires can be broken down into two separate phases. In the first phase, the initially solid wires heat up due to energy input until parts of the wire explode [Gie02]. As can be seen in [Gie02], not all wires explode simultaneously and plasma instabilities can occur.

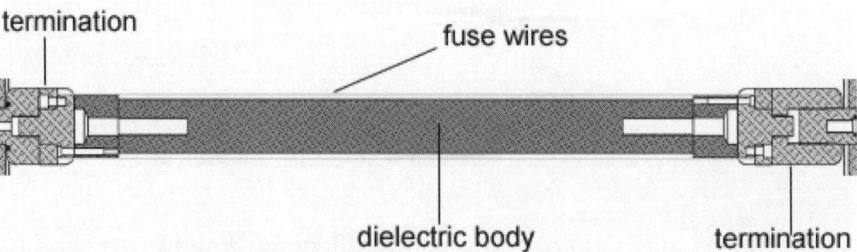

Fig. 6.8. Drawing of an exploding wire fuse used for experimental investigations of power conditioning circuits with an active length of 0.5m.

The explosion of the wires is a very complex process, which is described in detail in [Kim02]. Through exhaustive study of the literature and own experimental results, we generated a set of equations, which describe the dynamic behavior of the fuse as a circuit element very accurately for the parameter range of interest [Hee03]. The development of these equations is described in much more detail in Chap. 7.

With the material constants $A = 25$ and $C = 190$, we write:

$$\rho(t) = \begin{cases} \left[1 + A\left(\dfrac{\int_{0\mu s}^{t} J_{wire}(t)^2 \, dt}{BlowLimit_{Cu}}\right)^{2.5}\right] & \text{if } \int_{0\mu s}^{t} J_{wire}(t)^2 \, dt \leq BlowLimit_{Cu} \quad \text{a)} \\ \left[A + \exp\left(\dfrac{\int_{0\mu s}^{t} J_{wire}(t)^2 \, dt - BlowLimit_{Cu}}{BlowLimit_{Cu}} \cdot C\right)\right] & \text{otherwise.} \quad \text{b)} \end{cases}$$

(6.10)

Equation (6.10) shows a definition for the relative resistivity, ρ, as a function of time for the two phases of fuse opening. The relative resistivity, ρ, is a dimensionless number. The starting value at $t = 0$ is unity. The relative resistivity, ρ, is a multiplier for the fuse resistance, if the total initial resistance of the array of fuse wires at room temperature is considered to be the starting value (1.0). The controlling quantity for the fuse resistance is the integral of the square of the current density in the wires. The integral of the square of the total current is often called the action integral and has a unit of Joules/Ω. The heating phase of the solid wires ends if the integral of the J^2 is equal to the so-called Blow Limit for the wire material. The value for copper for the parameter range in question is given in Eq. (6.11). The value is slightly higher for faster current rise-times, but the variations are small.

$$BlowLimit_{Cu} := 1.63 \; 10^9 \; \text{amp}^2 \; \text{sec cm}^{-4} \quad (6.11)$$

Refer to Table 7.2 for values of the blow limit for materials other than copper.

Considering the given values for A and C, the resistance of the wires has risen to 26 times the initial value at the beginning of the second (explosion) phase. Afterwards the resistivity is rising exponentially, cf. Eq. (6.10b). A representative plot of the relative resistivity of the fuse wires is depicted in Fig. 6.9.

To study power conditioning circuits and specifically the complex fuse opening process, it is often advantageous, to initially replace an explosively driven generator with a conventional pulsed power source. This can be in the simplest case, as outlined below, a high energy density capacitor with a single closing switch. Such an arrangement eliminates the safety issues required to fire explosively driven generators as well as reduces the cost and turnaround time per shot. A typical circuit used for the evaluation of a power conditioning system, which is based upon a

168   6 Generator Modeling

fuse opening switch, is shown in Fig. 6.10. This circuit concentrates on the major components, omitting second order parasitic elements.

The sub-circuit that represents the fuse model is shown in Fig. 6.11. The input parameters for the fuse are the number of wires in parallel, the gauge number of the wires (AWG number), the active length of the fuse section and the parasitic inductance of the fuse element.

$$\text{mil} \equiv \frac{\text{in}}{1000} \qquad \text{cmil} \equiv \frac{\pi}{4} \cdot \text{mil}^2 \qquad A_{\text{cmil}}(\text{AWG}) := 2^{\left(\frac{50-\text{AWG}}{3}\right)} \qquad (6.12)$$

Equation (6.12) can be used to calculate the cross section of a wire with a given gauge (AWG) number. The equation gives the relationship of the gauge number with the cross-section of the wire in circular mills (cmil). A mil is defined as $1/1000^{th}$ of an inch and 1 cmil is the area of a circle with 1 mil diameter.

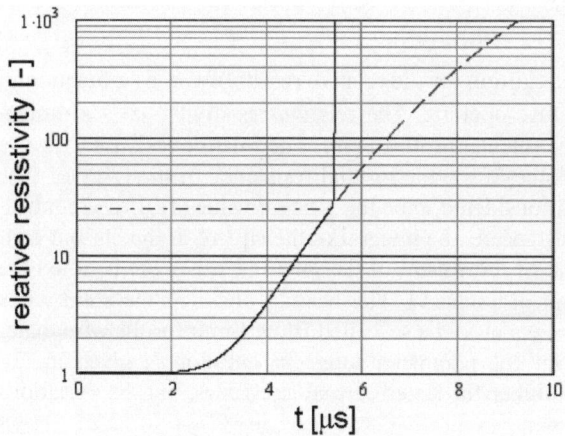

Fig. 6.9. Solid line: plot of the relative resistivity given in Eq. (6.10). Dashed line: Eq. (6.10a); Dash-dot line: Eq. (6.10b).

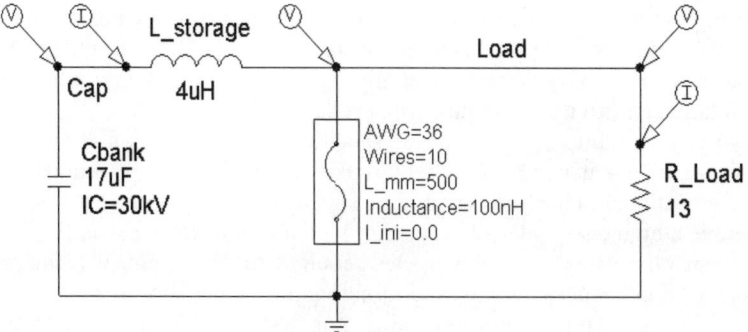

Fig. 6.10. Basic circuit for studying a power conditioning system with an opening switch.

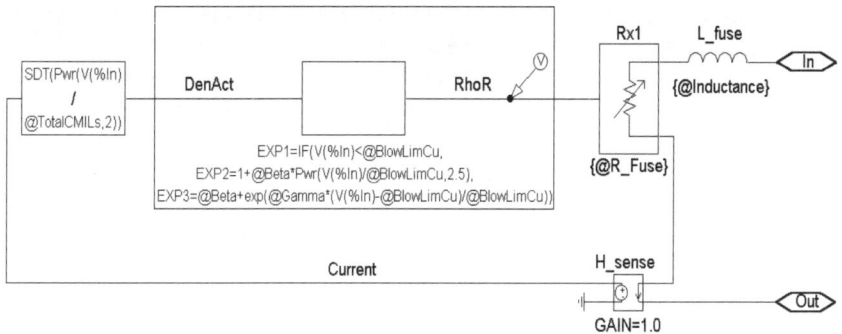

Fig. 6.11. Basic sub-circuit for the exploding wire fuse.

In order to evaluate the expressions given in Eq. (6.10) in PSpice©, appropriate scaling is necessary to assure that the value of the integral is not reaching values that would crash the simulator. Therefore the current density was scaled to amps/cmil and the limit value for the integral of the square of the current density is given by Eq. (6.13) for the new units (see Eq. (6.11) for MKS units).

$$BlowLimit_{Cu} := 4.185 \; 10^{-2} \; (amp/cmil)^2 \; sec \qquad (6.13)$$

The part with the frame around it, which is depicted in the center of Fig. 6.11, is an "ABM 1" device (Analog Behavioral Modeling with 1 input), which implements Eq. (6.10). The control functions expressed in "EXP1-3" correlate precisely with the definition for the relative resistance in Eq. (6.10) for the two regimes. The function block returns an output voltage starting with 1 volt to the control input of the element Rx1. Rx1 is a voltage controlled resistor which has a reference value that is equal to the cold resistance of the fuse wires. The actual value in the circuit is the reference value multiplied with the value of the control voltage in volts.

The sub-circuit for the voltage-controlled resistor is shown in Fig. 6.12. In this circuit the current-controlled current source "F_mirror" senses the current in the resistor and drives an equal current through the resistor "R_ref". The voltage drop across this resistor is measured and multiplied with the control voltage to generate the total voltage drop across the voltage controlled resistor, which is injected into the output loop by the voltage-controlled voltage source "E_Mult".

The result of the simulation for the circuit shown in Fig. 6.10 is given in Fig. 6.13. Shown in Fig. 6.13 are all the major parameters including the currents in the storage inductor and the fuse as well as the voltage and power in the load resistor. Also shown in the uppermost trace of Fig. 6.13 is the relative resistivity "RhoR" from the ABM device in the fuse model in Fig. 6.11, also cf. Fig. 6.9. For this plot, the vertical scale is logarithmic, whereas all other traces are plotted on linear scales.

The results of this simulation correspond well with experimental results we obtained from our experimental studies as shown in Fig. 6.14 and Fig. 6.15. Fig. 6.14 shows the actual scope image containing the currents in the storage inductor and the fuse from the experimental results reported in [Gie00].

The data has been imported into MathCAD® and plotted on a common vertical scale considering all scaling factors. This is shown in Fig. 6.15. If compared with Fig. 6.13, the experimental results in Fig. 6.15 show that the fuse model can be used to accurately predict the fuse performance observed in the laboratory.

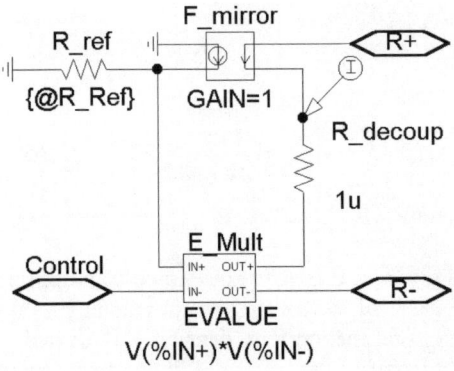

Fig. 6.12. Sub-circuit for the voltage controlled resistor in Fig. 6.11.

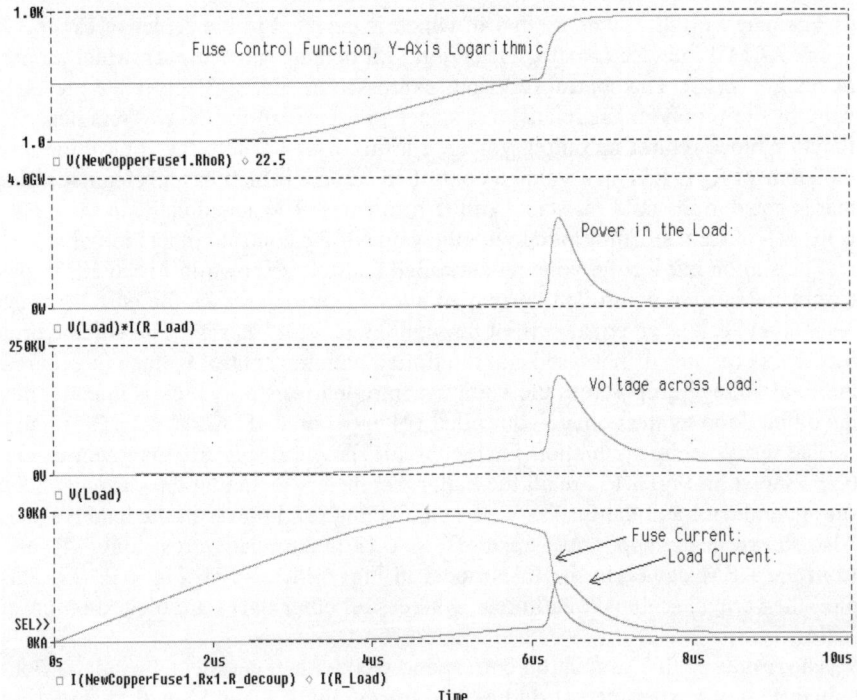

Fig. 6.13. Simulation results for the circuit shown in Fig. 6.10.

6.1 PSpice Simulation 171

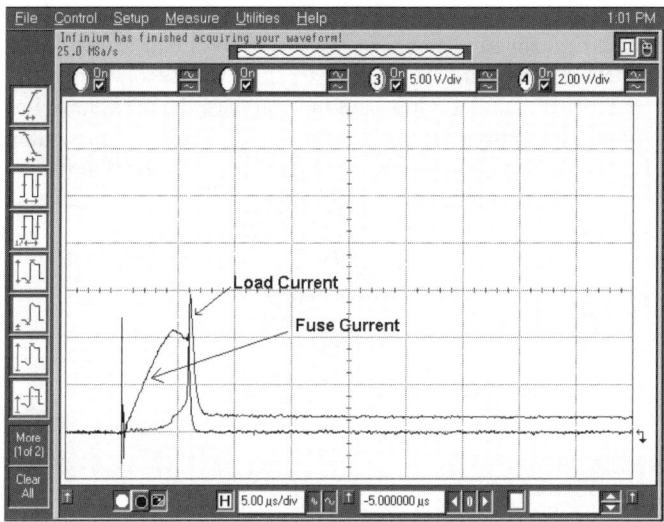

Fig. 6.14. Actual oscilloscope screen for experimental results for a system shown in Fig. 6.10.

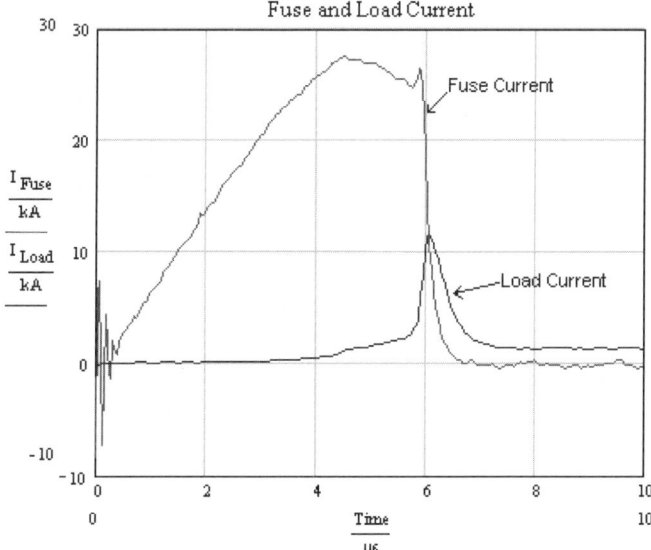

Fig. 6.15. MathCAD® plot of experimental results from Fig. 6.14.

6.1.4 Combining the Power Conditioning Circuit with the FCG Model

Now that the fuse model is developed and validated, we employ it to predict the performance of a power conditioning circuit that is used in combination with a flux compression generator, see Fig. 6.16. Here the attributes "L_ref" and "I_ini" have been added to the explosive generator model. The sub-circuit shown in Fig. 6.17 refers to these parameters using the "@" character. The parameters for the fuse, i.e. wire size, number of parallel wires, and the length of the wires have been adjusted for this example for appropriate performance. The results of the simulation are shown in Fig. 6.18. The bottom traces in Fig. 6.18 show that the fuse explodes right at the peak of the current in the FCG generator. In this example, the power conditioning circuit generates a peak voltage of 400 kV at the load.

We note that the results so far omit the intrinsic flux loss, cf. Chap. 5.4. In the following Chap. 0, we include the intrinsic flux loss and discuss its implementation into the Pspice FCG modeling.

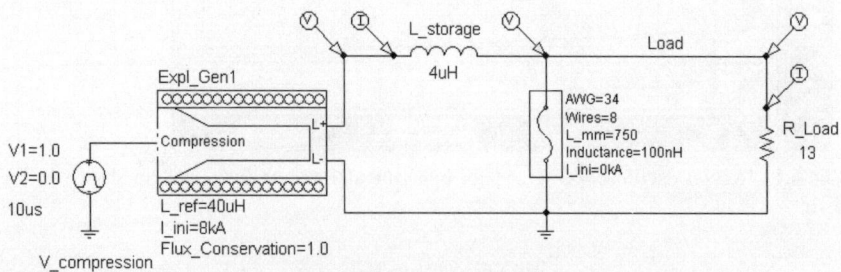

Fig. 6.16. Top level schematic for power conditioning circuit with FCG.

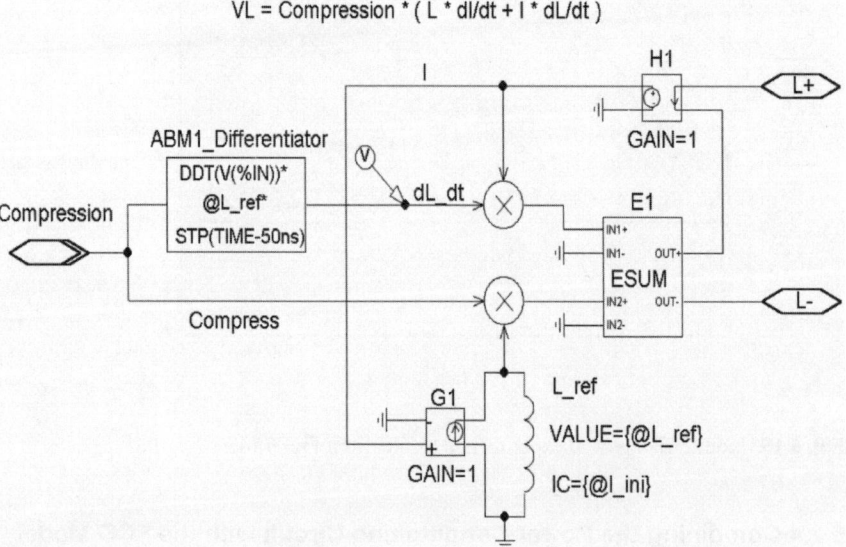

Fig. 6.17. Sub-circuit for FCG Generator.

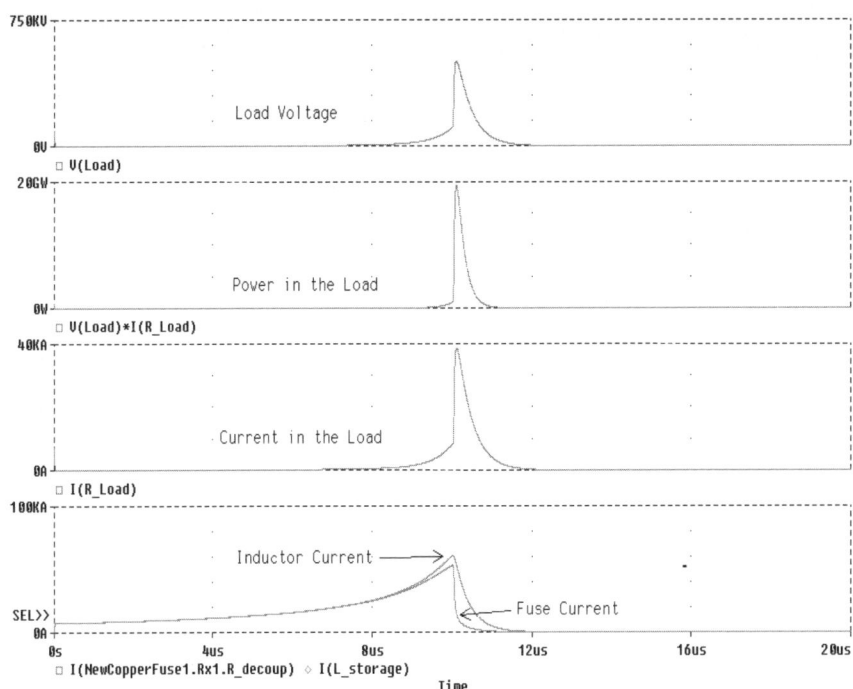

Fig. 6.18. Performance plot for power conditioning circuit with FCG.

6.1.5 Including Intrinsic Flux Loss into the FCG Model

During many of our own tests of magnetic flux compression generators and comparison of the results with model performance, it was found that not all of the magnetic flux, that is generated by the initial "seed" current is preserved and available for compression. During the compression process flux losses of various kinds occur with have been documented using very elaborate diagnostic techniques and extensive, detailed studies with electromagnetic codes, cf. Chap. 5. However, for performance prediction of systems with FCG generators it is important to have models that incorporate the effects of the various forms of flux losses and not be overwhelmingly complicated so as to demand too much computing power and time to generate a run of the model.

It was found, that using an intrinsic flux loss parameter in the second term of Eq. (6.2) yields very accurate results while preserving the elegant simplicity of the FCG model, cf. Eq. (2.13). For the next example an intrinsic flux loss of $\alpha = 0.85$ was used, which is typical for a generator with approximately 100 grams of high explosives. The circuit for this example is shown in Fig. 6.19. To utilize this flux loss parameter, an additional attribute, Flux_conservation, was added to the FCG model. Its value was set to 0.85. Fig. 6.20 shows the modified sub-circuit with incorporated intrinsic flux loss parameter, α.

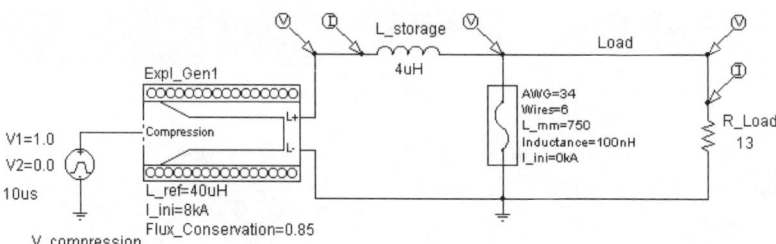

Fig. 6.19. Top level schematic for power conditioning circuit with modified FCG; Flux_Conservation = α = 0.85.

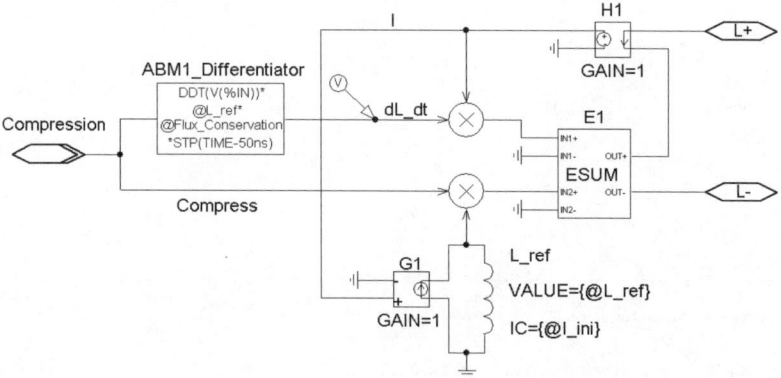

Fig. 6.20. Voltage controlled inductor sub-circuit for FCG generator modified to account for flux loss, cf. Eq. (2.13).

As a result the peak load voltage has dropped from about 400 kV to 240 kV, see Fig. 6.21, showing that the flux losses are very critical for the performance and an accurate model is needed for circuit performance prediction.

6.1.6 More accurate PSpice® model for the FCG

In the following, we present a more elaborate, yet PSpice® based, model for the FCG that will include current diffusion into the conductors in detail. The model is dividing the current conducting layers of the inner cylinder and the outer solenoid into many layers, which are all modeled separately. We compute the flux linkage of all the layers as well as the resistance of each layer individually. A drawing of the case studied here is shown in Fig. 6.22. The dynamic change of resistance is computed based on the integral heat deposited into the layer. By calculating the stray inductance of the individual layers and using the momentary resistance of the layer, we model the magnetic diffusion of the field into the conductors. Fig. 6.23 shows the flux linkage of the main flux with individual turns of the outer solenoid as a function of the current penetration depth into a wire and the geometrical relation of the penetration depth with the cross section of the wire.

6.1 PSpice Simulation

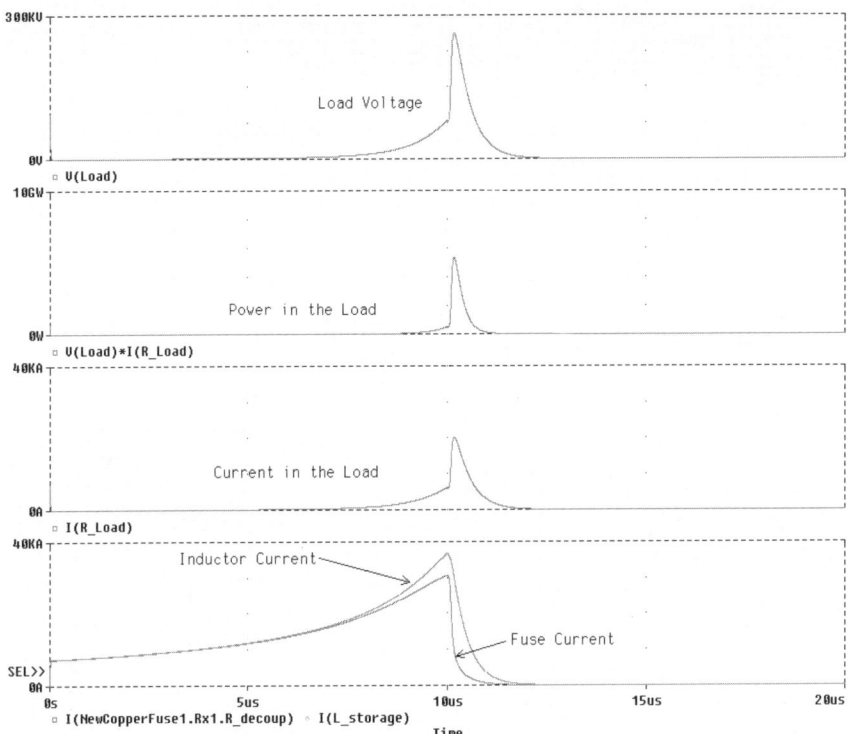

Fig. 6.21. Performance plot for power conditioning circuit with the FCG model incorporating intrinsic flux loss.

Fig. 6.24 shows the top level of a hierarchical PSpice® circuit that contains the generator and a resistive load. In his model, the value of the main inductor, the value of the nominal resistance of the remaining winding and the value of the leakage flux can be individually controlled. For this example, almost identical control functions for the three parameters have been chosen. All of the control functions are similar with the voltage "V_compression", for instance, in Fig. 6.1. These values represent the decrease in main inductance as well as the associated decrease of the resistance and leakage inductance of the remaining outer winding and armature (inner cylinder) during burnout. In addition to the nominal decrease in resistance, the resistivity of the outer winding and the armature is modeled in 8 different layers taking into account the accumulated heat from the current flow history and the associated increase in resistance. The general structure of the model underlying the FCG depicted in Fig. 6.24 is shown in the associated subcircuit in Fig. 6.25. The main variable inductor as well as hierarchical blocks for the other copper winding as well as the aluminum cylinder are visible. Also shown is a model of the explosively operated switch that crowbars the seed source and all the connections between the elements.

176 6 Generator Modeling

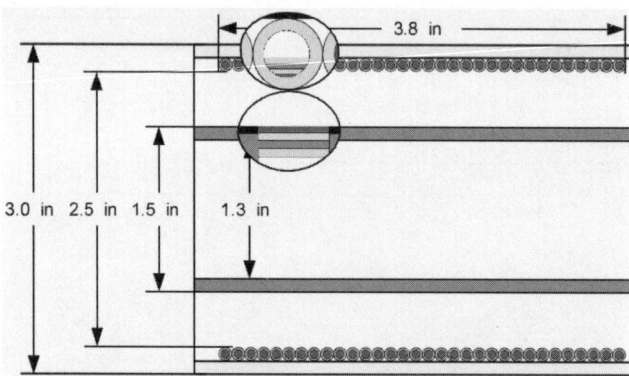

Fig. 6.22. Drawing of 32-turn FCG with detail of current layers in conductors.

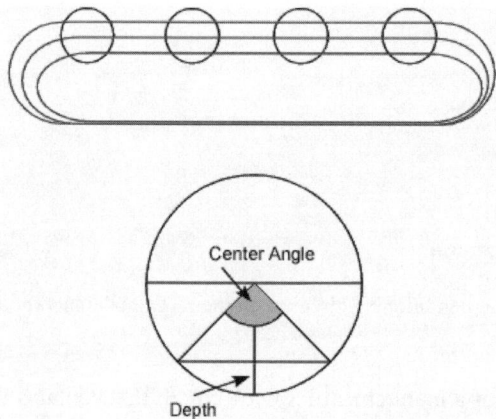

Fig. 6.23. Flux linkage with conductors of outer solenoid as a function of current penetration depth.

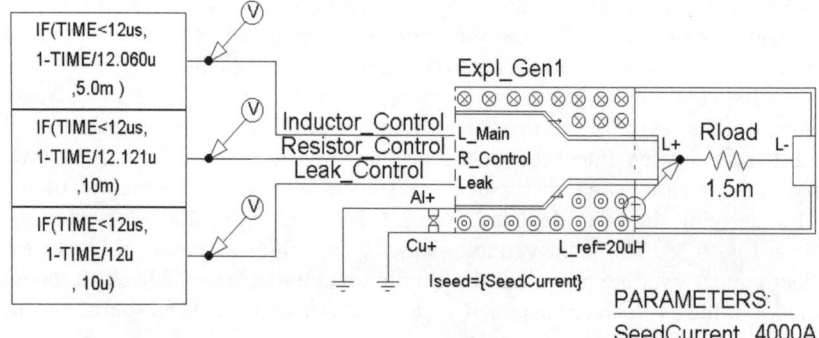

Fig. 6.24. Top level schematic for advanced FCG model.

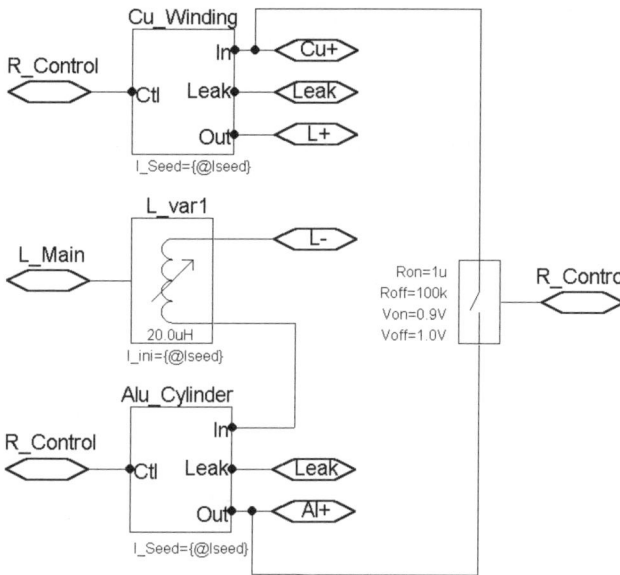

Fig. 6.25. Sub-Circuit for advanced FCG model.

Fig. 6.26 a and b depict the expansions of the hierarchical blocks for the copper winding and the aluminum cylinder of Fig. 6.25. These sub-circuits model the resistance of the 8 separate current layers and the leakage inductance between these layers in the copper winding and the aluminum armature. Eight layers with progressively increasing spacing were chosen after careful consideration of the overall geometry. The layers are more densely spaced on the outer edges where more current flows initially. The thickness of adjacent layers was chosen to increase by $\sqrt{2}$. All of the resistance elements are given their values at room temperature and mass from geometrical considerations. The geometry shown in the lower part of Fig. 6.23 was used to calculate the mass and resistance of the different layers of the copper winding. Also shown in Fig. 6.26 a/b are the initial values of the leakage inductances between the different conduction layers.

Fig. 6.27 a and b show the sub-circuits that are associated with the individual resistance elements in the models for the copper winding and the aluminum cylinder. In the lower part of these sub-circuits we calculate the dissipated energy, which is used to increase the relative resistivity of the layer during burnout. At the same time, the length of the remaining part of the layer and its remaining mass is reduced as the inner cylinder expands towards the end of the winding. Fig. 6.28 shows the model for the controlled leakage inductance between the layers. This model represents a voltage-controlled inductor without an energy conversion component.

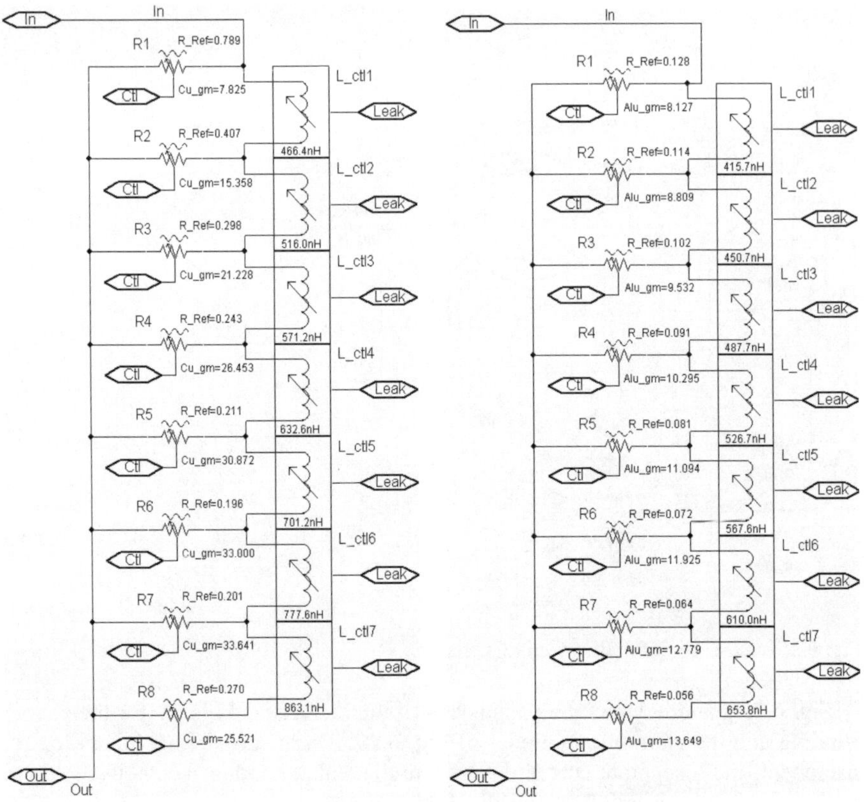

178 6 Generator Modeling

Fig. 6.26. Models for **a)** "Cu_Winding" and **b)** "Alu_Cylinder" in Fig. 6.25.

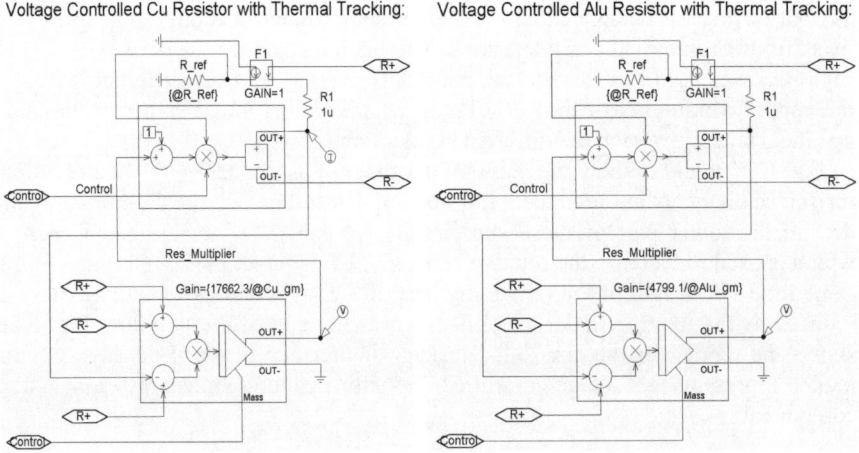

Fig. 6.27. Models for Cu and Al resistors with thermal tracking in Fig. 6.26 a/b.

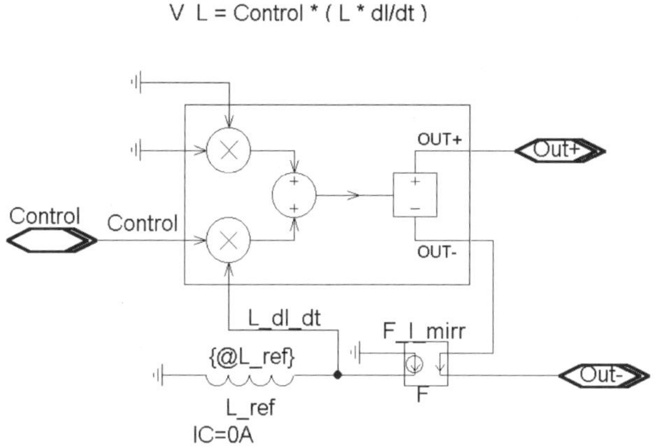

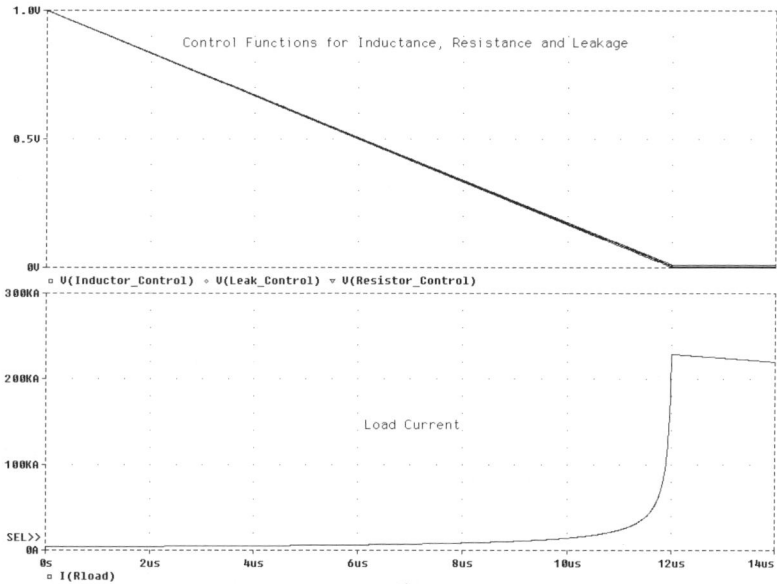

Fig. 6.28. Model for variable leakage inductance between current layers.

Fig. 6.29. Results from the advanced FCG model showing the load current and the control functions.

Fig. 6.29 shows results from the advanced model described above. The traces in the upper plot show the almost identical control functions, whereas the trace in the lower plot represents the load current. Fig. 6.30 shows further results of the simulation concentrating on the currents in the different layers of the copper winding. It is evident from this graph, that the outer layers carry most of the current in the final stages of the collapse of the inductor.

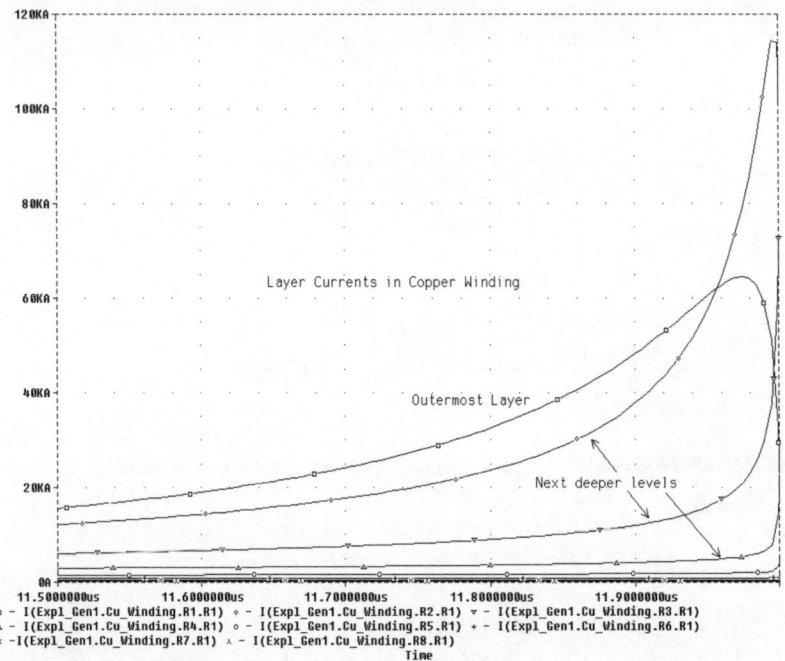

Fig. 6.30. Results from the advanced FCG model showing currents in the different layers of the copper winding.

6.1.7 PSpice® Model Conclusions

Thus far, in Chap. 6.1, PSpice® models for magnetic flux compression generators as well as models for critical elements of associated power conditioning circuits have been described. In particular, the behavior of exploding wire fuses has been modeled very accurately. We note that the provided numerical examples utilize a linear control function for the FCG inductance, cf. Fig. 6.29. In Chap. 9, we utilize the model developed in Chap. 6.1.5 to simulate and compare the experimental performance of an FCG in detail. In order to do so, we will use more realistic, nonlinear control functions for the inductance collapse. See, for instance, Fig. 2.6 for the inductance collapse of a single stage FCG.

Initial models describe the basic behavior of magnetic flux compression generators, while subsequent models incorporate more details of loss mechanisms in these devices to arrive at more accurate results. The complexity of the models increases rapidly with increasing level of modeling detail. Comparisons of the results of the simulations show, that even moderately complex models of magnetic flux compression generators and fuses yield good results. In the following, we will present an overview of a 2-D generator model, which was developed at the University of Loughborough, UK and goes beyond the lumped circuit description of Chap. 6.1.

6.2 Two-Dimensional (2D) Numerical Modeling

Many questions that relate to certain aspects of the performance of a helical flux-compression generator, FCG, excited either by a capacitor bank or a battery, or to the functioning of an FCG primed by an externally produced magnetic field, clearly require multidimensional considerations.

More specifically, the most important region in an FCG lies between the expanding armature cone and the helical stator coil. Here the kinetic energy of the armature is transformed into electrical energy as work is done by the armature in compressing the magnetic field. *This local magnetic flux-compression is manifested as a strong inductive effect, with increased azimuthal (θ) currents being induced in the expanding armature cone.* Although obviously important, this effect was not taken into account in previous models, most probably due to the complexity of the problem. As reported elsewhere [Nov03], *no description or calculation of this localized effect appears in previously published literature.*

The solution described below and used for advanced numerical modeling is based on decomposing the helical FCG into a collection of filamentary circuits. Although a number of other models have also decomposed the armature and the stator coil into rings, their azimuthal currents were then all considered to be equal, with the generator being described by only one circuit. The inherent flux-compression inductive effect is therefore neglected.

An immediate consequence of this is that local phenomena, such as coil-armature electrical breakdown under the very high electric fields developed in the region near the contact point, are not properly quantified, as their precise calculation is not possible without an accurate prediction of the magnetic field distribution. This in turn proves to be an extremely onerous task as the distribution is obviously 3-dimensional. Although the model presented here is 2D, it nevertheless takes fully into account the 3-dimensional distribution of the magnetic field.

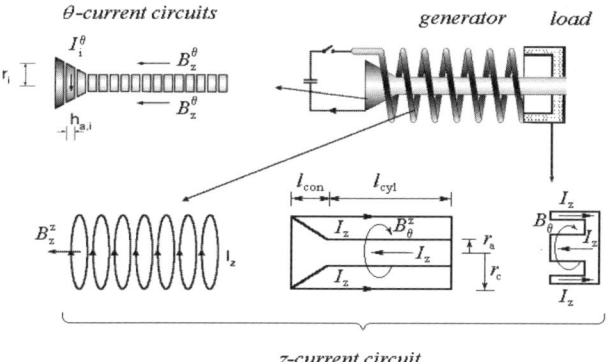

Fig. 6.31. Decomposition of a single-pitch helical FCG into a single z-circuit and multiple θ-current circuits.

A number of non-filamentary 2D codes described in the literature are very detailed [Fre77], [Mcg80], [Tip87] and need long run times on large and fast computers. Despite their complexity, they usually still require the introduction of adjustable parameters to match their predictions to the experimental evidence. The model described below overcomes a number of these disadvantages, and also includes not only such global features as phase transition in the conductors but also various local effects. Its main advantage is that *the basic concepts involved are simple.* The resulting code provides all the information that is required, even when the generator is fed with an input current much higher than the normal levels at which the same stable ratios of output to input current and energy are obtained for different input currents.

The Loughborough code is presented in detail and thoroughly benchmarked elsewhere [Nov97], and it is used here to identify and highlight a number of important features of the operation of a simple, single-pitch, single-stage FCGs.

6.2.1 A Filamentary 2D Model

A 2D set of equations that describe a helical FCG can be formulated on the basis of the resultant 3-dimensional magnetic field produced by the helical stator coil and armature. Considerations are based on the superimposed B_θ and (B_z, B_r) magnetic flux densities that are produced respectively by the z- and θ-directed currents, a technique first mentioned in [Cow84] and that has also proved to be useful in helical FCG inductance calculations [Fow89].

Fig. 6.31 shows how a helical FCG can conveniently be decomposed into equivalent z- (load) and θ-current carrying circuits. The first of these comprises the helical coil and the armature, and is closed by, for example, a coaxial load. The coil can be further decomposed into N rings, through which the same load current, I_z, flows azimuthally producing B_r^z and B_z^z fields and, together with the armature, into a two-part (conical plus cylindrical) coaxial structure, producing a B_θ^z field. Although Fig. 6.31 shows a constant pitch, single-section coil, with the same constant diameter along its length, an extension to multi-section variable diameter coils is straightforward. The initial armature geometry is also assumed to be cylindrical, although any alternative with an axially symmetric shape can readily be implemented.

The global armature θ-current induced by a time-dependent helical coil current I_z (the load current) may have a pronounced axial distribution. An adequate description of this involves the assumption of a number of separate θ-circuits (see Fig. 6.31), with each current flowing through a separate ring. Fig. 6.32 illustrates a typical case. This result demonstrates clearly an important difference from previous decomposed models in which, because magnetic diffusion through the armature wall and the inductive effects were neglected, only one circuit was used to represent the generator.

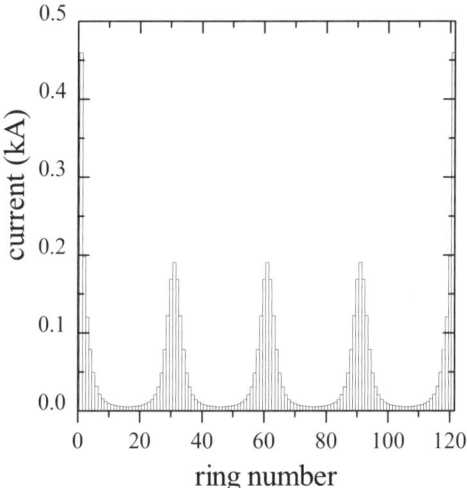

Fig. 6.32. Example of a continuous θ-current axial distribution in the armature induced by a sinusoidal stator (or outer feed coil) current. The axial position of the helical coil conductors corresponds to the maximum θ-currents.

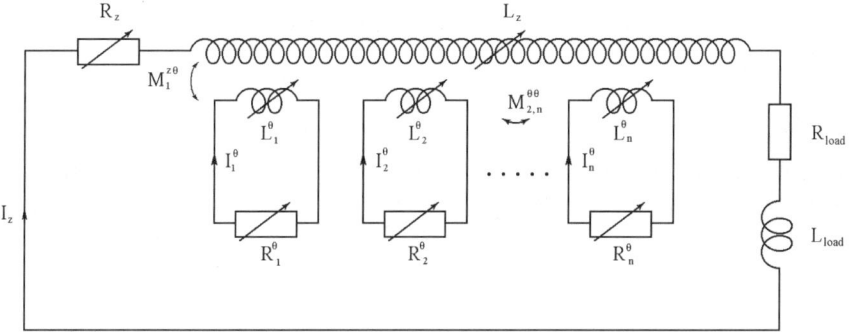

Fig. 6.33. Equivalent z- and θ-circuits for a helical FCG

A useful simplification of the resolved circuit is to assume that the number of armature circuits (or rings) equals the number of helical rings (equal to the number of helical coil turns N). This is however certainly not essential and it could be widely different (as in Fig. 6.32), either to provide a more precise description of the magnetic field distribution or to model special cases in which the generator is primed by an outer coil with a different number of turns. In this case, the θ-current in the armature will mirror that in the turns of the outer coil before the generator is crowbarred.

From the considerations above, the $(N + 1)$ equivalent circuits needed to describe the helical FCG are as given in Fig. 6.33.

The equation describing the z-(load) circuit then follows as

$$\frac{d}{dt}\left(L_z I_z + \sum_{j=1}^{N} M_j^{z,\theta} I_j^{\theta}\right) = -R_z I_z \qquad (6.14)$$

and the N equations required for the θ-circuits as

$$\frac{d}{dt}\left(M_i^{z,\theta} I_z + \sum_{j=1}^{N} M_j^{\theta,\theta} I_j^{\theta}\right) = -R_i^{\theta} I_i^{\theta} \qquad i = 1...N \ . \qquad (6.15)$$

6.2.1.1 Inductance Calculations

On the basis of the decomposed model for the generator, the total z-circuit inductance $L_z = L_z^z + L_\theta^z + L_{load}$ has components L_z^z and L_θ^z that relate, respectively, to the B_z^z and B_θ^z fields.

Thus for the B_z^z field

$$L_z^z = \sum_{i=1}^{N}\sum_{j=1}^{N} M_{i,j}^{z,z} \qquad (6.16)$$

where $M_{i,i}^{z,z}$ represents the self inductance of a ring (L_i^z). The mutual inductance between two coaxial rings of radii r_i and r_j separated by an axial distance, d_{ij}, is:

$$M_{i,j}(r_i,r_j,d_{ij}) = \mu_0[-\sqrt{(r_i+r_j)^2 + d_{ij}^2} E(x_{ij}) + \frac{(r_i^2 + r_j^2 + d_{ij}^2)K(x_{ij})}{\sqrt{(r_i+r_j)^2 + d_{ij}^2}}] \qquad (6.17)$$

where E and K are respectively the complete elliptic integrals of the first and second kind of modulus

$$x_{ij}^2 = \frac{4r_i r_j}{(r_i+r_j)^2 + d_{ij}^2} \qquad (6.18)$$

evaluated by fast subroutines [Pre89]. The time rate of change of mutual inductance is calculated with the aid of the expressions

$$\frac{\partial E}{\partial x}\frac{\partial x}{\partial t} \text{ and } \frac{\partial K}{\partial x}\frac{\partial x}{\partial t} \ , \qquad (6.19)$$

in which

$$\frac{\partial K}{\partial x} = -\frac{K}{x} + \frac{E}{x(1-x^2)} \text{ and } \frac{\partial E}{\partial x} = \frac{E}{x} - \frac{K}{x} \qquad (6.20)$$

More information on these numerical techniques can be found in [Nov98] and [Ena99].

For the inductance associated with the B_θ^Z field $L_\theta^Z = L_{cyl} + L_{con}$, the following needs to be considered:

$$L_{cyl} = \frac{\mu_0 l_{cyl}}{2\pi}\ln\left(\frac{r_c}{r_a}\right) \qquad (6.21)$$

and

$$L_{con} = \frac{\mu_0}{2\pi}[l_{con} - \frac{1}{tg(\alpha_c)}(r_a + l_{con}tg(\alpha_c))\ln(\frac{r_a + l_{con}tg(\alpha_c)}{r_a})] \quad (6.22)$$

The inductances L_{cyl} and L_{con} correspond, respectively, to the cylindrical and conical parts of the coaxial structure (α_c is the half-angle of the expanding cone). The inductances associated with the θ-circuits are obtained using similar techniques, with a detailed presentation being given in [Nov97] and [Nov98].

6.2.1.2 Resistance Calculations

The program uses a simple ohmic resistance model based on a variable one-dimensional skin depth presented elsewhere [Nov95]. Complete descriptions of equivalent resistances for modeling the various specific losses are also available elsewhere, for both high-energy [Nov95] and small-size [Gov80] generators. Using the notation of the earlier sections, R_z of Eq. (6.14) is given by

$$R_z = R_z^\theta + R_z^z + R_{load} \quad (6.23)$$

where R_z^θ is the combined resistance of the cylindrical and conical sections of the armature to the z-load current and R_z^z is the sum of the helical coil ring resistances. This term is calculated as $R_z^z = R_{DC} f_{skin} f_{proxy}$, where R_{DC} is the sum of the direct current resistances and the two correction factors f_{skin} and f_{proxy} take into account skin and proximity effects, with the techniques to calculate these being presented in [But25] and [Arn51].

The term $R_i^{\theta_i}$ in Eq. (6.15) is simply the resistance of a single armature ring.

6.2.1.3 Ring Dynamics

Armature ring

If the detonation front of the explosive makes contact with the first armature ring at $t = 0$, the nth ring begins to expand after a time given by

$$t_n = \sum_{i=1}^{n-1} \frac{h_{ai}}{D} \quad (6.24)$$

where h_{ai} is the width of the armature rings (Fig. 6.31) and D is the detonation velocity. It is assumed in the code that this movement is a constant radial expansion with a velocity $v_r = D \tan(\alpha_c)$, (α_c is also known as Gurney angle, see Chap.4) although a more accurate model that includes the initial acceleration phase could be based on either X-ray or photographic data [Fel84], cf. Chap. 4.

The effect of the magnetic field can be important for high current generators, when a more general approach that includes the corresponding magnetic deceleration term may be necessary.

Helical coil ring movement

It is assumed in the code that the helical coil is moved only by magnetic forces, and it is clearly possible to take into account both the independent axial and radial movements of each ring of the helical coil. Radial movement, which is of considerable importance in any generator design, is minimized by adding inertial mass, and a model can be developed for assessing this movement for an entire coil section [Nov95]. Adapting this model to include ring movement is a straightforward application of the laws of motion, with the driving magnetic force calculated from the local magnetic fields. The mass of a ring is regarded as supplemented by an appropriate portion of the inertial mass.

6.2.1.4 A Swicthing Problem

Solving the system of Eqs. in (6.14) and (6.15) at any time yields the corresponding generator currents. The magnetic field distribution (B_z, B_r) produced by the ring currents is then calculated using established techniques [Nov98] and the accompanying B_θ field distribution, using the approach detailed in [Ena99].

Although the process may appear straightforward, as indeed it is for many simple filamentary models developed for electromagnetic launchers, flux compression generators, etc, this is not the case here. The radial movement of the armature rings is considered initially as a continuous process, but this ceases to be so immediately after the first ring touches the helical coil. In effect the ring is then removed from the circuit, which immediately results in the process becoming discontinuous.

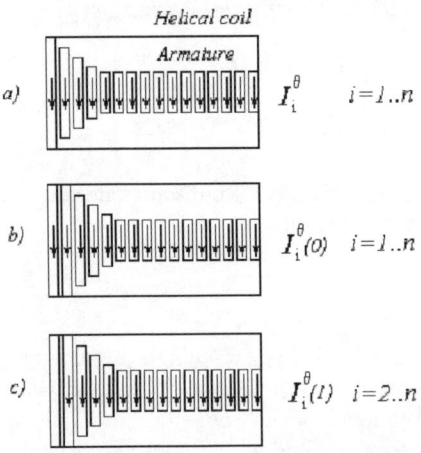

Fig. 6.34. Two-step process for currents following a switching process:
a) initial ring positions and currents
b) first step – intermediate currents due to ring movement as given by Eqs. (6.14) and (6.15)
c) second step – removal of first ring; final currents as given by Eqs. (6.25) and (6.26).

In one approach to the solution of circuits undergoing an inductance change due to a switching sequence [Lon87], it is assumed that the switching is a continuous process. This is certainly acceptable when the turns of a helical current carry the same current, and with the coil inductance being reduced progressively finite differences can be replaced by differentials. It is however not applicable in the present situation, where the different currents carried by the independent armature rings prevent the changes in a ring inductance from being made infinitely small. Unlike a helical coil (which can be regarded as having a fractional number of turns) a fraction of a ring has no meaning, and finite differences cannot be replaced by differentials.

It is clearly inevitable that the present model must involve a ***time quantization of the switching process***. To this end, the different currents that follow a circuit change can be calculated on the basis of the two-step process illustrated by Fig. 6.34, where n is the number of rings remaining in the circuit. The first stage of the quantization process requires the currents during the continuous movement of the armature rings from the position of Fig. 6.34a to be calculated by solving the set of $n + 1$ first-order differential Eqs. (6.14) and (6.15). This is done with the aid of the block-Newton iteration process [Pre89], followed by use of the LU decomposition method to solve the resulting set of algebraic equations [Pre89]. At the end of this step, the first two armature rings both make contact with the coil; see Fig. 6.34b. The first coil is then removed by the second step of the process, but while doing so the magnetic flux linkages must be conserved [Lon87]. Equations (6.25) and (6.26) then apply, and these can again be solved using the LU decomposition method to determine the new currents that then flow.

$$L_z \Delta I_z + \sum_{i=2}^{n} M_i^{z,\theta} \Delta I_i^{\theta} + M_1^{z,\theta} I_1^{\theta}(0) = 0 \qquad (6.25)$$

$$M_i^{z,\theta} \Delta I_z + \sum_{j=2}^{n} M_{i,j}^{\theta,\theta} \Delta I_j^{\theta} + M_{i,1}^{\theta,\theta} I_1^{\theta}(0) = 0 \qquad i = 2...n \qquad (6.26)$$

where $\Delta I_z = I_z(0) - I_z(1)$ and $\Delta I_i^{\theta} = I_i^{\theta}(0) - I_i^{\theta}(1)$, with 0 and 1 indicating the values immediately before and after the ring $i = 1$ is removed from the circuit as in Fig. 6.34c. This quantization process is repeated every time that an armature ring is removed from the circuit, and it generates no numerical noise.

6.2.2 Single-Pitch, Single-Stage Helical FCG

The arrangement of our simple single stage helical FCGs is straightforward and its design is described in Chap. 2, e.g. Fig. 2.9. The unsectioned helical coil is wound with a constant pitch using insulated conductors, with the armature expansion ratio for the generator considered here being 1.68. The generator length is three times that of the fully expanded armature cone, with two-thirds of the length being a helical coil and the remaining one-third a conical coaxial coil.

By including a very small coaxial component, the load inductance of the generator was conservatively estimated as 57 nH. The load geometry is however unusual, and frequency dependent changes that occur in the current distribution through the load will lead to a corresponding variation in this figure that makes precise theoretical predictions extremely difficult.

The FCG global resistance and inductance were determined from energy considerations, with 1-dimensional diffusion being included in the loss calculations. Proximity effects are used in both the helical coil calculations and in those involving interactions between the currents in the coil and those in the armature. Internal electric breakdown, 2π-clocking and other losses were all initially neglected. Since the generator was operating in an extreme regime, account was taken of any solid-to-liquid and liquid-to-vapor phase transitions in both the electrical conductivity and the equations of state for the conductors.

Since the diameter of the copper conductor used for the helical coil is only 2 mm (AWG12), the final linear current density for high initial (priming) currents is likely to exceed the characteristic figure of 0.34 MA/cm [Kno70] that corresponds to load currents above 220 kA. Under these conditions the energy multiplication of the generator (i.e. its gain) will be much reduced, due to the onset of non-linear diffusion and other effects such as electrical breakdown. Initially a saturation effect will appear, with a given increase in the input current producing a progressively reduced increase of the output current. It will be shown, however, that the code predicts that the conductors are eventually extremely heated to the point of melting and only the short operation time of the small generators keeps the conductors from being vaporized.

Fig. 6.35 shows an example of such high-current/high-energy calculations made *a priori* [Nov00] using the model described above, together with corresponding experimental data obtained much later [Neu02].

It is seen that the predicted energy gain varies little for an input current of up to about 100 A, but as this current is increased towards 1 kA an almost linear reduction in the gain is predicted. As the input current is further increased towards 10 kA there is a dramatic reduction in the gain that can be seen to be confirmed by the experimental findings.

Above 10 kA, the load current continues to increase, although there is undoubted evidence of saturation reported in [Neu02]. At even higher values of input current the generator energy gain will be less than unity, with the resulting energy in the load being less then that initial injected by the priming source i.e., the generator will fail eventually.

Overall, the a-priori model predicts the gain as a function of seed current quite accurately. The fact that the code over predicts at small current amplitudes and under predicts at large current amplitudes indicates that other mechanisms, presently not included in the model have to be considered (for a discussion of flux losses see, for instance, Chap. 5). Nevertheless, the present model manages without any adjustable, empirical parameter that is common to all other computer codes know to the authors.

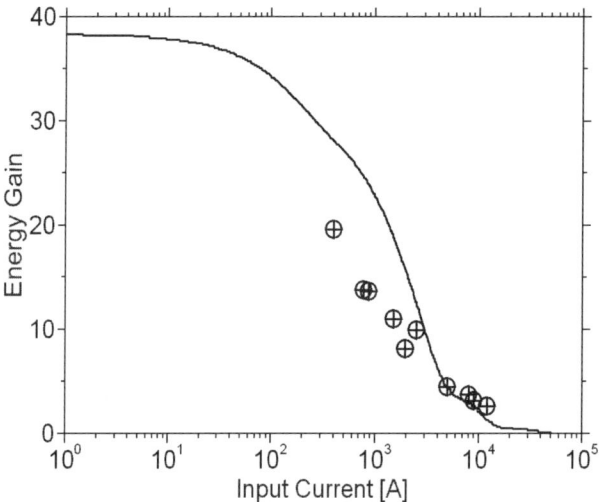

Fig. 6.35. Variation of energy multiplication (gain) with input current: (———) a priori theoretical prediction [Nov00], (●) experimental results [Neu02]. Error bars only take into account inaccuracies introduced by the load current measurements, as estimated in [Neu02].

6.2.3 Operation of the Simple, Single-Stage FCG

The different operating phases of the single-stage FCG primarily designed for the study of FCG basics, cf. Fig. 2.9., are examined below, to indicate the significant differences that exist between them. A short analysis of the most important phenomena is also given.

6.2.3.1 Phase Zero: Initial Current Injection

Although this is regarded in the model as a static phase, with no armature movement, expansion does occur outside the helical coil following activation of the detonator towards the end of the phase. Any corresponding minor inductive effects are however neglected in this analysis.

Phase zero begins with the discharge into the helical coil of an (assumed) 242 µF capacitor bank with an (assumed) inductance and resistance of 8 µH and 3 mΩ. With the bank charged to 4.5 kV, Fig. 6.36 shows that a current of 8.7 kA is produced after 125µs, when the detonation front reaches the plane of the crowbar and phase one begins. Since the actual power source is more complex, and includes more than one capacitor in a pulse-forming network, it is not surprising that some discrepancy occurs between the predicted result and the experimental result also given in Fig. 6.36. The differences are not however significant, and both the peak current and the rise time are identical.

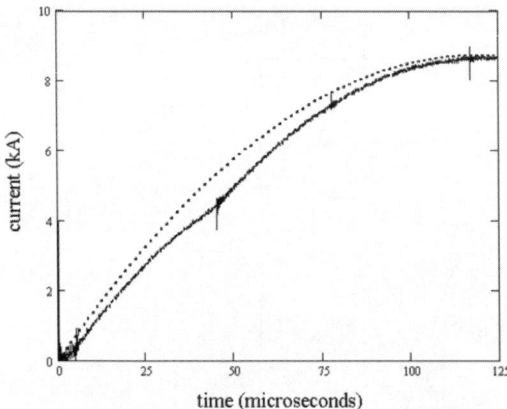

Fig. 6.36. Current during phase zero: (———) measured and (........) predicted.

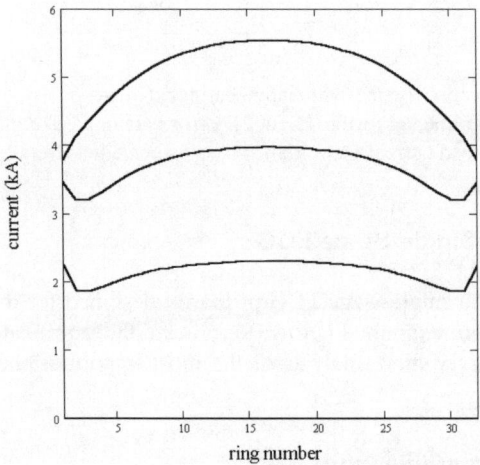

Fig. 6.37. Axial distribution of θ–current along armature (divided into 32 rings) during phase zero. From bottom to top: after 20 μs, 40 μs and 100 μs.

As the main current increases, inductive effects generate circulating θ-currents in the armature, and the resulting growth in the axial distribution of these is presented in Fig. 6.37. Calculated values of the generator inductance and resistance throughout phase zero are presented in Fig. 6.38, and despite the lack of any armature movement it can be seen that the changing current distribution has led to small variations in both parameters. *The authors are unaware of this effect having been reported previously in the literature on helical FCGs*, although variations in the inductances of coils and corresponding changes in the dL/dt due to current redistribution during the first moments of a capacitor discharge have been investigated [Lia98].

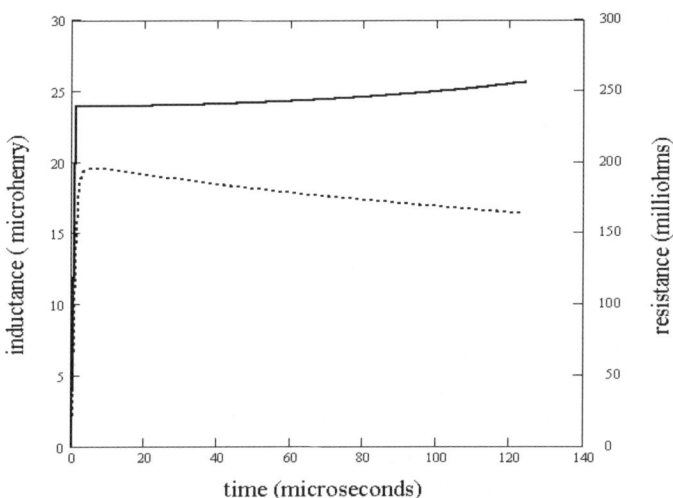

Fig. 6.38. Unexpected variations in generator (———) inductance and (.......) resistance during phase zero.

6.2.3.2 Phase One: Capture of Magnetic Flux and Preliminary Compression

Phase one operation begins with the fast expansion of the armature within the coil until it makes contact with the crowbar, during a pre-compression time of 0.56 µs. Although short, this phase in undoubtedly critical, and previous work [Neu02] has highlighted the possible losses that can occur if the design is not optimized, cf. Chap. 3.2. The capacitor bank remains in the circuit throughout this period, and considerable changes occur in the armature current distribution throughout the expanding region of the armature.

The power source is shorted as soon as the expanding armature first makes contact with the armature. Preliminary compression occurs as the cone builds up to its full length, and Fig. 6.39 shows the corresponding armature θ-current distribution. The 'valley' corresponds to the limits of the armature cone expansion under shock loading.

Only ohmic losses are included in the theoretical results of Fig. 6.39, with the close agreement between the predicted axial (load) current and the measured current confirming the very efficient action of the crowbar. Phase one lasts from 125.0 µs and 131.1 µs and ends when the expanding armature makes contact with the helical coil.

6.2.3.3 Phase Two: Main Flux Compression

Throughout phase two, the armature/coil contact point is traveling along the coil with a velocity of $D = 8,200$ m/s and the system inductance is decreasingly rapidly.

a)

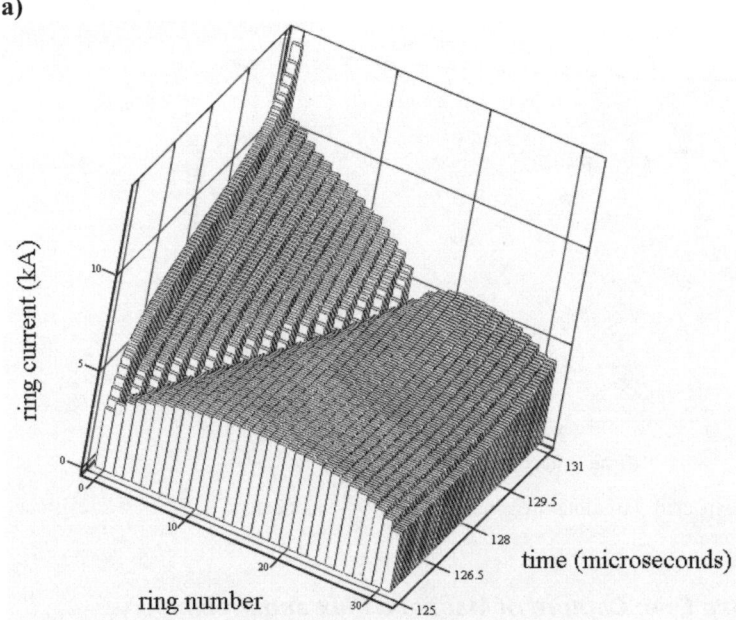

b)

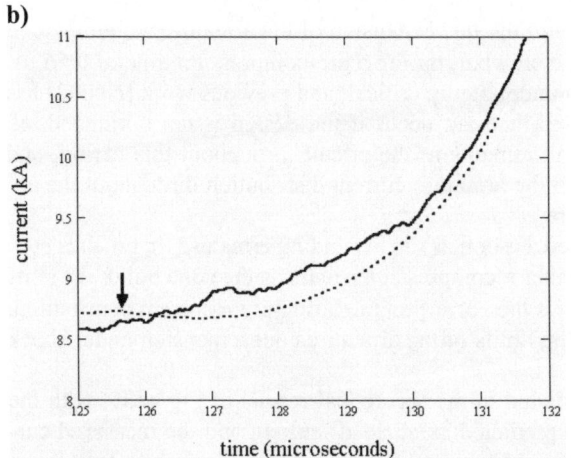

Fig. 6.39. a) Armature θ–current distribution during phase one. **b)** Load current during phase one: (———) integrated from magnetic probe signal and (........) predicted (↓ indicates armature/crowbar contact).

As a consequence of the high-intensity electric fields that are present in the contact region (see later), the code assumes that for the single-stage FCG considered here the contact is established immediately after the armature/coil separation falls to the thickness of the coil insulation.

Fig. 6.40b and Fig. 6.41 show both predicted and measured variations of the load current and of its time rate-of-change. Corresponding variations in the armature θ-currents and the inductively generated internal voltage ($I_{load}\, dL_{gen}/dt$) are given in Fig. 6.40a and Fig. 6.42 respectively.

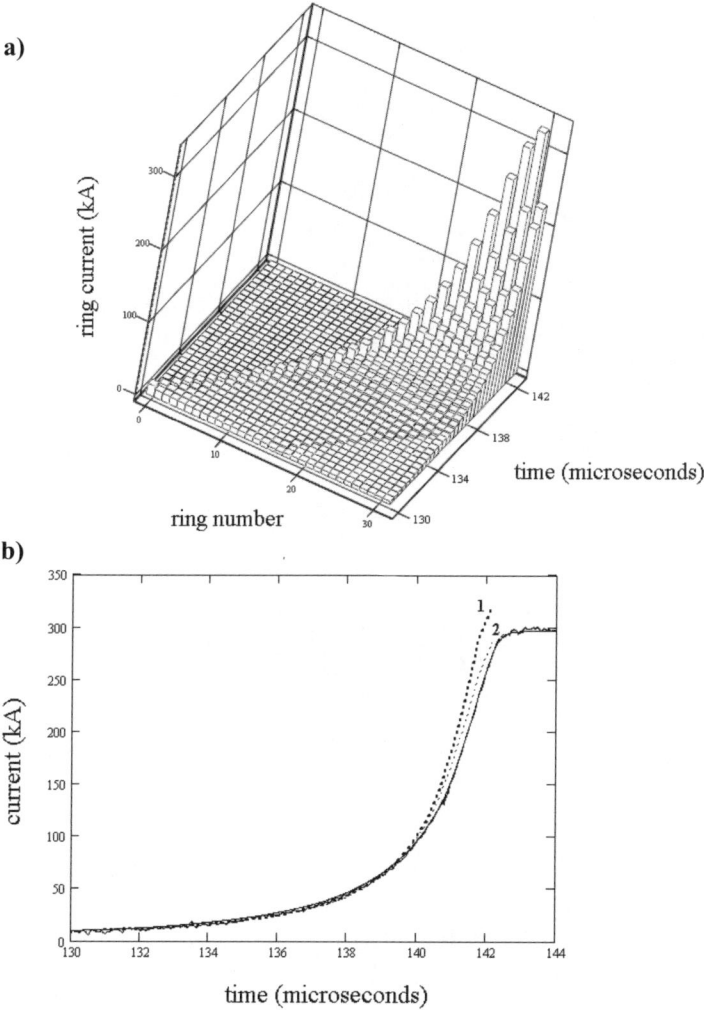

Fig. 6.40. a) Armature θ–current distribution during phase two. **b)** Load current during phase two: (———) integrated from both magnetic probe signal and from Rogowski coil and (……..) two theoretical predictions. 1-no supplementary losses, 2-electrical breakdowns included.

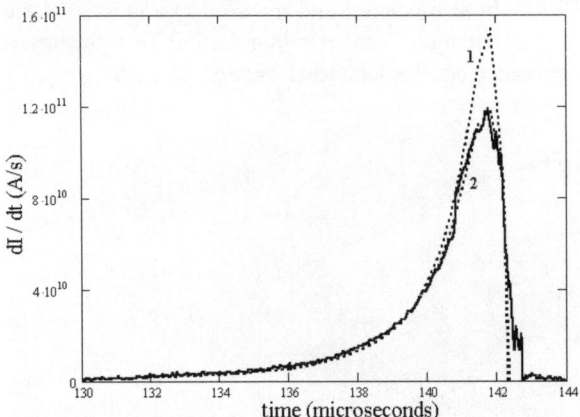

Fig. 6.41. Time rate of change of load current during phase two (———) measured (magnetic probe) and (……..) two theoretical predictions 1-no supplementary losses, 2-electrical breakdowns included.

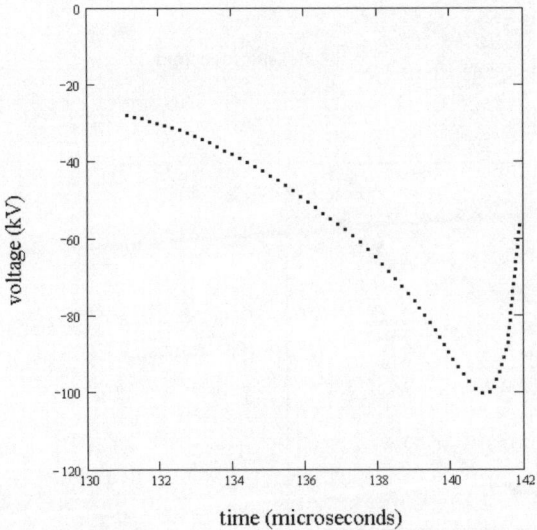

Fig. 6.42. Inductively generated internal voltage during phase two.

The conditions within the extremely small space that remains between the coil and the armature during the final half microsecond of compression differ markedly from those normally encountered, with the axial magnetic flux density exceeding 280 T and the radial electric field about 100 kV/cm. This very harsh environment arises from the superimposed effects of the last few coil turns and the final θ-rings, some carrying currents well in excess of 300 kA.

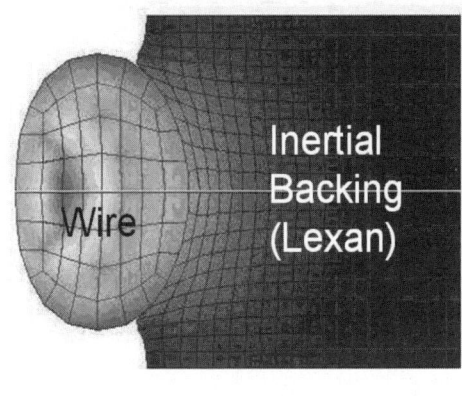

FCG axis

Fig. 6.43. Deformation of AWG12 copper wire due to magnetic pressure in last helix turn for 170 kA peak current amplitude. Peak local temperature of the copper wire due to ohmic heating is calculated to 1081 K if modeled with temperature dependent resistivity (T_{melt} = 1358 K), [Ben01].

The coil conductors are softened, may have partially melted and the magnetic forces are capable of moving the coil structure by more than 0.3 mm in less than 1 μs, see Fig. 6.43. This will certainly produce major effects during the final moments of compression, and is undoubtedly largely responsible for the 10% difference evident between the predicted and measured values of the peak current.

When electrical breakdown losses are included, the predicted results in Fig. 6.40b and Fig. 6.41 are brought even closer to the experimental data. In previously helical FCG investigations [Nov97], the electrical breakdown stress in air was calculated from the value at STP. In the present SF_6-filled generator, the figure calculated this way is multiplied by 2.7, which is the breakdown strength of SF_6 relative to that of air. Using this approach has always led to good agreement between Loughborough predictions and experimental data, although this has been regarded as fortuitous in view of the abnormal conditions inside the generator i.e., much higher than normal temperatures and ionization, both due to shock loading. Nevertheless, measurements [Neu01] have confirmed that the electrical breakdown stress of gases within MURI generators are indeed close to their STP values, see Chap. 5.2.

Fig. 6.44 illustrates the temporal variation of the magnetic streamlines at four instants during phase two. In Fig. 6.44b the outline of the armature cone can clearly be seen at the instant that it makes contact with the coil, and in all four figures the presence of streamlines within the armature indicates magnetic diffusion through its wall. The azimuthal component of the magnetic field is not shown, as it is confined to the interior of the generator with a substantially uniform distribution throughout.

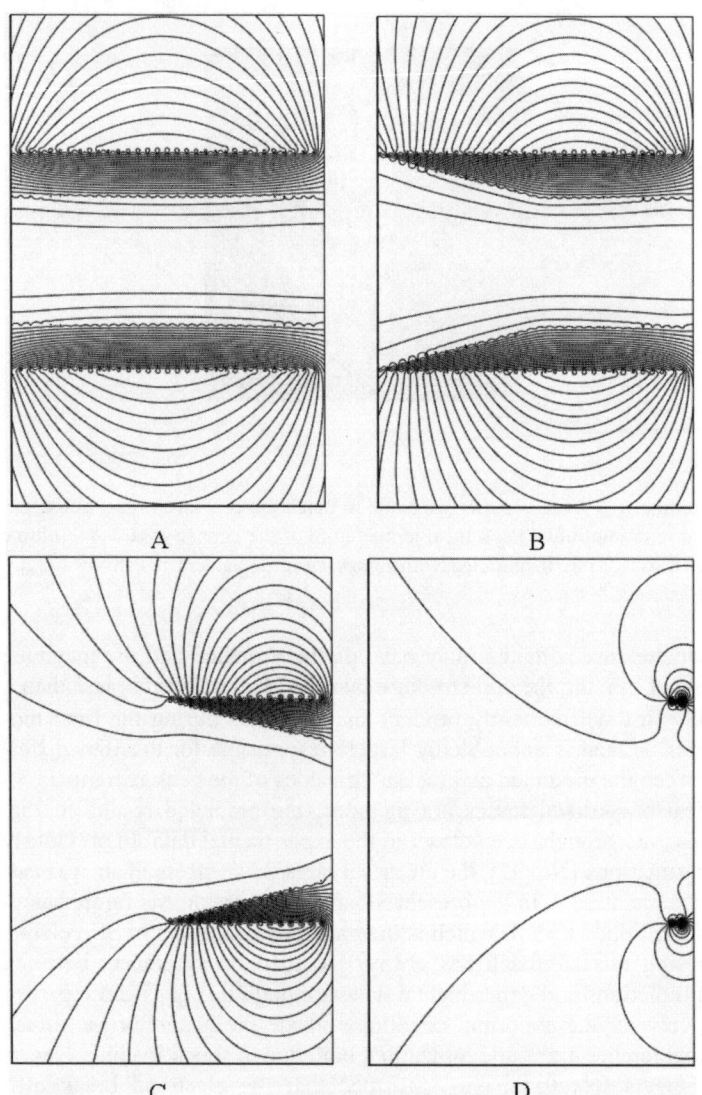

Fig. 6.44. Magnetic stream lines (stills from a WINDOWS AVI-type file): A. phase zero, B. end of phase one, C. middle of phase two and D. end of phase two

The occurrence of electrical breakdowns within the generator is immediately apparent from Fig. 6.45, which compares the dI/dt signals for similar single-stage FCG filled with either SF_6 or with air. There is generally no electrical breakdown through the gas in the SF_6- filled FCG, and the very small 2π-clocking effect is due to a minor armature/coil misalignment. As the two helical FCGs are made under the same closely controlled conditions [Neu02], they can be regarded as almost identical.

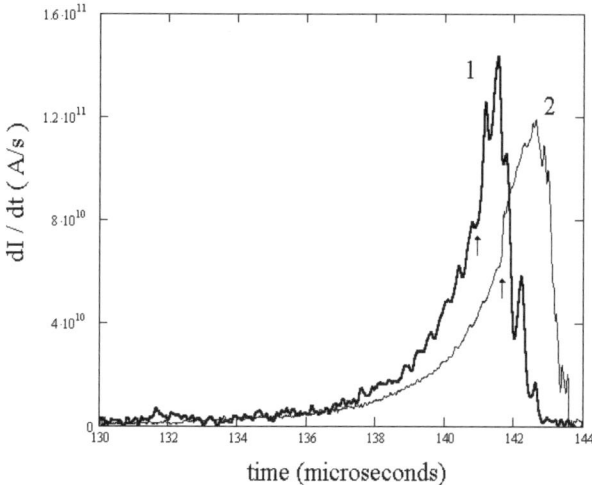

Fig. 6.45. Time rate of change of load current during phase two measured in experiments with air (1) and SF_6 (2). (for clarity the SF_6 data had been shifted by 0.85 µs). Arrows indicate time at which (according to predictions) the coil conductors melt.

Only one parameter is changed between the different results in Fig. 6.45 (the breakdown strength) and the (high) internal electric fields and the armature/coil eccentricity are about the same for the two FCGs. It is evident therefore that the enhanced 2π-clocking in the air-filled FCG is due to internal breakdown ahead of the coil/armature contact point, a conclusion that is well supported by voltage measurements made inside the helical FCG [Neu00]. Even so, with the inductively generated voltage exceeding 90 kV in the last microsecond of phase two operation, breakdowns are predicted even in the SF_6-filled helical FCG through a rather large path of approximately 0.15 mm. Fig. 6.45 confirms that these do occur in practice, limiting the final value of the load current.

Another interesting effect is related to the melting of the stator coil conductors. During the main flux-compression, the temperature and therefore the electrical resistivity of these conductors are both increasing with the corresponding higher values of the resistance limiting the increase in the dI/dt.

However, during the very short time of melting, the temperature and electrical resistivity are both constant and therefore a jump is predicted to appear in the dI/dt signals. This is confirmed by the experimental data presented in Fig. 6.45, where the arrows indicate the characteristic high jumps in the dI/dt signals corresponding to the conductor solid-liquid phase transition.

6.2.4 Summary 2D Model

The considerations above have shown the wide extent of information that a properly structured 2D code can provide on the performance of a single-stage FCG even when it is operated under an abnormally high input current. By identifying the sources of the losses that are encountered during an experiment (i.e., electrical breakdowns, phase transitions in coil conductors, etc.) the code provides an extremely powerful tool for use both in designing helical FCGs and in improving the performance of an existing design.

References

[Arn51] Arnold AHM (1951) The resistance of round-wire single-layer inductance coils. Proc. IEE 98: 94-100
[Ben01] Benton T, Hsieh KT, Stefani F, Neuber A, Kristiansen M (2001) Transient Analysis of Copper Stator Turns in MFCG. Paper No. H.06, International Conference on Pulsed Power Applications, March 29-30, 2001, Gelsenkirchen, Germany.
[But25] Butterworth S (1925) Alternating current resistance of solenoidal coils. Proc. Roy. Soc 107: 693-717
[Cow84] Cowan M, Kaye RJ (1984) Finite-element circuit model of helical explosive generators. In: Titov VM, Shvetsov GA (eds) Ultrahigh Magnetic Fields: Physics, Techniques, Applications. Nauka, Moscow, pp 241-245
[Ena99] Enache MC (1999) Numerical modeling and optical measurements for pulsed power. PhD Thesis, Loughborough University, UK
[Fel84] Felber FS, Caird RS, Fowler CM, Erickson DJ, Freeman BL, Goforth JH (1984) Design of a 20 MJ coaxial generator. In: Titov VM, Shvetsov GA (eds) Ultrahigh Magnetic Fields: Physics, Techniques, Applications. Nauka, Moscow, pp 321-329
[Fre77] Freeman JR, Thomson SL (1977) Numerical methods for studying compressed magnetic field generators. J. Computational Phys. 25: 332-352
[Fow89] Fowler CM, Caird RS (1989) The MARK IX generator. In: White R, Bernstain BH (eds) Digest of Technical Papers of the 7th IEEE Pulsed Power Conf., IEEE press, New York NY, pp 475-478
[Gie00] Giesselmann M, Heeren T, Kristiansen E, Kim J, Dickens J, Kristiansen M (2000) Experimental and analytical investigation of a pulsed power conditioning system for magnetic flux compression generators. IEEE Transactions on Plasma Science 28: 1368-1376
[Gie01] Giesselmann M, Heeren T, Neuber A, Kristiansen M (2001) Advanced modeling of an exploding flux compression generator using lumped element models of magnetic diffusion. Reinovsky R, Newton M (eds) Digest of technical papers of the 2001 IEEE Pulsed Power Plasma Science Conference, IEEE Press, Piscataway NJ, pp 162-165
[Gie02] Giesselmann M, Heeren T, Neuber A, Walter J, Kristiansen M (2002) High-speed optical diagnostic of an exploding wire fuse. IEEE Transactions on Plasma Science 30: pp 100-101
[Gov80] Gover JE, Stuetzer OM, Johnson JL (1980) Small helical flux compression amplifiers. In: P.J.Turchi, (ed) Megagauss physics and technology. Plenum Press, N.Y., pp 163-180
[Hee03] Heeren T (2003) Power conditioning for high voltage pulse applications. Ph.D. Thesis, Texas Tech University
[Kim02] Kim JG (2002) On the Resistance Modeling of the Fuse Opening Switches after Onset of Explosion, Ph.D. dissertation, Texas Tech University

[Kno70] Knoepfel H (1970) Pulsed High Magnetic Fields. North Holland Publ. Co., Amsterdam and London

[Lia98] Li L, Herlach F (1998) Magnetic and thermal diffusion in pulsed high-field magnets. J. Phys D: Appl. Phys. 31: 1320-1328

[Lon87] Long J, Lindner K, Zucker O (1987) Analysis and comparison of circuits undergoing a change of inductance via continuous sequential switching and/or geometrical change. In: Fowler CM, Caird RJ, Erikson DJ (eds) Megagauss technology and pulsed power applications. Plenum Press, N.Y., pp 593-607

[Mcg80] McGlaun JM, Thompson SL, Freeman JR (1980) COMAG-III: A 2-D MHD code for helical CMF generators. In: Turchi PJ (ed) Megagauss physics and technology. Plenum Press, N.Y., pp 193-203

[Neu00] Neuber AA, Dickens JC, Krompholz H, Schmidt MF, Baird J, Worsey PN, Kristiansen M (2000) Optical diagnostics on helical flux-compression generators. IEEE Trans. Plasma Science 28: 1445-1450

[Neu01] Neuber A, Holt TA, Dickens JC, Kristiansen, M (2001) Thermodynamic state of the magnetic flux-compression generator volume. Reinovsky R, Newton M (eds) Digest of technical papers of the 2001 IEEE Pulsed Power Plasma Science Conference, IEEE Press, Piscataway NJ, pp 98-101

[Neu02] Neuber A, Dickens J, Cornette JB, Jamison K, Parkinson ER, Giesselmann M, Worsey P, Baird J, Schmidt M, Kristiansen M (2002) Electrical behaviour of a simple helical flux compression generator for code benchmarking. IEEE Trans. Plasma Science 29: 575-581

[Nov95] Novac BM, Smith IR, Stewardson HR, Senior P, Vadher VV, Enache MC (1995) Design, construction and testing of explosively-driven helical generators. J. Phys. D: Appl. Phys. 28: 807-823

[Nov97] Novac BM, Smith IR, Enache MC, Stewardson HR (1997) Simple 2D model for helical flux-compression generators. Laser and Particle Beams 15: 379-395

[Nov98] Novac BM, Tudorache VG, Smith IR, Gregory K (1998) Two-dimensional filamentary modelling. Rom. J. Phys. 43: 265-287

[Nov00] Novac BM, Smith IR (2000) Extended notes for MURI programme (Loughborough, unpublished).

[Nov03] Novac BM, Smith IR (2003) Loughborough 2-D simulation of MURI flux-compression generators. IEEE Trans. Plasma Science 30: 1654-1658

[Pre89] Press WH, Flannery BP, Teukolsky SA, Vetterling WT (1989) Numerical Recipes in Pascal: The Art of Scientific Computing. Cambridge University Press, Cambridge

[Tip87] Tipton RE (1987) A 2D Lagrange MHD code. In: Fowler CM, Caird RJ, Erikson DJ (eds) Megagauss technology and pulsed power applications. Plenum Press, N.Y., pp 299-306

[Tui88] Tuinenga PW (1988) SPICE, a guide to circuit simulation & analysis using Pspice®. Prentice Hall, Engelwood Cliffs, NJ 07632

7 Power Conditioning

Tammo Heeren, Michael Giesselmann, and Andreas A. Neuber

7.1 Fuse Opening Switch

7.1.1 Historical Overview

The use of a fuse as an opening switch was first reported in scientific literature in the late 1950s and early 1960s. They were used in conjunction with controlled nuclear fusion experiments, fast pinches, or hypervelocity guns. These types of experiments require a typical current rise of the order of 10^{13} A/sec. Conventional practice had been to use high voltage capacitors in series with a closing gap (spark gap). However, it should be emphasized that we can only achieve a limited rate of current rise with this standard approach.

Starting with the basic voltage-current relationship for inductors

$$\left.\frac{dI}{dt}\right|_{max} = \frac{V_{peak}}{L_{total}}, \qquad (7.1)$$

and assuming ideal conditions where the capacitor is the sole provider of inductance, also assuming modern high voltage capacitors with a typical self-inductance in the order of $> 10^{-8}$ H, the maximum dI/dt is limited to $< 10^{12}$ A/sec for a peak voltage of 100 kV. In order to increase peak current rise times H. C. Early and F. J. Martin devised a system that makes use of a "fast opening fuse" in a pulse shaping system to deliver a large dI/dt to a load. The fuse consisted of a copper foil sandwiched between copper terminals, arranged in a parallel plate transmission line setup. Fiber glass cloth was used as a quenching medium and thin Mylar sheets as additional insulators. With this inductive energy storage system with fuse opening switch they reported a peak rate of current change of 1.5×10^{13} A/sec [Ear65].

7.1.2 Fuse Designs

Various fuse designs are presented in the literature. These designs essentially follow two main approaches. They either utilize a thin, wide, and long exploding metallic foil or thin metallic wire(s). The foil fuse either comes in a parallel plate design [Rei82, Ear65, DiM70, Bue85, Par85, Mai66, Wil85] or in a cylindrical arrangement [Mai66, Wil85, Rei89a] where the cylindrical fuse can also be arranged in layers with insulating material between the layers to yield a wider fuse with smaller diameter. The exploding wire fuse typically comes as a single wire or a wire array that may or may not be placed in a cylindrical conductor. Usually this outer conductor serves as a current return path as well [Nov98, Nas61, Vit81, Sal74, Gie99, Gie99a, Gie00]. Other designs include explosives or other means to interrupt the current path. However, these systems are not discussed here. Rather, the emphasis will be on exploding wire fuses. Fig. 7.1 shows a typical parallel plate exploding foil fuse. The fuse simply replaces a section of a parallel plate transmission line. The foil fuse is clamped between two opposite terminal blocks and surrounded by a quenching medium. This design also enables subjecting the fuse to large pressures from top and bottom. If long fuses are required an S-shape with fuses stacked about each other could be used. Multiple fuses can be stacked in parallel to achieve a larger fuse cross-sectional area. The fuse presented in Fig. 7.2 is beneficial if even wider fuses are necessary. One design could incorporate multiple cylindrical fuses separated by insulating sheets, or could simply be one large sheet of fuse foil placed on an insulating sheet, and rolled to a cylinder. Terminal blocks can be placed on both ends to allow for good current conduction into the foil(s).

A typical fuse arrangement for wire fuses is depicted in Fig. 7.3. The illustration on the left shows a multi-wire setup with fuse support structure, terminal blocks, and quenching medium. Wires can be arranged in multiple layers to increase cross-sectional area. In some applications the number of wires might be as low as 1 wire, and as high as 200 or more. Research applications that study wires explosions or material properties under extreme conditions often use a single wire to eliminate effects of other wires and to better control experimental conditions. In real opening switch applications mostly multiple wire fuses are used, since higher currents have to be interrupted.

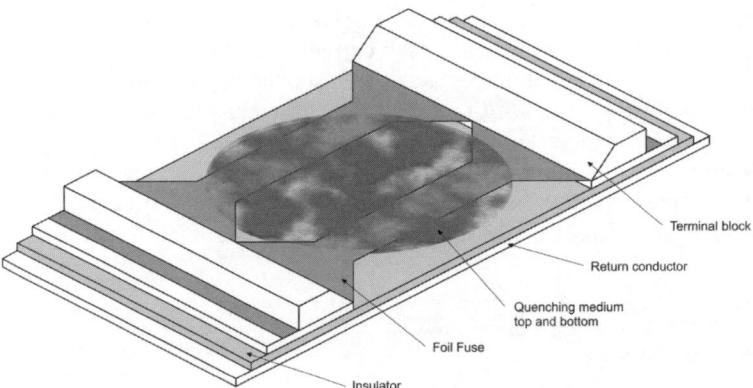

Fig. 7.1. Generic parallel plate exploding foil fuse with insulator, quenching medium, terminal blocks, and return conductor.

7.1 Fuse Opening Switch   203

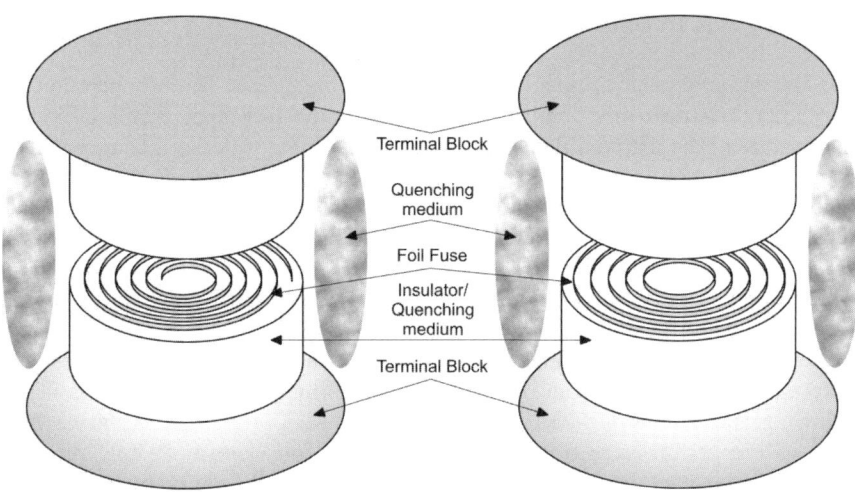

Fig. 7.2. Exploding cylindrical layered foil fuse with fuse support structure, quenching medium, terminal block, and insulating material

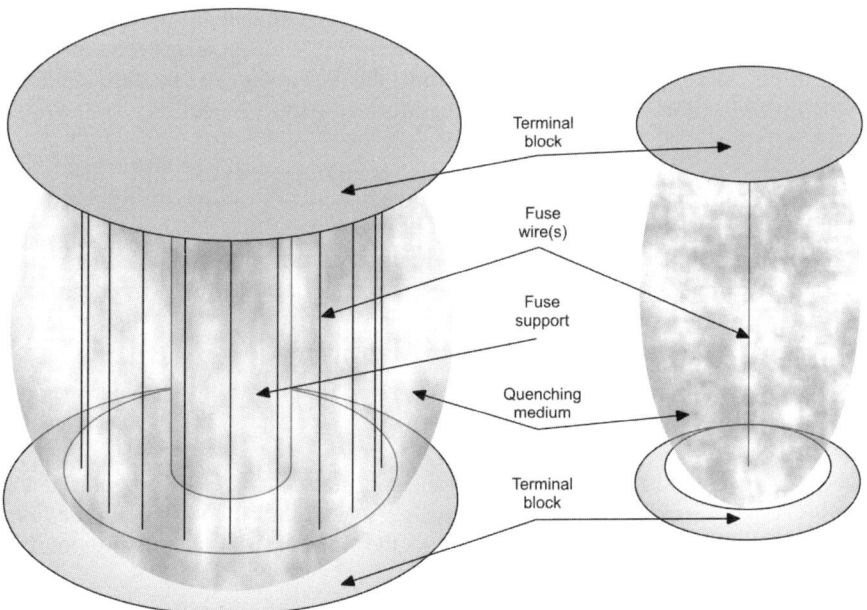

Fig. 7.3. Exploding wire fuse arrangement for single and multiple wires. Wires are mounted on the terminal block and surrounded by the quenching medium. A support structure might be employed to ease assembly.

7.1.3 Methods to Determine Fuse Parameters

The fuse initially behaves as a simple resistor in the circuit. During operation, as more energy is deposited into the fuse material, the resistance increases as the temperature of the material increases. At some point the material undergoes melting and eventually vaporizes. At this point the resistance increases much more rapidly than previously and the fuse is considered to have opened.

Typically fuses, as opening switches, are used in some type of inductive energy storage device, thus it is desirable to interrupt the fuse close to current maximum to deliver the highest possible energy, or voltage to the load. Therefore vaporization of the entire fuse material should take place at this point in time. Fuse parameters such as cross sectional area, length, fuse material, number of wires, and surrounding medium, have an impact on the performance. We will discuss in the following subsections how all parameters are chosen, and what influence they have on the performance of the fuse. The discussion starts with how the fuse cross-sectional area and length are determined, and is followed by how to choose the correct fuse material. The impact of the surrounding material and other performance influencing parameters are followed by a discussion of what types of explosions can occur in fuses. We will further introduce an equivalent action time-scale and its derivation for capacitive and flux compression generator systems. This is followed by a discussion of an equivalent system that allows the representation of an arbitrary current source by a capacitive source requiring the same fuse for current interruption. We will conclude the discussion of the fuse opening switch with the introduction of a simplified fuse-resistivity model.

7.1.4 Optimal Fuse Cross-Sectional Area

C. Maisonnier in [Mai66] established the standard method to compute the preferable cross sectional area and length of the fuse material. The following derivation is closely related to the one presented in [Mai66].

The power dissipated in the fuse material is proportional to

$$P_{diss}(t) = \frac{\rho \cdot h}{s} \cdot I_{fuse}(t)^2 \qquad (7.2)$$

where ρ is the resistivity of the material, constants h, s are the height and the cross-sectional area of the fuse, respectively, and I_{fuse} the current through the fuse. If e is the internal energy per unit mass of the fuse material then at any point in time

$$P_{diss}(t) = m \cdot de/dt \qquad (7.3)$$

or

$$\frac{\rho \cdot h}{s} \cdot I_{fuse}(t)^2 = m \cdot de/dt \qquad (7.4)$$

7.1 Fuse Opening Switch

where m is the mass of the fuse. Typically, the resistivity is a function of deposited energy. Thus, reorganizing Eq. (7.4) above and introducing the energy dependence of ρ yields

$$\frac{h}{s \cdot m} \cdot I_{fuse}(t)^2 \, dt = \frac{1}{\rho(e)} de \tag{7.5}$$

If also the density $\gamma = \dfrac{m}{s \cdot h}$ of the material is introduced into the equation then

$$\frac{I_{fuse}(t)^2}{s^2 \cdot \gamma} dt = \frac{1}{\rho(e)} de \cdot \tag{7.6}$$

As mentioned above, the fuse should be driven to vaporization energy, e_v, of the total fuse material and should reach this point at t_0. Thus, if both sides of the equation are integrated and the equation is reorganized, then

$$\frac{1}{s^2} \int_0^{t_0} I_{fuse}(t)^2 dt = \gamma \int_{e_0}^{e_v} \frac{1}{\rho(e)} de = a \cdot \tag{7.7}$$

The right hand side of the equation is solely dependent on material properties and is readily evaluated by using tables of physical constants giving the required values. [Mai66] denotes this value a. In older literature this value is typically multiplied by a factor of $\sqrt{2}/\pi$, which is due to the heritage of the exploding wire experiments. Mostly, these values are taken using slow adiabatic heating. Adiabatic in the sense that the time required to heat the wires to vaporization temperatures, t_{Vap}, is less than the thermal time constant of the surrounding medium, and slow in the sense that the electric time constant of the fuse material is smaller than t_{Vap}, such that uniform heating can occur. However, this is often not the case in exploding wires experiments and a needs to be multiplied by a factor k_1, where $1 < k_1 < 3$.

In cases where the circuit impedance is much larger then the resistance of the fuse, the left hand side of the equation is, to a great degree, only dependent on circuit parameters and the cross-sectional area of the fuse.

$$h = \int_0^{t_0} I_{fuse}(t)^2 dt \tag{7.8}$$

is called the integral of current action and is sometimes denoted h. As Reinovsky in [Rei89] presents, it is convenient to normalize the integral of current action by the peak current action at t_0, thus resulting in

$$\frac{h}{I_{fuse}(t_0)^2} = \frac{1}{I_{fuse}(t_0)^2} \cdot \int_0^{t_0} I_{fuse}(t)^2 dt = t_{eq} \cdot \tag{7.9}$$

The value t_{eq} can be understood as an "equivalent action timescale" [Rei89]. It makes it possible to compare systems with similar actions but different current waveforms. This simplifies computation of cross-sectional area and length for different systems. Thus using Eqs. (7.7), (7.8), and (7.9) the cross-sectional area of the fuse can be evaluated by

$$s_{fuse} = \sqrt{\frac{h}{a}} = \alpha \cdot \sqrt{t_{eq}} \cdot I_{final}. \tag{7.10}$$

Where the constant $\alpha = 1/\sqrt{a}$ and is similar to a unit length for fuses and I_{final} is the current at which the vaporization of the wires/ foil should take place. The scaling factor k_1 is set to unity.

7.1.5 Optimal Fuse Length

The optimal fuse length will inherently depend on what is considered to be an optimum, and this, in turn, will depend on the type of load that is used. For an inductive load, the inherent losses in the switching process have a lower limit that is determined by the fact that flux conservation has to be achieved. The ratio of dissipated energy over total energy is defined by

$$v = \frac{L_{load}}{L_{load} + L_{storage}} \tag{7.11}$$

and has a minimum of 50% if maximum energy is to be transferred into the load, i.e. if load inductance and storage inductance are equal. This also means that at least 50% of the energy will dissipate in the switch during the switching process. For the case that the load has a resistive component the efficiency becomes dependent on the switching process and on how fast the switch impedance increases. For the case of a purely resistive load an energy efficiency of $\geq 90\%$ can be achieved if the switch resistance increases faster than $37 \cdot R_{load}^2 / L_{storage}$ [Hee03]. A more detailed analysis of this can also be found in [Pai95].

To enable us to determine an optimal fuse length independent of the load and the detailed characteristics of the switching process consider the inductive energy storage system shown in Fig. 7.4. A similar system is discussed by Reinovsky in [Rei89]. The current is provided by an arbitrary current source. Initially the arbitrary load is isolated from the system by the output switch. During the time when the current through the inductor L and the fuse is increased, the resistance of the fuse is considered negligible. When peak current is reached the fuse is considered to open with a linearly increasing resistance. Upon reaching peak voltage across the fuse, the output switch closes and connects the load to the system.

Reinovsky notes the interesting point that the energy dissipated into the fuse after vaporization and until the time of peak voltage is independent of the rate of resistance rise, and is always 63% of the energy stored in the inductor. This is true for any arbitrary load, if it is isolated from the system until the time of peak voltage.

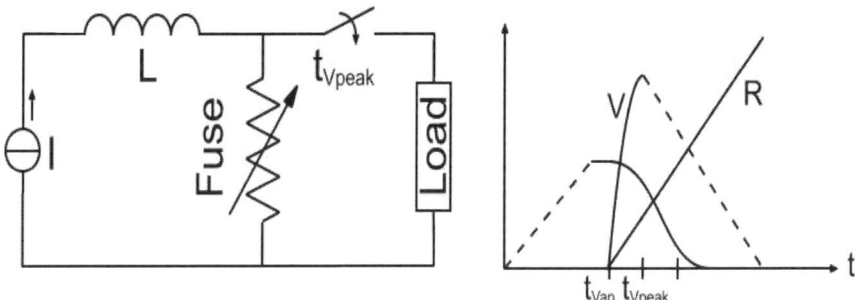

Fig. 7.4. Schematic of an arbitrary inductive energy storage system with storage inductance L, output switch triggered at t_{Vpeak}, arbitrary load, and arbitrary current source. The fuse is represented by a linearly increasing resistance. The right hand graph qualitatively shows the temporal behavior of current, I, through L and the fuse, voltage, V, across the fuse, and resistance, R, of the fuse. Dashed lines indicate arbitrary behavior.

It also means that the maximum amount of energy (under ideal conditions) that could be coupled into the load is 37% of the energy stored in the inductor, and is independent of the load. Real fuses however, require a finite amount of energy to vaporize, thus reducing the amount of energy available for deposition into the load. However, in many cases energy efficiency is not of main interest, but maximum voltage.

To discuss real applications we need to deviate from the assumptions previously made. We can separate the operation into three time intervals. The interval from start to t_{Vap}, from t_{Vap} to t_{Vpeak}, and from t_{Vpeak} onwards. The latter intervals were already discussed above and we consider the assumptions to stay valid. In real applications with real fuses, the fuse will have some finite resistance $R_0 \neq 0$ at the time of vaporization t_{Vap}, and the resistance at any time after vaporization is proportional to R_0. Thus, the larger R_0, the larger the resistance at the voltage maximum, $R(t_{Vpeak})$, and the larger the voltage at voltage maximum $V(t_{Vpeak})$.

One way of changing R_0 is by changing the length of the fuse (we can not change the cross sectional area of the fuse since that will change the time of vaporization t_{Vap}, we could use a different fuse material, but that will greatly impact the fuse behavior after vaporization). However, we can not increase the length of the fuse indefinitely, since we are limited by the amount of energy required to vaporize the material. Since we will dissipate 63% of the energy after the fuse has vaporized, we only have 37% of energy available to actually drive the fuse to vaporization temperatures. Thus, if we make the fuse as short as possible, we could achieve maximum energy transfer, but minimum voltage. But, if we choose the length of the fuse such that it will dissipate 37% of the energy before vaporization, R_0, $R(t_{Vpeak})$, and $V(t_{Vpeak})$ will be larger.

Thus for the case where an arbitrary load is switched into the circuit at voltage maximum across the fuse $v_{max} = 37\%$. This will lead to maximum voltage across the fuse, and load. This, however, is contrary to [Rei89], who states that v has to be between 63% and 100%, because he includes the energy deposited into the fuse material after vaporization into the energy available to vaporize the material However, our discussion above focused on the fact that the material has vaporized before it begins to show a significant increase in resistance. Thus, the 63% of energy deposited into the fuse material during this time are clearly not available for vaporization.

The energy dissipated in the fuse material is proportional to

$$W_{fuse} = v \cdot s \cdot h \cdot \gamma \cdot w, \qquad (7.12)$$

where w is the maximum deposited energy per unit mass at normal conditions. [Mai66] only includes the latent heat of vaporization and includes an additional multiplier k_2 in the range $1 < k_2 < 3$. Larsson in [Lar03] also takes into account the energy necessary to bring the material up to vaporization temperature.

$$w_{total} \cong c_s \cdot \Delta T_s + h_m + c_f \cdot \Delta T_f + h_v \qquad (7.13)$$

$$\Delta T_s = T_M - T_0$$
$$\Delta T_f = T_b - T_m$$

where c_s is the specific heat of the solid, h_m the heat of fusion or melting, c_f the specific heat of the fluid, h_v the heat of vaporization, T_m the melting temperature, T_b the boiling temperature, and T_0 the room temperature. Typically, the heat capacitance of the solid and the liquid material is considered equal. Since the cross-sectional area is determined by the integral of the current action or its equivalent action timescale, the length of the fuse would determine its total vaporization energy. Considering the above-mentioned points the optimal fuse length is given by

$$l_{fuse} = \frac{v \cdot W_{source}}{w_{total} \cdot \gamma \cdot s_{fuse}} = \frac{1}{w_{total} \cdot \gamma \cdot \alpha} \cdot \frac{1}{\sqrt{t_{eq}}} \cdot \frac{v \cdot W_{source}}{I_{final}} \qquad (7.14)$$

where s_{fuse} was replaced by its equivalent from Eq. (7.10). At the time of wire explosion the energy stored in the source has ideally commutated into the storage inductor, thus

$$W_{source} = \frac{1}{2} \cdot L_{store} \cdot I_{final}^2. \qquad (7.15)$$

Defining $\beta = (2 \cdot w_{total} \cdot \gamma \cdot \alpha)^{-1}$ and assuming $k_1 = 1$ then the desired length of the fuse is

$$l_{fuse} = v \cdot \beta \cdot \sqrt{t_{eq}}^{-1} \cdot L_{store} \cdot I_{final}. \qquad (7.16)$$

[Rei87] arrives at a similar result for β, however, he only considers the energy necessary to drive the entire material through vaporization and does not include the energy necessary to bring the material up to the point of vaporization. Thus, the numbers presented here differ slightly from the numbers published in [Rei87].

In any case, the length calculation as well as the area calculation are nothing more than good approximations, since the material constants involved in the calculation sometimes vary greatly between sources. Also, the value α, which is the integral of current action until the time of explosion, increases with increasing current density as [Hee03] shows for various materials in a literature survey.

Nonetheless, even if the fuse length is not optimal, opening will most likely still occur, this observation is supported by [Nas61] who states that even partial vaporization can account for a considerable increase in resistance.

The discussion above focuses on a system in which the (arbitrary) load is isolated from the system until the time of peak voltage across the fuse. If the load is not isolated, or connected to the system at an earlier time, the calculations become much more complex, since they will greatly depend on the how and how fast the impedance of the opening switch changes. Since less energy is available for vaporization the fuse will become shorter.

7.1.6 Optimal Fuse Material

The choice of the ideal material is not an easy one. From a systems design point of view a material with a low β seems attractive since it means the shortest fuse. However, from a performance point of view, the metal properties are important. The initial resistivity at room temperature will contribute to an early energy loss and will increase the time-constant of the system. The percentage of energy dissipated in the fuse material until the point of vaporization is, within a few percent, equal for the materials discussed here. As Table 7.1 shows, the energy to vaporize the materials is typically about 80% of the total energy necessary to drive the material from a solid state to vapor. Since the fuse is designed to dissipate a certain amount of energy independent of the material, the energy dissipated to bring the material to the boiling point will be (almost) equal for all (4) materials. Ideally all the energy should be deposited into the material during the liquid and vaporization phase since this will result in a higher dR/dt of the fuse. Thus, a fuse with low resistance at room temperature would be interesting. This is not necessarily the material with the lowest resistivity at room temperature, since the optimal fuse has different cross-sectional areas and lengths for different materials. Here, we define a figure of merit, R_N, as the normalized resistance of the fuse, normalized to system parameters and the explosion at current maximum. Since the length-proportionality constant, β, can be regarded as a normalized length and the width-proportionality constant, α, as a normalized width, we write for the normalized resistance, R_N,

$$R_N = \rho_S \cdot \frac{\beta}{\alpha}, \qquad (7.17)$$

which should be as low as possible. From the traditional 4 metals (Ag, Au, Al, and Cu) Aluminum is the most attractive, followed by Silver, Copper, and Gold. On the other hand, a material with a large increase in resistance for a small increment in energy is most desirable since it will yield the highest dR/dt. We introduce a figure of merit, dR_N as the (normalized) ratio of the temperature coefficient of the resistivity of the liquid metal α_ρ over the heat capacitance of the liquid metal c_l. Thus dR_N is given by

$$dR_N = \frac{\alpha_\rho \cdot \beta/\alpha}{c_l \cdot \beta \cdot \alpha} = \frac{\alpha_\rho}{c_l \cdot \alpha^2}. \tag{7.18}$$

The nominator results in the resistance change of the fuse as a function of temperature and the denominator reflects the temperature change (of the entire fuse material) as a function of deposited energy. Therefore, dR_N is the normalized resistance change as a function of energy input. Since dR_N heavily relies on approximations it should be understood as a guideline to choosing the right material only. Here, gold is clearly the metal of choice, followed by silver, copper, and aluminum. The same ranking for materials silver, copper, and aluminum where observed as well by Heeren and by Reinovsky in [Hee03] and [Rei87], respectively.

7.1.7 Fuse Quenching Material

All parameters such as wire length, total fuse cross-sectional area, and fuse material will influence at what time during the process the fuse will explode, how much energy it will dissipate, and how much of an impact it will have on the overall system performance. However, the behavior of the fuse after explosion, that is the rise (or fall) of fuse resistivity and the speed at which it will happen depend on another parameter of the fuse, the surrounding material, also called "quenching medium". In general gaseous, liquid, or solid media can be used. The fuse performance after explosion and the interaction with the load will greatly depend on the right choice. Discussion of the fuse quenching materials are among others given in [Vit81], [Nas61], and [Rei87], and results depending on the quenching medium are presented in [Par85], [Sal74], and [Bue85].

To understand the impact of the quenching material and why it is needed, a better understanding of the mechanisms of the fuse explosion is necessary. Early investigations into exploding wire fuses concentrated on an explosion in gas (air included), at atmospheric pressure, vacuum, or pressurized. The phenomenon that is used to utilize exploding wire fuses as an opening switch is in much of the early literature called "dark pause" or "dwell time". It essentially meant the interruption of current flow though the fuse or, in turn, a high resistivity phase of the fuse. However, an additional current spike through the fuse (after the first deliberate current flow though the wire) after the "dark pause" was also observed, which is clearly not desired in an opening switch arrangement.

It has been well established that the rise in resistivity is a consequence of the vaporization of the fuse material and its trajectory through the density-temperature space [Rei87].

Table 7.1 Table of material constants for Silver, Gold, Aluminum, and Copper. Data take from [Mai66], [DiM70], [Kno70], [Tka03], [Rei87], and [Cus63]. Parameters with marked with an asterisk are estimated. A more extensive list can be found in [Hee03].

Material	Descripton	Ag (Silver)	Au (Gold)	Al (Aluminum)	Cu (Copper)
a $[10^{16}\ A^2 \cdot s/m^4]$	Integral of current action	8.7	6.7	4.4	13.1
α $[10^{-9}\ m^2/A \cdot s^{0.5}]$	Length proportionality constant	3.4	3.9	4.5	2.8
β $[10^{-3}\ A \cdot s^{2.5}/kg \cdot m]$	Size prop. constant	4.8	3.16	3.04	3.43
R_N	Normalized resistance	0.116	0.111	0.074	0.126
dR_N $[10^4\ kg/m^3]$	Normalized energy coefficient	3.5	9	1.2	2
w_{vap} [kJ/gm]	Energy of vaporization	2.3	1.7	10.9	4.7
w_{total} [kJ/gm]	Total energy of vaporization	2.9	2.1	13.5	5.9
w_{vap}/w_{total}		79%	81%	81%	80%
γ $[10^3\ kg/m^3]$	Density of the solid	10.50	19.32	2.70	8.96
ρ_s [$\mu\Omega$cm]	Resistivity of the solid	8.2	13.68	10.9	10.3
ρ_l [$\mu\Omega$cm]	Resistivity of the liquid	17.2	31.2	24.2	21.1
ρ_v [$\mu\Omega$cm]	Resistivity of the vapor	28*	56*	50*	34*
T_v [K]	Temp. of vaporization	2428	3081	2740	2843
c_l $[10^3\ J/kg \cdot K]$	Heat capacitance liquid	0.26*	0.14*	1.08	0.51
α_ρ $[10^{-10}\ \Omega\ m/K]$	Resistivity temperature coefficient	0.9	1.4	1.45	0.89

During the process of fuse vaporization, a vaporization wave is moving inward into the molten layers of fuse material, and a wave of vaporized material is moving outward. As the wave moves away from the fuse, its density decreases. At some point in time the density of the vapor will have decreased to a density at which the mean-free path for electrons will be long enough to cause, in conjunction with the voltage across the fuse, avalanche breakdown.

The purpose of the quenching medium is to impede this low density front long enough to allow all necessary action to take place before the hydrodynamic pressure drives the quenching medium away and the density drops. If no impeding medium is present the right conditions for avalanche breakdown will be present almost immediately and no current interruption will occur [Nas61]. The breakdown can be shifted to higher voltages/ later times by increasing the mass that is impeding the expansion. This is achieved by using higher gas pressure and/or a heavier gas [Nas61, Vit81]. Not only does the quenching medium serve as inertial confinement for the expanding fuse vapor, but also aids in guiding the fuse resistivity on an optimal path through the density-temperate space, and absorbs free electrons. However, some of the quenching material will be exposed to liquid metal temperatures and might ionize if the ionization temperature is too low. This, of course, is counter-productive, since it will add electrons to the vapor and lower the breakdown voltage.

The following is a list of quenching media used in research. In most cases quartz sand or glass beads are employed.

- Solid Media
    - Granular medium
        - Glass beads [Rei87, Hee00, Hee03]
        - Quartz sand/ power [Rei87]
        - Alumina Al_2O_3 [Rei87]
        - Silicon carbide SiC [Rei87]
    - Solid medium
        - Polyethlene [Rei87, Nov95]
        - Paraffin [Rei87]
        - Mylar [Rei87, Nov95]
- Liquid
    - Water [Rei87]
    - N_2 [Shi89]
- Gas
    - Air [Rei87, Nas61, Shi97]
    - O_2 [Nas61]
    - N_2 [Rei87]
    - He [Nas61]
    - Ar [Nas61]
    - $CClF_3$ (Chlorotrifluoromethane) [Nas61]
- Vacuum [Vit81], [Nas61]

Gas as a quenching medium has recently become more popular for fuses in small explosive systems [Meg04], however, any possible superiority of this approach over, for instance, granular media has yet to be shown.

7.1.8 Other Performance Influencing Factors

Besides the right choice of fuse material, length, cross-sectional area, and surrounding medium, other factors can influence the performance of the fuse. Reinovsky in [Rei87] illustrates that heat sinking or cooled quenching material can greatly increase peak current and voltage at the load. Cooled quenching medium can result in 3 fold increase in peak current, and cooled and heat sinking quenching medium might result in as much as a 6 fold increase in peak current.

We have shown [Hee00] that, while keeping the total cross-sectional area constant, a larger number of thinner wires is beneficial with regard to higher peak currents. Typically, a larger number of thinner wires, while keeping the optimal cross-sectional area, will perform better. A reasonable limit for a minimum wire size is found by whatever small wire diameter is still mechanically stable enough to survive the fuse manufacturing process.

The rate of energy input into the wires also has an impact on the performance. Not only can the type of explosion change (see next subsection), higher rates of energy input can shift the explosion to higher total energies. Higher rates of energy input, i.e. higher dI/dt as typically experienced in an FCG based system over a capacitive system, will cause the boiling point of the metal to shift to higher temperatures, similar to the situation in a pressure cooker. Higher rates of energy input, or higher current densities, will cause the wires to explode at higher integrals of current action, effectively increasing the value of the blow limit, h_e. Primarily, the fuse's number of wires should be adjusted, i.e. reduced, to that. We performed a survey of various sources showing the dependency of the action integral at the time of explosion/burst as a function of peak current density [Hee03]. One order of magnitude increase in current density can cause a 100% increase of the action integral required to explode the wires. We note, however, that the fuse's cross sectional area, s, shows a rather weak dependence on the action integral, i.e. $s \propto 1/\sqrt{h}$.

7.1.9 Types of Fuse Explosions

How a fuse explodes does not only depend on extrinsic parameters such as wire size and length, on intrinsic parameters such as resistivity, vaporization temperature, or on the quenching medium, but to a large degree also on the amount of energy that is available and delivered to the wires. Chace and Levine in [Cha60] summarize the important aspects and define four types of explosions.

If the energy delivered to the wires is entirely too small ($W_{Storage} < W_{Vaporization}$), the wires, at most, will melt but not vaporize. This, of course, is undesirable, since the vaporization phase is responsible for the steep increase in resistance.

However, assuming the available energy is sufficient to vaporize the wires, the rate of energy input into the wires becomes important. Obviously, if the rate of energy input is in the order of the thermal time constant of the surrounding medium, the heat will dissipated before sufficient heating can occur. Provided that the rate of energy input is sufficiently large such that only very little energy is deposited into the environment (adiabatic heating), we can classify three different types of explosion, the slow explosion, the fast explosion, and the explosive ablation.

At some point after the fuse material has melted, instabilities will occur. These instabilities will lead to localized changes in resistivity, which in turn will cause faster heating of the material at these points. These localized points with higher resistivity will, due the larger energy input, vaporize before other volumes of liquid fuse material. Since these points will cause an increase in resistance, the energy input into the fuse will decrease and not all fuse material will vaporize. Thus, if t_{Vap}, the time to deposit $W_{Vaporization}$ into the wires, is larger than the time to develop instabilities, t_{Inst}, not all material will vaporize, and we speak about slow explosion. However, if $t_{Vap} < t_{Inst}$ then nearly all fuse material will vaporize, and fast explosion will occur. Tkachenko in [Tka03] provides the following equation to calculate t_{Inst}:

$$t_{Inst} = \frac{\pi \cdot R^2 \cdot \sqrt{\gamma_l / \mu_0}}{I} = \sqrt{\frac{\gamma_l}{\mu_0}} \cdot \frac{1}{J} \cdot \qquad (7.19)$$

Where J is the current density, R the radius of the wire, and γ_l and μ_0 the density of the liquid metal and the permeability, respectively. Typical values for t_{Inst} are in the range of 10^{-6} sec.

If we further increase the rate of energy deposition into the wires, we will encounter explosive ablation, because the current will not have sufficient time to diffuse into the entire wire cross-section, cf. Chap. 3.1, before the first layer is evaporated. This is also called onion-peel-effect and can be achieved if t_{Vap} is in the range of 10^{-9} to 10^{-8} sec [Cha60].

7.1.10 Equivalent Action Timescales for Capacitive Systems

The "equivalent action timescale" already mentioned above and initially introduced by R. E. Reinovsky in [Rei89] provides a convenient way to calculate the fuse cross sectional area and length for systems with different (time-varying) source impedances, e.g. for systems that provide a different time-varying current to the fuse. It not only allows comparing systems with the same general current shape, e.g. sinusoidal for capacitive systems, but also comparing systems with generally different current shapes. Although only two sources are discussed here, the general approach can be extended to all types of energy sources and thus is not limited to the discussion here. In the first case the energy is provided by a capacitor, resulting in a sinusoidal current shape. In the second case a magnetic flux compression generator provides the energy, resulting in an exponential current shape.

In general, the "equivalent action timescale" is defined by

$$t_{eq} = \frac{1}{i_{fuse}(t_{peak})^2} \int_0^{t_{peak}} \left[i_{fuse}(t)\right]^2 dt \cdot \quad (7.20)$$

Thus, it is simply the integral of the current squared until the time of maximum current and divided by the maximum current squared. This calculation can very well be applied to numerical data, if no analytical equation is available. If a new fuse has to be designed for a system where the interaction between fuse impedance (until the time of explosion) and the energy-source is unknown, measuring the current trace of the energy-source discharging directly into an inductor without fuse is sufficient, since the fuse impedance should be small for most of the cycle from start to peak current. In this case the inductance should be similar to the storage inductance, cf. Chap. 9.2, found in the final system. Of course, adjustment to the fuse parameters will always be necessary, but this should be a good first approximation.

The equation for the time-varying current in an inductive energy storage system with a capacitive primary energy source, essentially a RLC circuit, is given by

$$i_{fuse}(t) = \frac{V}{\omega \cdot L} \cdot e^{-\delta \cdot t} \cdot \sin(\omega \cdot t), \quad (7.21)$$

where

$$\delta = \frac{R}{2 \cdot L}, \quad \omega = \sqrt{\omega_0 - \delta^2}, \text{ and } \omega_0 = \frac{1}{\sqrt{L \cdot C}}. \quad (7.22)$$

Substituting i_{fuse} in Eq. (7.20) with i_{fuse} given in Eq. (7.21), and performing equivalent transformations yields

$$t_{eq}(\xi) = \frac{\pi}{4 \cdot \omega} \cdot \left[1 + \left(\frac{\pi}{2} - 2\right) \cdot \xi\right], \quad \xi = \frac{\delta}{\omega}. \quad (7.23)$$

Eq. (7.23) is valid for a range from 0 to 1 for ξ. The case of $\xi = 0$ represents no damping and $\xi = 1$ critical damping of the RLC circuit. Since in most cases a maximum current in the inductor is desired, the damping should be kept to a minimum ($\xi \to 0$). This results in a somewhat simplified equation for t_{eq}, given by

$$t_{eq} = \frac{T}{8}, \quad T = 2 \cdot \pi \sqrt{L \cdot C}, \quad (7.24)$$

where T is simply the period time of the undamped oscillation.

7.1.11 Equivalent Action Timescales for FCG Systems

The primary application for fuse opening switches in the context of this book is to use them in conjunction with a magnetic flux compression generator. A number of different compression generators have been presented in the literature. Despite their different designs their output current typically follows the same generic exponential rise with an initial value of I_{seed}, a runtime t_{run}, a current gain of G_l, and an exponential factor ε. Thus, we approximate the generic generator current by

$$i(t) = I_{seed} \cdot \exp\left\{\ln(G_l) \cdot \left(\frac{t}{t_{run}}\right)^{\varepsilon}\right\} = I_{seed} \cdot G_l^{\left(\frac{t}{t_{run}}\right)^{\varepsilon}}. \qquad (7.25)$$

If Eq. (7.25) is included in the equation for the integral of current action, cf. Eq. (7.20), the equivalent action timescale for a generic flux compression generator is

$$t_{eq} = \frac{1}{G_l^2} \cdot \int_0^{t_{run}} \left[G_l^{\left(\frac{t}{t_{run}}\right)^{\varepsilon}}\right]^2, \qquad (7.26)$$

which only has an analytical solution for $\varepsilon = 1$, and requires numerical evaluation for any other case. Setting $\varepsilon = 1$ yields

$$t_{eq} = \frac{1}{2} \cdot t_{run} \cdot \frac{G_l^2 - 1}{G_l^2 \cdot \ln(G_l)} \cong \frac{1}{2} \cdot t_{run} \cdot \frac{1}{\ln(G_l)}\bigg|_{G_l \gg 1}, \qquad (7.27)$$

for a current gain of $G_l \gg 1$.

7.1.12 System Equivalency

In many cases it is desired to optimize the fuse before real tests with the final energy source are made. This could be due to issues such as cost, repetition rate, safety, and various other reasons. For instance, if an explosive flux compression generator is used as source: It is expensive to fire, the repetition rate is low, and many safety regulations apply which make it undesirable to use the flux compression generators to optimize the fuse. In many cases it is possible to replace the generator with a capacitor for fuse optimization purposes. Three parameters are important in this consideration, I_{peak}, t_{eq}, and L. The fuse cross-sectional area, s_{fuse}, given by Eq. (7.10), and the fuse length, l_{fuse}, are proportional to $t_{eq}^{0.5}$ and to $I_{final} = I_{fuse}(t_{peak})$. To have a fuse that has the same s_{fuse} and l_{fuse} both numbers, as well as the fuse material and the inductance, need to be the same. The equivalent action timescale for an LC oscillation is given by Eq. (7.24) and needs to be equivalent to t_{eq} of the final system with the real energy source, thus the capacitance of the replacement system is given by

$$C = \frac{16}{L} \cdot \left(\frac{t_{eq}}{\pi}\right)^2. \qquad (7.28)$$

Since the capacitance is now known, the equation for the peak current in an LC oscillation, given by

$$I_{finalLC} = V \cdot \sqrt{C/L}, \tag{7.29}$$

can be used to solve for the capacitor voltage V, which evaluates to

$$V = I_{final} \cdot \sqrt{\frac{L}{C}} = \frac{4}{\pi} \cdot I_{final} \cdot \frac{L}{t_{eq}}. \tag{7.30}$$

Both equations can be used to design an equivalent system that has the same peak current and equivalent action timescale as the real system does. From the fuse's point of view both systems look similar and thus can be used to optimize the fuse with regard to explosion time and energy. Following optimization in a non-explosive test bed, the fuse needs to be fine tuned by operating it with the explosive driven FCG. This will still take a considerable number of shots, depending on the experience of the designer and the repeatability of the explosive system. Nevertheless, far less explosive tests will be needed if the fuse is optimized as described above.

7.1.13 Fuse Resistivity Modeling

In many cases it is desirable to be able to model the resistivity of the fuse as a function of some input parameter such as time, energy, temperature, or current action integral. This is especially the case if the "cost" of operating the entire inductive energy storage system (source, fuse, and load) is high. Costs in the sense of labor, time, monetary expenses, or safety. Often all parts of the systems are tested and evaluated independent of each other and interactions are uncertain. Simulation can fill the gap and provide a tool to evaluate the complete system beforehand. This will help to identify problem areas and might help to avoid surprises. True magneto-hydrodynamic modeling can become rather difficult and complex since all interactions of the liquid/vaporized material with itself and the surrounding medium, including fluid-dynamics and equations-of-state, which requires extensive density-temperature-resistivity tables [Lin85, Lin01], need to be modeled. In many cases this is far too complex, complicated, and time consuming to be useful in every-day simulations.

Since the resistance of the fuse is not a function of time, but rather a function of deposited energy, it is not useful to express the resistivity as a function of time. It is far from simple to obtain the resistivity as a function of energy, and a more convenient way is to express the resistivity as a function of current action. The usefulness of the "Action Integral" is discussed, for instance, in [And59].

We present a set of relatively simple equations that express the resistivity of the fuse as a function of the current action integral [Hee03]. They are based on empirical observations and the result of extensive curve-fitting efforts of experimental fuse data. The equations are valid for a wide variety of fuse materials and are only dependent on four material specific parameters. The resistivity during the heating and melting phase can be expressed by

$$\rho(h) = \rho_0 \cdot \left(1 + A \cdot \left(\frac{h}{h_e}\right)^B\right) \quad \text{for } h \leq h_e. \tag{7.31}$$

Parameters ρ_0 and h are material resistivity at room temperature, and the integral of current action, respectively. The integral of current action, necessary to vaporize a unit mass of material, the blow limit, is denoted by h_e. Parameters A and B are results of curve-fitting efforts, and also material dependent. After vaporization takes place the resistivity is better modeled by

$$\rho(h) = \rho_0 \cdot \left(A + \exp\left[\frac{h - h_e}{h_e} \cdot C\right]\right) \quad \text{for } h > h_e. \tag{7.32}$$

The parameter C is, again, a result of curve fitting efforts. The following Fig. 7.5 shows the normalized resistivities (ρ/ρ_0) as a function of normalized integral of current action (h/h_e) for three different materials (copper, silver, and aluminum). A list of modeling parameters for the three metals is given in Table 7.2. In many cases these parameters need to be adjusted to reflect actual experimental conditions such as surrounding medium, actual wire material used, and peak current density. However, this should be relatively easy if measurements are accompanied by simulations. If the equations are included in a simulation the resulting currents accurately followed our measured currents from an equivalent experimental setup, see, for instance, Fig. 7.6. The thin trace shows the actual measurement of the load current, whereas the bold trace shows the simulated current.

An example implementation of a PSpice® model that simulates an exploding wire fuse is shown in Fig. 7.7, cf. Chap. 6.1. It shows the overall inductive energy storage system with the fuse model in the middle, and the actual fuse model. All important parameters can be accessed from the top level and no modifications of the model are necessary. Important simulation input parameters such as integral of current action, resistivity, and resistance and are accessible on the top level.

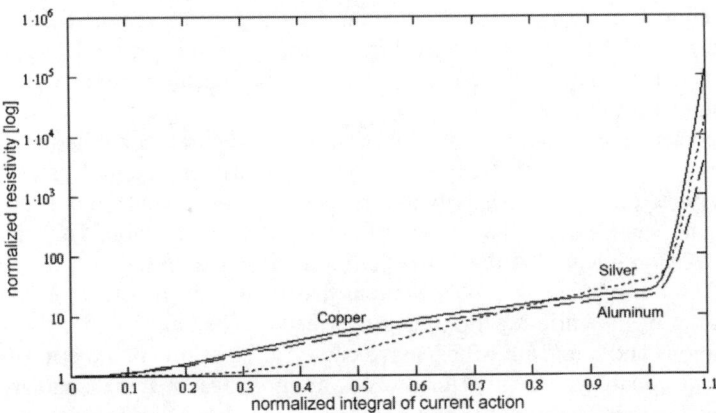

Fig. 7.5. Normalized resistivity (ρ/ρ_0) as a function of normalized integral of current action (h/h_e) for copper (solid), silver (dotted), and aluminum (dashed), cf. Eqs. (7.31), (7.32).

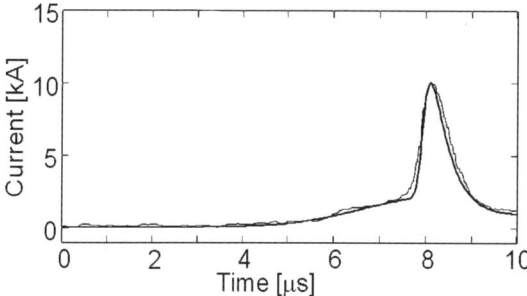

Fig. 7.6. Load current as a function of time. Measured data (thin line) and simulated data (bold line). It shows the ability to accurately simulate the heating and liquid phase, as well as the high resistivity vapor phase with the equations given in Eqs. (7.31) and (7.32). Peak current and pulse width are also simulated quite accurately.

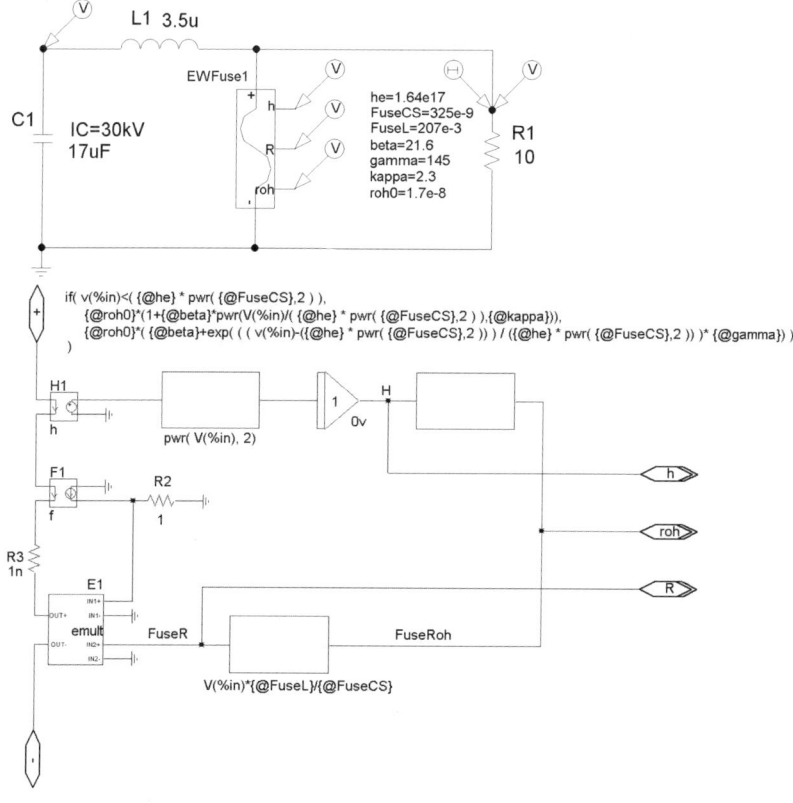

Fig. 7.7. Example implementation of an exploding wire fuse model. The top schematic shows the overall simulation of the inductive energy storage system with the fuse in the middle. All important fuse parameters are accessible from the top level. The bottom schematic shows the actual implementation of the fuse.

Table 7.2 List of modeling constants for copper, silver, and aluminum for wires surrounded by impact beads, and a maximum current density of 10^{11} A/m² [Hee03].

	Description	Copper	Silver	Aluminum
h_e [10^{17} A²s/m⁴]	Blow limit	1.6	1.03	0.59
A	Scalar constant	23.9	36	19
B	Exponential parameter before vaporization	2.3	4.5	2.3
C	Exponential parameter after vaporization	118	100	85

7.1.14 Exploding Wire Fuse Calculations for a Staged FCG

As an example, we perform in the following the exploding wire fuse calculations for the small dual-stage FCG presented in Chap. 9. We approximate the current produced by the output stage of this generator by

$$i(t) = \frac{I_{Final}}{e^1 - 1} \cdot \left[e^{\left(\frac{t}{t_{run}}\right)^\varepsilon} - 1 \right], \quad (7.33)$$

where I_{Final} is the current at the end of the generator runtime t_{run}. A single point along the current trace is sufficient for determining the exponential factor ε for currents with approximately e^x rise. Thus, ε is determined by

$$\varepsilon = \frac{\ln(\ln(1 + a \cdot e^1 - a))}{\ln(b)}, \quad (7.34)$$

with $a = \frac{i(t_1)}{I_{Final}}$ and $b = \frac{t_1}{t_{run}}$.

We will use Fig. 7.8 for our example. Typically a point in the middle of the current trace yields a good exponential factor. Including Eq. (7.33) into the equation for the equivalent action timescale, Eq. (7.20), yields

$$t_{eq} = \int_0^{t_{run}} \left[\frac{1}{e^1 - 1} \cdot \left(e^{\left(\frac{t}{t_{run}}\right)^\varepsilon} - 1 \right) \right]^2 dt, \quad (7.35)$$

which has no analytical solution. However, the normalized solution for t_{eq},

$$t_{eqN} = \frac{t_{eq}}{t_{run}} = \int_0^1 \left(\frac{e^{t^\varepsilon}-1}{e^1-1}\right)^2 dt \qquad (7.36)$$

can be approximated by

$$t_{eqN} \approx 0.257 - 0.079 \cdot \ln(2.877 \cdot \varepsilon - 1.877) \qquad (7.37)$$

with $1 \leq \varepsilon \leq 4$.

See Fig. 7.8 for an comparison between the integral equation and the approximation. These equations can now be used to design a fuse for example for an MFCG producing the current shown in Fig. 9.12. In this example the following parameters are assumed:

- generator runtime: $t_{run} = 7$ μs
- system inductance: $L_{system} = 3$ μH
- final output current: $I_{final} = 15$ kA
- wire diameter: $\varnothing_{wire} = 125$ μm
- wire cross-sectional area: $s_{wire} = 0.012$ mm^2
- wire material: silver

From the wire material properties and Table 7.1 α and β are obtained:

- $\alpha = 3.4 \times 10^{-9}$ m^2/A·s$^{0.5}$
- $\beta = 4.8 \times 10^{-3}$ A·s$^{2.5}$/kg·m.

To calculate the exponential factor ε, a point along the current trace half way through the run time is chosen, thus (using $b = 3.5/7 = 0.5$ and $a = 2.9/15 = 0.193$ in Eq. (7.34)) yielding:

- $\varepsilon = 1.802$
- $t_{eqN} = 0.163$
- $t_{eq} = 0.163\ t_{run} = 1.138$ μs

These parameters can now be used to calculate the total fuse cross sectional area, s_{fuse}, (from Eq. (7.10)), and maximum fuse length, l_{fuse}, (from Eq. (7.16) with $n = 37\%$) and result in

- $s_{fuse} = 0.054$ mm^2
- $l_{fuse} = 7.5$ cm
- $wires = s_{fuse} / s_{wire} \approx 4$

If a similar temporal current behavior is assumed, but a 45 kA peak output current, then fuse cross-sectional are and fuse length turn out to be

- $s_{fuse} = 0.163$ mm^2
- $l_{fuse} = 22.5$ cm
- $wires = s_{fuse} / s_{wire} \approx 13$

Whereas the number of wires determines the time at which the wires evaporate, the wire length determines the amount of energy consumed. For an inductive energy storage system where the load is isolated from the system until the time of peak voltage, wires with the maximum length will consume close to the total energy available, but will in turn produce the highest voltage. Shorter wires will consume less energy, but will also produce less voltage.

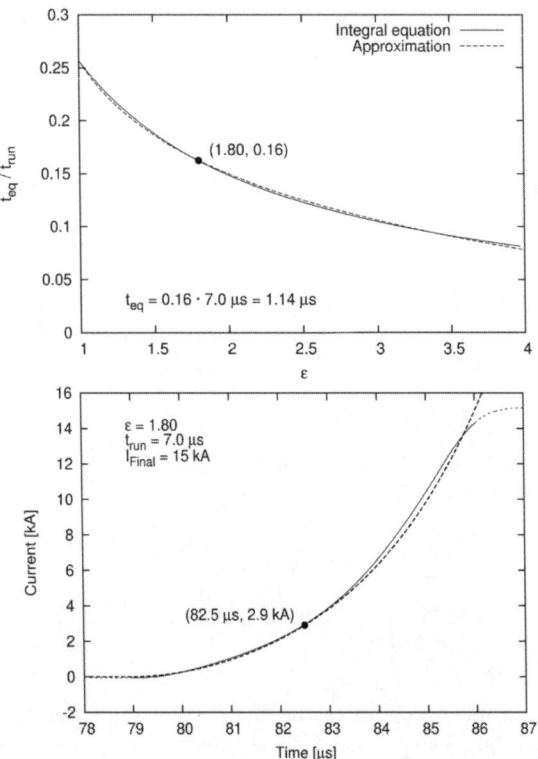

Fig. 7.8. Top: Plot of Eq. (7.36) (solid line) and the approximation from Eq. (7.37) (dashed line). Indicated in the graph is the point used to calculate t_{eq} in the example. Bottom: Plot of the MFCG current from Fig. 9.12 (thin line), the portion of the MFCG current used curve-fit the approximation function (thin solid line), and the approximation function from Eq. (7.33) (thick dashed line). Indicated is the point to calculate ε using Eq. (7.34).

These calculations can only yield a good first approximation of the fuse dimensions and adjustments especially to the parameter α need to be made once measurements of the actual system are available.

7.1.15 Fuse Summary

Fuse opening switches have been employed for the last 50 years. They are typically used in inductive energy storage systems to deliver a large dI/dt to a load that is parallel to the fuse. Fuses as opening switches come in a wide variety of forms and shapes. Some fuses use metallic foils in a parallel plate arrangement of rolled up foils in a cylindrical arrangement. Others come as wire fuses with one or more wires, typically in a cylindrical setup. Most commonly copper, silver, gold, and aluminum are used as fuse material. The performance of the fuse is determined by parameters such as cross-sectional area, length, material, surrounding medium, peak current density, and arrangement.

Some basic calculations are available to determine optimal length and size of the fuse. The most important system parameter is the equivalent action timescale t_{eq}, defined in Chap. 7.1.10. It makes the fuse design largely independent of source characteristics.

The most important fuse parameters are the length proportionality constant, β, and the size proportionality constant, α. Both allow calculating fuse length and size from system constants such as peak current, inductance, and equivalent action timescale. Constants α and β are discussed in Chaps. 7.1.4 and 7.1.5, respectively.

Out of the four most common materials used for fuse opening switches, copper, silver, gold, and aluminum, gold is probably the best choice with regard to current rise time.

The main purpose of the quenching medium is to delay the expansion of the plasma that develops during fuse explosion. It can be gaseous, liquid, or solid. Quenching media with increased heat sinking properties, or cooled quenching media are beneficial, and media that allow electron attachment can also improve performance. Due to their good performance and easy application quartz sand and glass beads are used quite often as quenching medium.

A factor that can influence the performance are higher peak current densities as they tend to shift the time of explosion to later times. In the case of wire fuses, the fuse will usually perform better if a larger number of thinner wires is used. Despite some difficulty with variations of parameters that determine fuse size and length, fuse opening switches, and especially exploding wire fuse opening switches, are used quite frequently and successfully.

7.2 Energy Extraction with Transformer

As outlined before, cf. Chap. 2, Magnetic Flux Compression generators, FCGs, typically produce currents in the range of hundreds of kA at low voltages. Typical Pulsed Power loads, however, require voltages in the range of hundreds of kV. In order to feed these loads from magnetic flux compression generators, power conditioning, for instance, by utilizing a pulse transformer, is required. While it is difficult to achieve a tight transformer coupling with an air core transformer, we note that a successful air core design, driven by a fast plate generator, has been demonstrated [Eri84]. To decide on a specific design, the tradeoff between saturation (if magnetic material is used) and poor coupling (if air core is used) has to be considered.

The method discussed in the following section is using a tightly coupled transformer with a large step-up ratio to transfer the energy from the FCG into an energy storage capacitor of high voltage levels. The initial discussion will describe the theory behind this approach including some numerical examples followed by experimental results and more refined models [Gie01, Dix04].

7.2.1 Theoretical Background

A magnetic flux compression generator of moderate size (50-200 grams of explosives), the kind of which was used in the studies reported about in this handbook, generates a fast-rising (µs) current with amplitudes of up to 400 kA. To capture the energy and generate high voltages, a tightly coupled transformer with a large step-up ratio can be used. The output current of the magnetic flux compression generator would be the primary current of the transformer. A capacitor on the secondary could be used to capture and hold the energy. An example of a readily available tightly coupled transformer with a large turns-ratio is current transformers (CT) used by utilities to measure large AC currents. Such a transformer is normally operated with a short circuit in place for utility applications and can in fact produce a large voltage if the short is removed by accident. We have shown that CTs can be used with a capacitive load in pulsed power applications and can produce high voltages. The behavior of these transformers is sufficiently ideal to initially consider ideal transformers with the understanding of the limits of flux densities in real magnetic materials.

A transformer is a device with at least two windings that are linked by a mutual magnetic flux. If the mutual flux linkage is changing as a function of time, voltages are induced in both windings according to Eq. (7.38).

$$V_1 = N_1 \frac{d}{dt} \Phi_m \qquad V_2 = N_2 \frac{d}{dt} \Phi_m \qquad (7.38)$$

The basic equation linking the voltages, currents and turns on each winding is given by:

$$\frac{V_1}{V_2} = \frac{N_1}{N_2} = \frac{I_2}{I_1} . \qquad (7.39)$$

Rearranging this equation yields a fundamental relation that is often called "Ampere-Turns Balance".

$$N_1 I_1 = N_2 I_2 , \qquad (7.40)$$

which simply means that the magnetic flux excitation produced by the secondary load will balance the magnetic flux excitation from the primary.

Since the voltages on the windings are determined by the derivative of the mutual flux, the value of the mutual flux is determined by the time-integral of the voltages seen on the windings. Eq. (7.41) shows the relationship between the secondary voltage and the mutual flux.

$$\frac{1}{N_2} \int V_2(t) dt = \Delta \Phi \qquad (7.41)$$

The flux can be expressed as the product of the flux density and the cross sectional area of the transformer core resulting in Eq. (7.42) that relates the change in flux density, B, to the time integral of the voltage on the secondary winding. The maximum change of the magnetic flux density, B, is given for ferromagnetic cores as a change from positive to negative saturation. The core can be pre-magnetized in the appropriate direction before firing the magnetic flux compression generator. Beyond saturation, the transformer coupling is very weak and not suitable for energy extraction. This represents the fundamental limitation of iron core transformers. In the following example some numbers will be given to appreciate the performance of such a system.

We assume that a magnetic flux compression generator after the burn phase is delivering 400 kA for 10 µs. We also assume that we have a strongly coupled transformer with a turns-ratio of 1:100 connected to the output of the magnetic flux compression generator. In this case, the output current of the MFC generator will flow in the single turn primary of the transformer. The core of the transformer shall have a cross section of 80 cm². We further assume that the output of the transformer is connected to a 100 nF capacitor. The voltage on the capacitor is then given by

$$V_C(t) = \frac{1}{C} \int_{0\mu s}^{t} I_2(t') dt' \cdot \quad (7.42)$$

As shown in Fig. 7.9, the final capacitor voltage after 10 µs is 400 kV and the stored energy is 8 kJ. For this example, Eq. (7.43) yields a flux density change of 2.5 Tesla in the core with the above cross section.

$$\frac{1}{N_2 A} \int_{0\mu s}^{10\mu s} V_C(t) dt = 2.5 \, \text{T} \quad (7.38)$$

7.2.2 Experimental Work

To validate the aforementioned concept, we used a commercial current transformer with a 1:100 turns ratio. The maximum continuous primary current of this transformer was 500A.

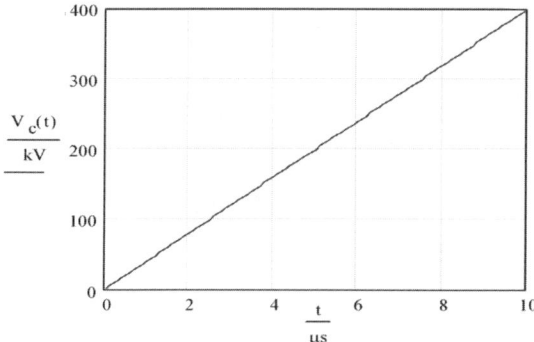

Fig. 7.9. Capacitor Voltage for given example, ideal case.

226   7 Power Conditioning

Fig. 7.10. Photgraph of the current transformer and a MFG for size comparison.

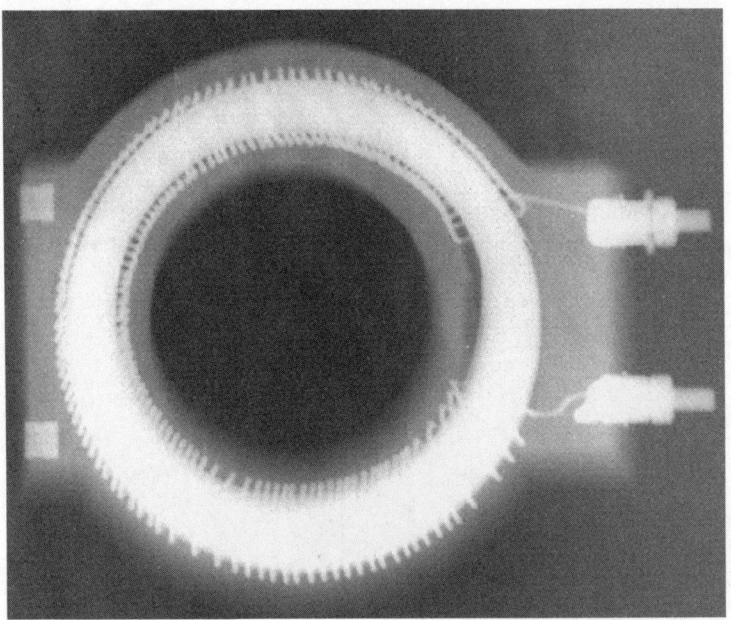

Fig. 7.11. X-Ray photograph of Current Transformer used in our experiments.

Fig. 7.10 shows a photograph of this current transformer and a magnetic flux compression generator (MFG) for size comparison. An x-ray photograph of the current transformer, showing the core as well as the secondary winding, is presented in Fig. 7.11. From this picture the cross-section of the magnetic core can be determined. Fig. 7.12 shows the schematic of our experimental setup to evaluate the energy extraction principle.

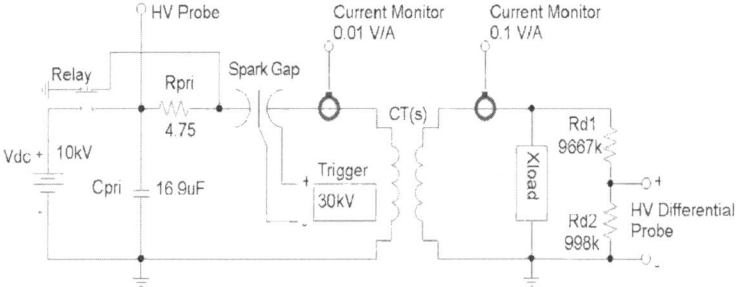

Fig. 7.12. Circuit diagram of experimental setup.

Fig. 7.13. Photograph of experimental setup with the circuit schematic given in Fig. 7.12.

We used a high-energy capacitor and a low impedance resistive load to generate a controlled primary current for the current transformer, see Fig. 7.13. The load depicted in Fig. 7.12 was typically a capacitor. In a final application, a diode could be used to prevent the capacitor from being discharged, when the primary current in the transformer decay's or the transformer core saturates; whichever occurs first.

Typical results from our experimental investigations are depicted in Fig. 7.14. The three graphs in Fig. 7.14 show the voltage across a capacitor (160 nF) on the secondary of the transformer for 3 different initial conditions of the residual magnetic flux in the transformer core. In the curve labeled "forward", the residual flux has the same polarity as the flux from the current pulse. In this case, the core will saturate soon after the onset of the primary current and the maximum secondary voltage on the capacitor is the smallest. In the curve labeled "reset", the residual flux has been set to zero before the current pulse, resulting in a significantly higher output voltage.

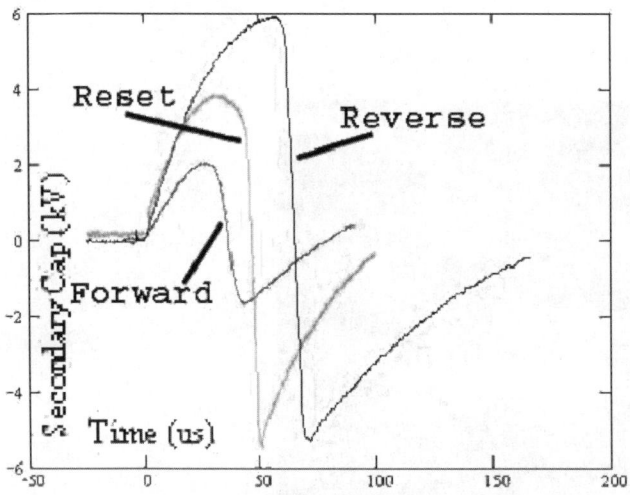

Fig. 7.14. Typical results from our experimental setup described in Fig. 7.12.

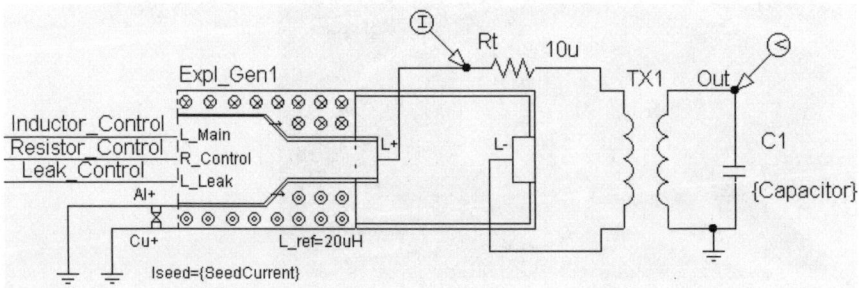

Fig. 7.15. Schematic using advanced MFG model and output Transformer.

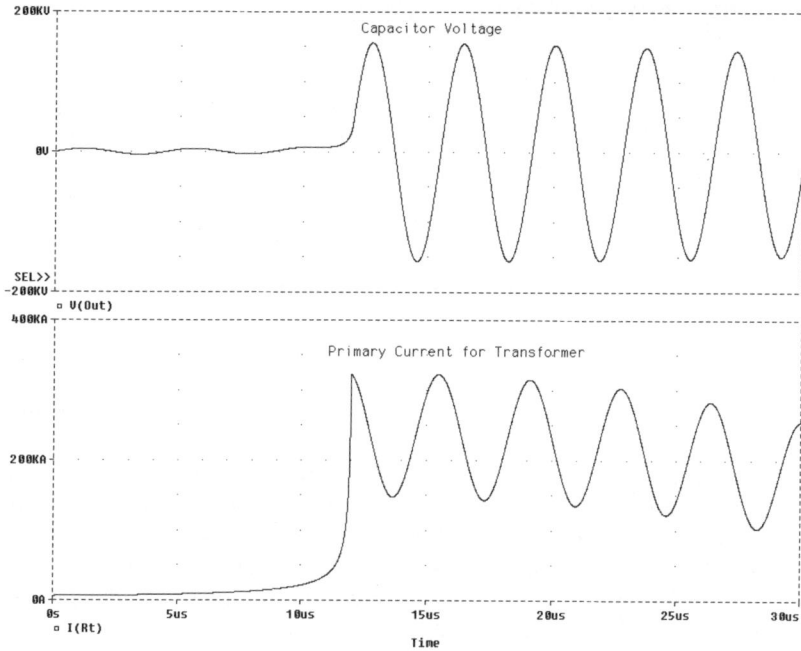

Fig. 7.16. Calculated results from the circuit shown in Fig. 7.15.

The best results are obtained, if the residual flux in the core is opposed to the flux from the current pulse. The easiest way to accomplish this is to physically reverse the transformer after every shot. In this case, labeled "reverse", the core will have the largest flux density swing before saturation and yield the largest output voltage.

7.2.3 Modeling of an FCG with Energy Extraction by Transformer

Fig. 7.15 shows a schematic for a system level study using an advanced FCG model, cf. Chap. 6.1 and an output transformer. A load capacitor of 100 nF is connected to the secondary of the energy extraction transformer. Fig. 7.16 shows the results obtained with a 6 kA seed current. The upper trace in Fig. 7.16 represents the voltage on the output capacitor, whereas the lower trace shows the primary input current into the capacitor. The voltage levels are slightly lower than for the ideal case shown in Fig. 7.9, but still very respectable.

In summary, the theoretical background as well as experimental results and PSpice® models for power conditioning systems for magnetic flux compression generators have been described. In particular, energy extraction by use of strongly coupled transformers with high turns ratios have been discussed. The approach has been shown to be feasible and comparison of models with experimental data yield good results.

7.2 Dynamic Transformer

One major limitation of single circuit or single stage FCGs for many practical applications is their limited energy gain (another limitation is that single stage FCGs run efficiently only into small inductive loads). This limitation is due to the fact that a single stage FCG, at best, tends merely to conserve its initial magnetic flux. Even worse, the flux conservation can be as low as several 10%, primarily depending on the specific generator design or particularly on the generator size for a specific design, cf. Chap. 5.

In the early 60's, concepts were developed for multiplying the magnetic flux with utilizing two or more FCGs coupled appropriately [Cow64]. The most successful methods employ a transformer to couple two successive FCG stages, with the transformer being dynamic (transformer is explosively changed) or passive (standard transformer). The basic circuits for these methods are depicted in Fig. 7.17.

Both methods rely on closing the switch, S, when the inductance L_G or L_{G1} of the first FCG is explosively reduced close to a minimum. Both transformers, the dynamic, TD, and the passive, TX, are designed as air-core transformers since any known magnetic material would saturate at the high magnetic field levels. The secondary of the TD serves as the second stage FCG. Meaning, that the primary (few turns) and secondary (many turns) inductance of the TD along with its mutual inductance will be reduced to a minimum, thus pushing the flux into the load inductance, L_L. Hence, with closing the switch, S, the secondary of TD effectively traps the flux that has been generated by the primary of TD.

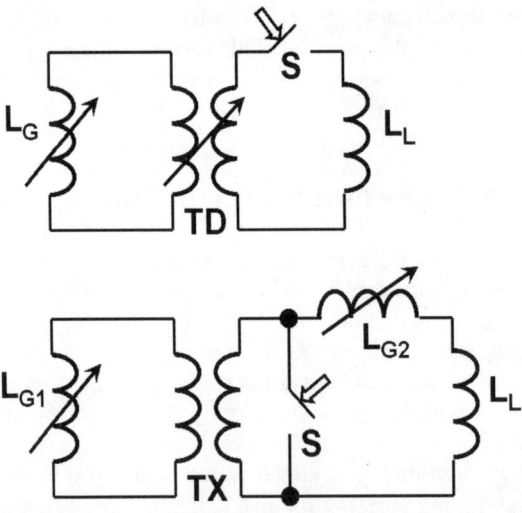

Fig. 7.17. Basic circuits for dynamic transformer coupling, TD, and passive transformer coupling, TX, of two FCG stages. Arrows indicate the explosive-driven reduction of the FCG inductance, L_G, and the TD inductance on the microsecond timescale.

In the passive transformer case, TX serves to step up the voltage, thus pushing current into the second FCG with the initial inductance, L_{G2}, typically larger than L_{G1}. With closing of the switch, S, the transformer is decoupled from the right hand of the circuit and the second stage, L_{G2}, runs like a single stage transformer. Both approaches will, if properly designed, deliver an output magnetic flux larger than the seed magnetic flux.

It should be noted that the initial flux can be generated either by directly seeding the first generator, L_G, with current from, for instance, a capacitor bank, or with the magnetic field from, for instance, a permanent magnet [Gol96] or a field coil energized by appropriate means. In all cases, a magnetic flux has to be established in the FCG volume, which, in the case of indirect seeding will be trapped by the FCG. Indirect seeding with flux trapping has the advantage that the FCG main inductor and a possible load are currentless during the seed phase, thus avoiding premature heating of the FCG conductors and other circuit elements.

It has been shown that, although the dynamic transformer or "flux-trapping" approach does not produce the highest gain per stage [Cna83], it is still likely to be favored in most cases as it uses less components, and the switch, S, can be quite easily integrated into the dynamic transformer. We discuss an implementation and experimental results of a dynamic transformer design in Chap. 9.

References

[And59] Anderson GW, Neilson FW (1959) Use of the "Action Integral" in exploding wire studies. Exploding Wires 1: 97-103

[Bue85] Bueck JC, Reinovsky RE (1985) High performance electrically exploded foil opening switches. In: Turchi PJ, Rose MF (eds) Digest of technical Papers, 5th IEEE Pulsed Power Conference, IEEE press, New York, NY, pp 287-290

[Cha60] Chace WG, Levine MA (1960) Classification of wire explosions. J. Appl. Phys. 31: 1298

[Cna83] Cnare EC, Kaye RJ, Cowan M (1984) An explosive generator of cascaded helical stages. In: Titov VM, Shvetsov GA (eds) Ultrahigh magnetic fields. Moscow, Nauka, pp. 50-56

[Cow64] Cowan M, Crawford JC (1964) Explosively driven, flux multiplying electrical generators. (Sandia Report SC-WD-64-691) and references therein

[Dim70] DiMarco JN, Burkhardt LC (1970), Characteristics of a Magnetic Energy Storage System Using Exploding Foils. J. Appl. Phys. 41: 3894-3899

[Dix04] Dixon B (2004) Pulsed Power Conditioning with strongly coupled Transformers. M.S. Thesis, Texas Tech University

[Ear65] Early HC, Martin FJ (1965) Method of producing a fast current rise from energy storage capacitors. Rev. Sci. Instrum. 36: 1000-1002

[Eri84] Erickson DJ, Caird RS, Fowler CM, Freeman BL, Garn WB, Goforth JH (1984) A megavolt pulsed transformer powered by a fast plate generator. In: Titov VM, Shvetsov GA (eds) Ultrahigh magnetic fields. Moscow, Nauka, pp 333-341

[Gie99] Giesselmann M, Zhang J, Heeren T, Kristiansen E (1999) Pulse power conditioning with a transformer for an inductive energy storage system. In: Stalling C, Kirbie H (eds) Digest of technical papers of the 12th IEEE International Pulsed Power Conference, IEEE press, Piscataway NJ, pp 1476-1479

[Gie99a] Giesselmann M, Heeren T, Kristiansen E. (1999) Simulation, design, and test of a MOV pulse shaping device for high power microwave generators. In: Stalling C, Kirbie H (eds) Digest of technical papers of the 12th IEEE International Pulsed Power Conference, IEEE press, Piscataway NJ, pp 1433-1436

[Gie00] Giesselmann M, Heeren T, Kristiansen E, Kim J (2000) Experimental and analytical investigation of a pulsed power conditioning system for magnetic flux compression generators. IEEE Transactions on Plasma Science 28: 1368-1376

[Gie01] Giesselmann M, Heeren T, Neuber A, Kristiansen M (2001) Advanced modeling of an exploding flux compression generator using lumped element models of magnetic diffusion. In: Reinovsky R, Newton M (eds) Digest of technical papers of the 2001 IEEE Pulsed Power Plasma Science Conference vol 2, IEEE Press, Piscataway NJ, pp 162-165

[Gol96] Golovina VV, Isakov VP, Lopatin MV, Mintsev VB, Ushnurtsev Aye (1996) Magnet-Powered Initial Stage of Magnetic Flux Compressor. In: Chernyshev VK, Selemir VD, Plyashkevich LN (eds) Megagauss and megaampere pulse technology and applications, part 1. RFNC-VNIIEF, Sarov, Russia, pp 333-335

[Hee00] Heeren T (2000) Power conditioning in inductive energy storage systems. M.S. Thesis, Texas Tech University

[Hee03] Heeren T (2003) Power conditioning for high voltage pulse applications. Ph.D. Thesis, Texas Tech University

[Kno70] H. Knoepfel (1970) Pulsed High Magnetic Fields. North-Holland Publishing Company, Amsterdam, London

[Lar03] Larsson A, Appelgren P, Bjarnholt G, Hultman T, Nyholm SE (2003) Energy flow in a pulsed-power conditioning system for high-power microwave applications. In: Giesselmann M. Neuber, A (eds) Digest of Technical Papers of the 14th IEEE International Conference vol 2, IEEE Press, Piscataway NJ, pp 935-938

[Lin85] Lindemuth IR, Brownell JH, Greene AE, Nickel GH, Oliphant TA, Weiss DL (1985) A computational model of exploding metallic fuses for multimegajoule switching. J. Appl. Phys. 57: 4447-4460

[Lin89] Lindemuth IR, Reinovsky RE, Goforth JH (1989) Exploding metallic fuse modeling at Los Alamos. In: Titov VM, Shvetsov GA (eds) Megagauss fields and pulse power systems, part 1. Nova Science Publishers, Commack, NY, pp 269-274

[Lin01] Lindemuth IR, Atchison WL, Faehl RJ, Goforth JH, Tasker DG (2001) Improved modeling of electrically exploded fuse opening switches in flux compression generator experiments. In: Reinovsky R, Newton M (eds) Digest of technical papers of the 2001 IEEE Pulsed Power Plasma Science Conference vol 2, IEEE Press, Piscataway NJ, pp. 909-912

[Mai66] Maisonnier C, Linhart JG, Gouilan C (1966) Rapid transfer of magnetic energy by means of exploding foils. Rev. Sci. Instrum. 37: 1380-1384

[Meg04] Megagauss X Conference, (2004) Berlin, Germany

[Nas61] Nash CP, McMillan WG (1961) On the mechanism of exploding wires. The Physics of Fluids 4: 911-917

[Nov95] Novac BM, Stewardson HR, Smith IR, Senior P (1995) Analysis of helical generator driven exploding foil opening switch experiments. In: Baker WL, Cooperstein G (eds) Digest of technical papers of the 10th IEEE International Pulsed Power Conference vol 2, IEEE Press, Piscataway NJ, pp 1126 - 1131

[Nov98] Novac BM, Magureanu M, Smith IR (1998) High electric fields sustained in fast EBW experiments. Rapid Communication, J. Phys. D: Appl. Phys. 31: L57-58
[Pai95] Pai ST, Zhang Q (1995) Introduction to high power pulse technology. Advanced Series in Electrical and Computer Engineering – Vol. 10, World Scientific, Singapore
[Par85] Parker JV, Parsons WM (1985) Foil Fuses as opening switches for slow discharging circuits. In: Turchi PJ, Rose MF (eds) Digest of technical Papers, 5th IEEE Pulsed Power Conference, IEEE press, New York, NY, pp 283-286
[Rei82] Reinovsky RE, Smith DL, Baker WL, Degnan JH, Henderson RP, Kohn RJ, Kloc DA, Roderick NF (1982) Inductive store pulse compression system for driving high speed plasma implosions. IEEE Transactions on Plasma Science 10: 73-81
[Rei87] Reinovsky RE (1987) Fuse opening switch for pulse power applications. Opening Switches. In: Guenther A, Kristiansen M. (eds) Opening switches, advances in pulsed power technology, vol. 1, Plenum Press, New York, London
[Rei89] Reinovsky RE (1989) High voltage power conditioning systems powered by flux compression generators. In: White R, Bernstain BH (eds) Digest of Technical Papers of the 7th IEEE Pulsed Power Conf., IEEE press, New York NY, pp 971-974.
[Rei89a] Reinovsky RE, Lindemuth IR, Goforth JH, Caird RS, Fowler CM (1990) High-performance, high-current fuses for flux compression generator driven inductive store power conditioning applications. In: Titov VM, Shvetsov GA (eds) Megagauss fields and pulse power systems, part 1. Nova Science Publishers, Commack, NY, pp 453-464
[Sal74] Salge J, Braunsberger U, Schwarz U (1974) Circuit breaking by exploding wires in magnetic energy storage systems. In: Proceedings of the International Conference on Energy Storage, Compression, and Switching, Asti-Torino, Italy, pp. 477-480
[Shi89] Shimomura N, Akiyama H, Maeda S (1989) Improvement of a Pulsed Power Generator by Fuses in the Liquid Nitrogen. Trans. of the Institute of Electrical Engineers of Japan 109-A: 323
[Shi97] Shimomura N, Nagata M, Akiyama H (1997) Compact pulsed power generator using a Marx circuit and an optimized exploding wire. In: Cooperstein G, Vitkovitsky I (eds) Digest of technical papers of the 12th IEEE International Pulsed Power Conference vol 2, IEEE press, Piscataway NJ, pp 1202 - 1207
[Tka03] Tkachenko SI, Vorob'ev VS, Malyshenko SP (2003) Parameters of wires during electric explosion. Appl. Phys. Letters 82: 4047-4049
[Vit81] Vitkovitsky IM, Scherrer VE (1981) Recovery characteristics of exploding wire fuses in air and vacuum. J. Appl. Phys. 52: 3012-3015
[Wil85] Wilkinson GM, Miller AR (1985) Generation of sub-microsecond current rise times into inductive load, using fuses as switching elements. In: Turchi PJ, Rose MF (eds) Digest of technical Papers, 5th IEEE Pulsed Power Conference, IEEE press, New York, NY, pp 280-282

8 Seed Sources

James C. Dickens and Andreas A. Neuber

8.1 Seed Sources Basics

FCGs act as energy amplifiers and as such require some initial energy source. In the case of helical FCGs, this initial energy is in the form of magnetic flux between the stator and the armature. Although generator performance is relatively independent of how the initial flux was generated, it is important to note that only the magnetic flux in this region can be utilized during the generator operation. For instance, the magnetic flux within the armature will not be compressed as it expands and in fact is lost. Because the location of the initial magnetic flux is so important, special care must be taken in the design of the seed source to ensure optimal performance. This includes not only spatial considerations but also time of application and duration of seeding flux.

One of the most common ways of seeding a flux compression generator is to apply a current pulse to the stator windings directly or alternately to an outer seed winding. This current pulse can be derived from a simple capacitor discharge into the windings, or from another explosively driven device such as a ferroelectric or ferromagnetic generator. Regardless of the seed generator used, the current risetime should be short relative to the magnetic flux diffusion into the armature to achieve optimal energy efficiency.

Another consideration in the seeding of FCGs is the timing of the main detonation wave. The main detonation must be time such that the armature expands to make contact with the crowbar at the same time the seed current has reached its maximum value. Again this ensures maximum energy efficiency.

An alternative to active seeding is the use of permanent magnets to achieve the requisite initial magnetic flux. This form of seeding typically uses some type of rare-earth magnets with a magnetic field shaping structure, to optimally form an axial magnetic field between the stator and armature. This type of seeding is limited to smaller energy devices due to the size and weight of the permanent magnets. Permanent magnet seeded FCGs can be used as the initial energy source for a larger FCG by supplying the current pulse to its external seeding coil.

8.2 Capacitor Driven

Capacitive energy storage is most common method of seeding FCGs and in fact is the most common energy storage scheme for most pulsed power systems. The reason for this is its relative simplicity compared to other storage methods. In this scheme, a capacitor is charged with a dc power supply to some predetermined voltage and then it is switched into the FCG's primary windings or an auxiliary seed winding, cf. Fig. 2.5. The size of the capacitor and its characteristics is determined from a system point of view with considerations such as generator size, maximum generator output current, generator gain, initial generator inductance and/or seed winding inductance. The most important consideration is of course the desired seed energy in the form of magnetic field within the generator volume. This initial energy is related to the magnetic field by

$$E_o \propto \oint B_o^2 \cdot dV, \qquad (8.1)$$

where E_o is the initial energy within the generator, B_o is the initial magnetic field within the generator, and V is the working volume of the generator.

Another important consideration in designing the seed source is the rise-time of the magnetic field within the operating volume. It is desirable to seed the magnetic field in as short of a time as practical. If the rise-time is too slow, the magnetic field will diffuse into the armature region and be lost for compression and essentially wasted. Long seed times can also cause premature heating of the FCG windings when the generator is seeded without an auxiliary winding (i.e. I^2R heating of the main windings). As discussed previously, this heating results in reduced performance.

Most FCG capacitive seeding systems can be modeled as a simple RLC circuit, with

$$R < 2\sqrt{\frac{L}{C}}, \qquad (8.2)$$

where R is the total system resistance (winding, switch, etc), L is the total seed system inductance (FCG winding inductance or seed winding inductance, cabling to/from the FCG to capacitor, etc), and C is the seed capacitor. Because the initial magnetic field within the generator is proportional to the seed current, we find that

$$B_0 \propto V_{c(initial)}, \qquad (8.3)$$

where $V_{c(initial)}$ is the initial capacitor voltage. We also find that the maximum B_0 occurs at time,

$$T_{B_o(max)} = \frac{\pi}{2}\sqrt{LC}, \qquad (8.4)$$

and the maximum current is proportional to

$$I_{max} \propto V^2 \sqrt{\frac{C}{L}} \ . \tag{8.5}$$

The initial magnetic flux within the generator is bounded by the stator winding on the outside and the solid armature on the inside. The solid armature resists magnetic flux diffusion by forming a mirror current on the surface that is equal but opposite to that of the stator winding. If the initial current is applied too slowly, the magnetic seed field will diffuse into the armature volume and be lost for compression. The rate of this diffusion and depth is related to the skin depth of the armature material and its radial resistance. For simple AC currents, the skin depth is proportional to the frequency of the AC current. Because a capacitive seed system can be modeled with as a simple RLC circuit, we can approximate the skin depth δ, cf. Chap. 3, of the seed flux with the simple expression:

$$\delta = \sqrt{\frac{2\rho}{2\pi f \mu_o}} \ , \tag{8.6}$$

where ρ is the bulk resistivity (ohm-meter), μ_o is the permeability constant $4\pi \times 10^{-7}$ Henries/meter and

$$f = \frac{2\pi}{\sqrt{LC}} \ . \tag{8.7}$$

Skin depth is the point at which the field has fallen by a factor of $1/e \cong 1/2.71$. For aluminum, the bulk resistivity is around 2.8 microOhm-cm, and thus with an armature thickness of 2-3 mm, the seeding time should be no longer than 50-100 μs.

The use of an auxiliary seed winding allows more freedom in the choice of capacitor size and voltage. For instance, in general, if a lower voltage seed circuit is desired, a single turn seed winding could be chosen that would result in a relatively small seed circuit R and L and correspondingly higher seed current I that could result in the same magnetic flux within the generator volume.

Solid-state switching is easily achievable with seed capacitor voltages under 3-5 kV. Beyond this voltage level, the complexity and cost of the solid state switches becomes impractical, particularly for single shot usage. The most common solid-state switching element for pulsed discharge seeding is the thyristor (SCR). While the steady state current rating of a thyristor may be only 10-100 A, the pulsed discharge capability of the device may be 10-100 kA. The primary failure mechanisms for thyristors due to pulsed discharge are over-voltage or localized melting due to excessive heating from I^2R losses. For short pulses, with high dI/dt, this localized heating is more common, due to the finite time of the conduction plasma spread across the device. To reduce device failure due to high dI/dt, a saturable reactor is often placed in series with the switch to allow the conduction plasma time to spread before the high current is applied to the switch.

To minimize device failures due to over-voltage, conservative device limits must be used in the circuit design or adequate snubbers must be used. For example, if 1 kV thyristors are to be used in a circuit designed for 2 kV, it would be advisable to place three devices in series, rather than two. A typical solid-state seed source used for laboratory testing is illustrated in Fig. 8.1a. This circuit utilizes a multi-tapped trigger transformer to simultaneously turn-on all three thyristors. The resistor and capacitor across each thyristor is to compensate for static leakage imbalance and for the dynamic turn-on differences of each device. This circuit has a voltage limit of 2-3 kV and a current limit of 20-30 kA. Obviously with larger thyristors, these limits would increase accordingly.

For higher seed source voltages (>5 kV), a triggered spark-gap switch is often used. These types of switches have voltage and current ratings > 100 kV and > 500 kA in a single switch. To achieve reliable triggering with minimal jitter, a high voltage (10-20 kV), low current trigger generator is required (a compact pulse transformer is often used).

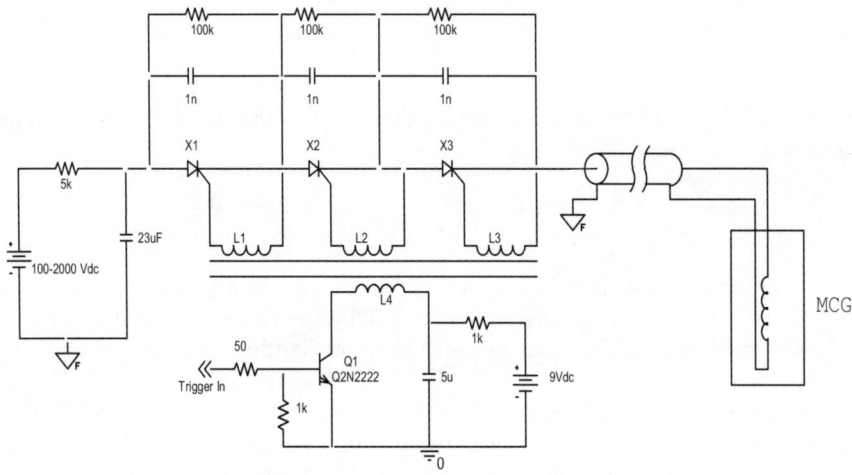

a. Solid-state laboratory seed source.

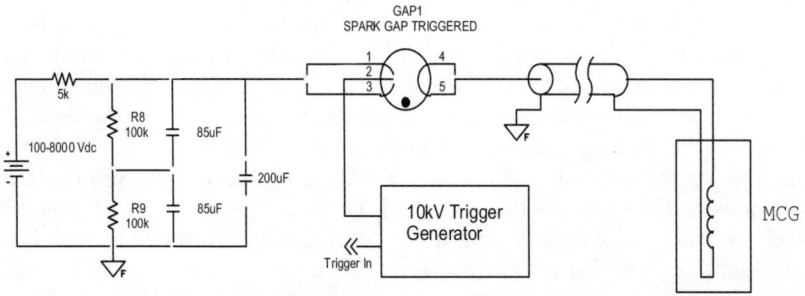

b. Spark-gap switched seed source.

Fig. 8.1. Capacitor based seed sources

8.3 Explosively Driven Seed Sources

8.3.1. Ferroelectric Seed Sources

An explosive ferroelectric generator (FEG) utilizes the unique properties of ferroelectric materials to use the energy stored in explosives to generate electrical energy. The active element in an FEG consists of a piece of ferroelectric material with metal deposited onto two opposite sides. An explosively generated pressure/shock wave is sent through the material, which generates a voltage between the metal plates. A simple circuit model of an FEG consists of a pulsed current source that drives charge into the capacitance between the two metal plates and the load. The capacitance across the device is significant (on the order of one to many nF) due to the high electromagnetic permittivity of the ferroelectric material. This simple model does not take into account such factors as the characteristics of the shock wave and the varying conductivity and permittivity of the ceramic material during generator operation [Tak02]. However, it is sufficient for use in the design of a seed current source. A simplified diagram and circuit model is shown in Fig. 8.2. A more complicated, but realistic ferroelectric generator is illustrated in Fig. 8.3 [Shk01].

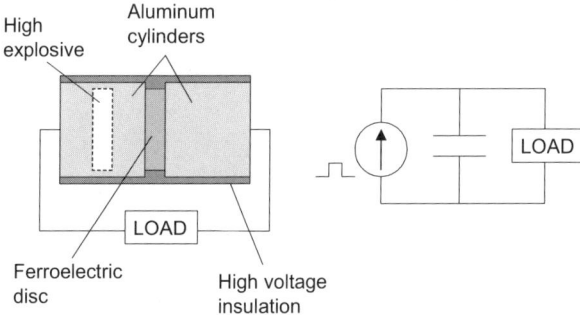

Fig. 8.2. Simple ferroelectric generator cross-section and equivalent circuit.

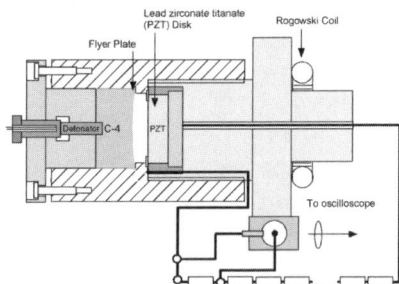

Fig. 8.3. Ferroelectric generator design

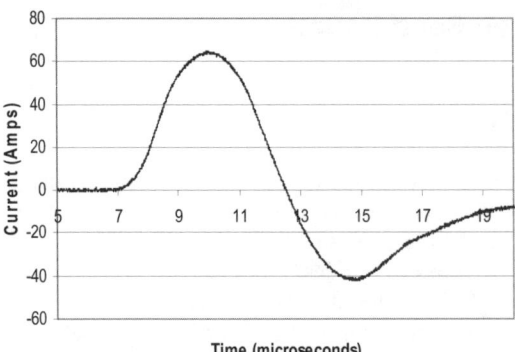

Fig. 8.4. Measured Output Current of 1.7 Joule seed source into 825 µH load.

The design of the seed source requires matching the total capacitance of the source with the inductance of the load to get the maximum energy transferred into the load before the energy is lost due to the conductivity of the shocked ceramic. The maximum amount of energy generated to date is 1.7 Joules into an 825 µH load. This device had two six-disc-in-series generators driving current into the load in parallel. This generator had a specific energy density of 2.25 J/kg, where the energy is defined as the instantaneous energy, $\frac{1}{2}LI^2$, in the inductor. The measured output waveform of this system is shown in Fig. 8.4. In general, ferroelectric generators perform best when operated in a high-voltage / high impendence mode. As a result, a high inductance seed winding is required for operation with ferroelectric seeding.

8.3.2. Ferromagnetic Seed Sources

Explosive ferromagnetic generators utilize the phenomena of shock wave demagnetization of hard ferri- and ferromagnets. A shock wave, initiated by high explosives passes along the magnet body and demagnetizes as it passes though the element. The initial flux Φ is coupled to a coil (with number of turns N), wound tightly around the magnetic element to maximize the coupling coefficient. After the shock wave passes, the change in magnetic flux $\Delta\Phi$, generates an electromotive force $E_g(t)$ in the coil according to Faraday's law:

$$E_g(t) = N \frac{-d\Phi(t)}{dt}. \tag{8.8}$$

From this simple equation, it would seem that the use of high field magnets, such as rare-earth $Nd_2Fe_{14}B$ would yield the highest $\Delta\Phi$ and thus the highest energy output. In general this is the case; however, the process of shock wave demagnetization is more complicated and is influenced by many factors such as induced eddy currents within the magnetic element and shock wave propagation and amplitude.

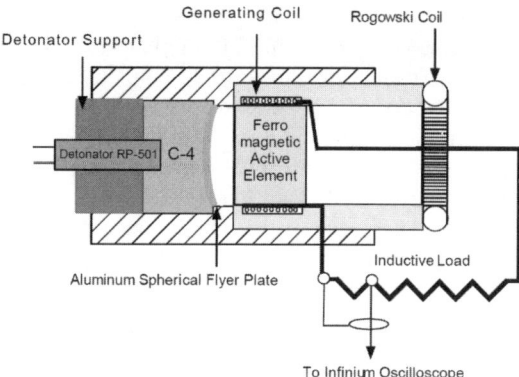

Fig. 8.5. Explosive ferromagnetic generator.

The greatest limitation to ferromagnetic generator performance is the problem of induced eddy currents. This is especially true for the rare-earth type magnets due to their good electrical conductivity [Shk02]. A simple ferromagnetic generator is illustrated in Fig. 8.5.

To overcome the eddy current limitations of shock induced demagnetization, we explored a variation of a ferromagnetic generator that uses flux compression. This type of device, called a constant magnet bus generator, utilizes two pairs of magnets to generate an initial magnetic field. This magnetic field is compressed by the expansion of a central aluminum tube between the magnets into two stationary copper strips on opposing sides. A current is then generated in the copper strips and transferred to the load, connected at the end opposite the detonation point [Sch03]. The major parameters for this design consist of the initial field intensity in the gap, H_o, the copper strip length, l, and the space between the copper strips and the aluminum tube, d. Knowledge of these values, as well as the size of the load, allows for the determination of the final current value given by:

$$I = \frac{l \cdot d \cdot \mu \cdot H_o \cdot \psi}{L}, \qquad (8.9)$$

where ψ is the magnetic flux conservation factor, determined empirically. The value of ψ was not explicitly stated in the literature, but was determined to be approximately 0.711 [Boj94].

The PMG design, shown in Fig. 8.6, uses four Neodymium-Iron-Boron magnets ($B_r = 1.23$ T), which generate an initial field strength of approximately 5.26 x 10^5 A/m in the gap. Each magnet had a width of 12.7 mm, height of 25.4 mm, and a length of 50.8 mm. The central aluminum tube, filled with C-4, had an inner radius of 4.45 mm and an outer radius of 6.35 mm. The copper strips were approximately 1.8 mm thick and 10 mm wide. The strips were cut from a copper pipe having inner and outer radius of 20.4 mm and 22.2 mm, respectively. Each strip was 190 mm in length before bending. The critical angle associated with the copper strips is seen on the right side of the magnets in Fig. 8.6. That angle is equal to the Gurney angle created by the aluminum tube, which is approximately 14°. The thickness of the mild steel flux paths was determined through use of the field simulation program Maxwell 2D by Ansoft Corp. In order to contain 95% of the flux lines in the metal plates, a thickness of 12.7 mm was necessary. The final flux-path-steel-plate dimensions were 63.5 mm x 50.8 mm x 6.35 mm, which contained about 80% of the flux lines. The simulation was also used to verify the initial magnetic field intensity, H_o, already stated. Two Lexan pieces were used as the support structure for the magnets. At the end of the tube, a piece of PVC was used to separate the copper strip from the aluminum tube in order to prevent any high voltage breakdown between the two pieces. An aluminum plug was placed at the end of the tube and served as one of the terminals for the load. The other terminals were at the end of each copper strip.

The load sizes used for testing ranged from 50 nH to 1 µH. Each coil was made from Litz wire in order to keep the load resistance to a minimum. The current provided by the copper strips was measured separately in order to verify that each side delivered equal current to the load. The dI/dt associated with each load was measured with a Rogowski coil. A typical output current waveform is shown in Fig. 8.7. The total energy generated was determined by the summation of the energy generated in each load, which is given by:

$$E = \frac{1}{2} \cdot L \cdot I^2. \tag{8.10}$$

Using a load size of 0.5 µH, the PMG generated 600 A in each copper strip, yielding a total energy of 0.2 J at 0.228 J/kg [Sch03].

Each of the devices described shows potential for use as a seed source for flux compression generators. The devices were demonstrated driving inductive loads with their efficiency strongly depending on the size of the load inductance. To become truly competitive to capacitor bank seeding, it would be necessary to increase the energy density of the explosive driven seed sources. Table 4.1. shows a comparison between the energy densities achieved by the various seeding methods. The PMG device was capable of supplying a maximum 0.94 J into 50 nH, the smallest load size tested. Elimination of the currently undetermined flux losses would yield even more substantial results.

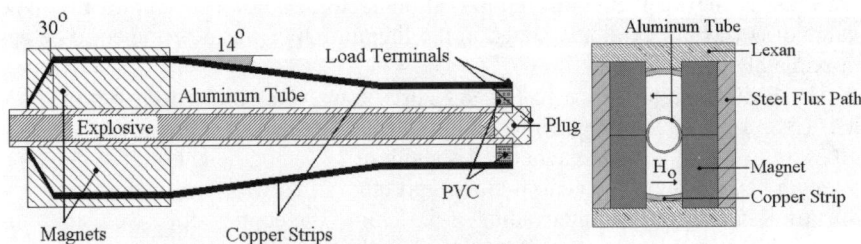

Fig. 8.6. Side view (cross-section) and front view of permanent magnet generator.

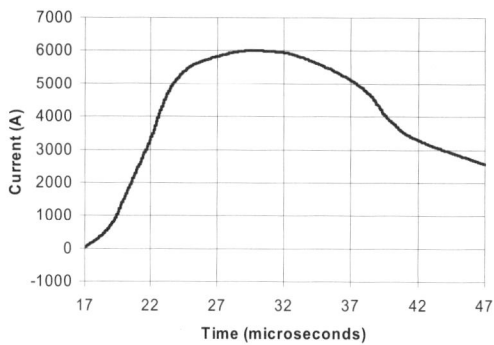

Fig. 8.7. Measured Output Current of 1.0 J/kg seed source into 50 nH load.

Table 4.1. List of a seed source devices and corresponding specific energies achieved.

Device	Specific Energy	Load
Capacitor Bank	~20 J/kg *	-
Shocked Magnet	2.3 J/kg	Low inductance
FEG	2.25 J/kg	825 µH
PMG	1.07 J/kg	50 nH

*up to ~ 400 J/kg for large high energy density capacitors

The ferroelectric generators constructed were capable of supplying energies up to 1.7 J into a very large inductive load (a peak of 63A into 825 µH). This is a low energy level for seeding an FCG, but the method shows promise for the future. Of the explosive designs presented in Table 4.1, the FEG method shows the most potential given that it can drive much larger load impedances than the other methods. Further development of this method will show if it possible to push the specific energy density to values on the order of capacitive seeding.

8.3.3. All Explosive FMG-FCG Example System

In this example system, a Ferromagnetic generator system is coupled to a FCG system via flux trapping from an external coil built into the FCG. In the example system described, we manufactured the FCG by winding the stator onto a mandrel with inserted armature and following application of two-component epoxy filler. We achieved intimate coupling with a calculated coupling coefficient of ~ 0.97 between the one-turn "primary winding", which is energized by the FMG. The FMG consists of a center-bored NdFeB-35 rare earth magnet and is loaded with approx. 1 g C-4. The FMG and FCG are coupled using a low inductance strip line that widens from initially 25 mm to 38 mm as it feeds into the primary of the FCG. We incorporated a low-inductance current viewing resistor into the strip-line that enables measuring the FCG seed current. This system is illustrated in Fig. 8.8 [Neu02].

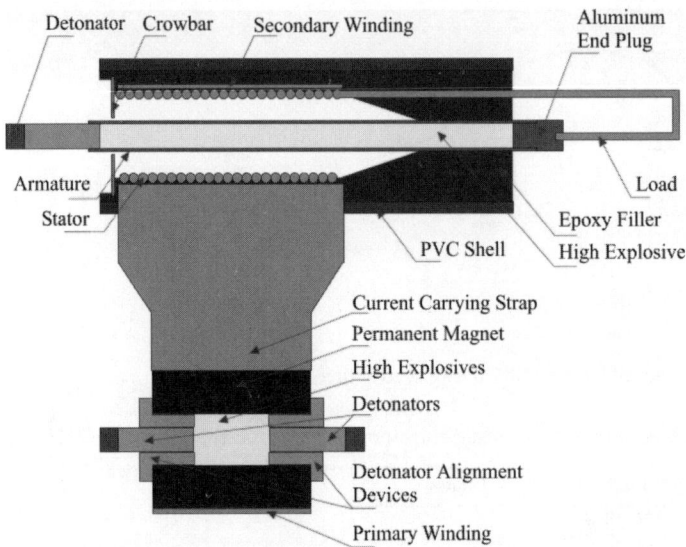

Fig. 8.8. Small all explosive FCG system (19 mm FCG stator diameter).

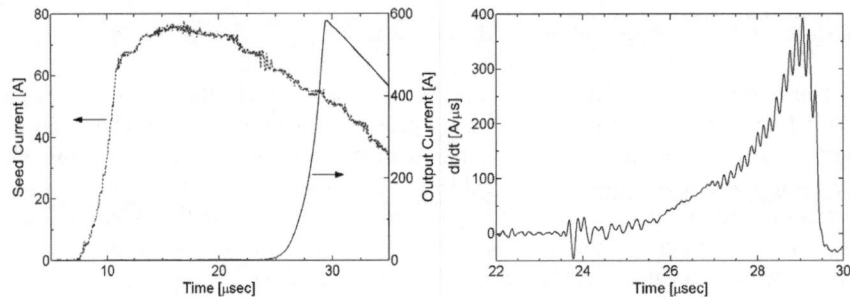

Fig. 8.9. Current and *dI/dt* waveform of FMG seeded mini-FCG. The seed current is scaled with the transformer ratio of 1:25 between the single turn primary and the secondary of the FCG.

It takes about 15 μs for the FMG output current to reach its maximum after detonation is initiated. Hence, the FCG detonator is timed so that the moment of crowbar approximately coincides with the seed-current maximum. The FCG carries no electrical current up to the moment of crowbar, thus premature resistive heating is avoided. The FCG's stator diameter was 19 mm, the armature diameter 9.5 mm, and the amount of C-4 used was 6 gram. As expected, no current prior to crowbar is noticed in the indirectly, via flux trapping, seeded FCG, see Fig. 8.9. For ease of comparison, the seed current is scaled with the turn-ratio of 1:25, the ratio of stator turns and the single turn seed coil.

Each of the devices described above shows potential for use as a seed source for flux compression generators. The devices were demonstrated driving inductive loads. However, further development would be required to construct practical explosive seed sources. Either source would require an impedance transformer to effectively drive the typical 10-100 µH loads presented by our small FCGs. Also, it would be necessary to increase the energy density of each.

References

[Tka02] Tkach Y, Shkuratov S, Talantsev E, Dickens J, Kristiansen M, Altgilbers L, Tracy P (2002) Theoretical treatment of explosive-driven ferroelectric generators. IEEE Trans. on Plasma Science 30: 1665-1673

[Shk01] Shkuratov SI, Kristiansen M, Dickens JC, Neuber AA, Altgilbers LL, Tracy PT, Tkach Y (2001) Experimental study of compact explosive driven shockwave ferroelectric generators. In: Reinovsky R, Newton M (eds) Digest of technical papers of the 2001 IEEE Pulsed Power Plasma Science Conference, IEEE Press, Piscataway NJ, pp 959-962

[Shk02] Shkuratov SI, Talantsev E, Dickens J, Kristiansen M (2002) Compact explosive-driven generator for primary power based on a longitudinal shock wave demagnetization of hard ferri- and ferromagnets. IEEE Trans. on Plasma Science 30: 1681-1691.

[Sch03] Schoeneberg N, Walter J, Neuber A, Dickens J, Kristiansen M (2003) Ferromagnetic and ferroelectric materials as seed sources for magnetic flux compressors. In: Giesselmann M. Neuber, A (eds) Digest of Technical Papers of the 14th IEEE International Conference, IEEE Press, Piscataway NJ, pp 1069 - 1072

[Boj94] Bojko BA, Gurin VE, Lyudaev RZ, Pavlovskii AI (1994) Autonomous Cascade MC-System with Constant Magnets. In: Cowan, M, Spielman RB (eds) Megagauss magnetic field generation and pulsed power applications, part 1. Nova Science Publishers, Commack, NY, pp 468-47

[Neu02] Neuber A, Shkuratov S, Holt T, Dickens J, Talantsev Y, Walter J, Kristiansen M (2002) All-explosive pulsed power generator system. In: Selemir VD, Plyashkevich LN (eds) 2002 Meggauss-IX. RFNC-VNIIEF, Sarov, Russia, pp 232-238

9 Practical FCG Pulsed Power System

Andreas A. Neuber, Juan-Carlos Hernandez

9.1 Dual Stage Generator

As we have outlined in Chap. 5 and also Chap. 7.2, the flux or energy multiplication of single stage FCGs is severely limited. This is in particular true if a load with rather high impedance, say several micro Henries or tens of Ohms is connected to the FCG output. The performance of a single stage FCG would be dismal if fired into such a load. Typically, a single stage FCG would exhibit only acceptable performance if fired into a load with an impedance of two orders of magnitude lower (several 10 nano Henries or hundreds of milli Ohms).

Experiments on the performance of dual stage FCGs with respect to utilizing them as a driver for inductive energy storage with a fuse open switch were carried out in the 90s, e.g. [Leo96], or more recently [Neu03]. We will present in Chap. 9.1 what it takes to drive 40 kA through a 3 µH load utilizing a dual stage FCG with a 25 mm armature diameter.

9.1.1 Performance by Stage

The smallest wire pitch used for the staged FCG was set to 1.25 mm. Following the relationship between the armature's expansion angle, θ, and the pitch, p,

$$\Delta a = \frac{p}{4} \tan \theta \qquad (9.1)$$

partial turn-skipping can be avoided [Che86] if the armature is round and centered with respect to the helix within $\Delta a \sim 0.08$ mm, cf. Chap. 4. We consider this required level of accuracy as only moderate so that we could utilize the manufacturing methods previously applied to our single stage, single pitch generators, which we primarily studied to gain insight in the basic physics of these generators [Neu02], cf. Chap. 2.

In brief, the helices L2, L4 followed by L1 and L3, see Fig. 9.1, are wound on a dielectric mandrel (we used Delrin) and held in place by thin layers of epoxy. We paid specific attention to the layer between primary and secondary helices, such as L3 and L4, so that breakdown prone air voids were almost completely avoided.

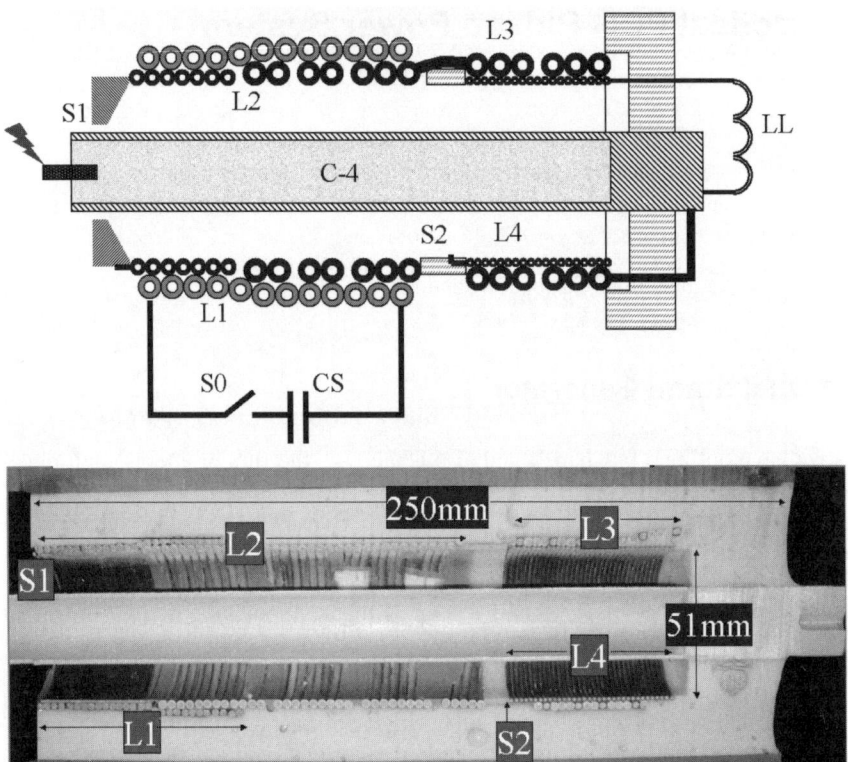

Fig. 9.1. Top, schematic of the dual staged MFCG (160 g HE). Bottom, staged MFCG connected to an inductive load, LL. Storage capacitor (50 μF) – CS, Closing switch – S0, field coil for first stage – L1, Crowbar for L2 – S1, first stage coil – L2, primary of dynamic transformer – L3, Crowbar for L4 – S2, secondary of dynamic transformer – L4. See Table 9.1 for specific helix dimensions.

Before removing the mandrel, we added mechanical strength to the dual stage FCG by inserting the generator into a PVC pipe and casting the space between generator and pipe with epoxy (overall thickness of outer layer ~ 25 mm). We made no attempt to minimize this layer thickness. One could think of heavier materials, such as concrete or similar, as inertial backing with smaller layer thickness. In any case, the inertial backing merely needs to contain the explosive and magnetic forces occurring during generator operation on the several 10 microsecond time frame. The generator as well as the backing typically disintegrates into small fragments on the millisecond timescale.

The seamless aluminum armature, 25 mm outer diameter, 2 mm wall, was partially annealed, thus ensuring proper expansion up to the helix diameter of 51 mm, cf. Chap. 4. We chose the initial generator pitch as small as reasonably possible in order to achieve a sufficiently large dL/dt to overcome the initially large resistance, $R(t)$ of L2. Starting with the lumped circuit equation for the FCG from Chap.2, Eq. (2.13),

$$L(t) \cdot \frac{dI(t)}{dt} + \alpha \cdot I(t) \cdot \frac{dL(t)}{dt} + I(t) \cdot R(t) = 0 \qquad (9.2)$$

it can be easily derived that one must have

$$\alpha \cdot \left| \frac{dL(t)}{dt} \right| / R(t) > 1 \qquad (9.3)$$

to have an instantaneous current gain > 0, cf. Eq. (2.17). The parameter α describing the intrinsic flux loss has a typical value of ~ 0.8 for our small generators (0 < α < 1 for realistic conditions, $\alpha = 1$ would mean no intrinsic loss). Or even more restricting, for an instantaneous energy gain > 0 the following has to be true, cf. Eq. (2.18):

$$(2\alpha - 1) \cdot \left| \frac{dL(t)}{dt} \right| / 2R(t) > 1 \qquad (9.4)$$

Table 9.1 MFCG helix dimensions and inductances. All wires off-the-shelf; L2, L3, and L4 with Teflon insulation (AWG12: ~ 2 mm core diameter, 3 mm total dia.)

Dimensions	Wire	# Turns	Inductance
L1	2 x AWG 12 [*]	15	10 µH
L2 [**]	AWG 22	40	52 µH
	2 x AWG 12	6	
	3 x AWG 12	3	
	4 x AWG 12	2	
	5 x AWG 12	1	
L3 [**]	7 x AWG 12	~ 2.5	~ 300 nH
L4 [**]	AWG 20	30	29 µH
	AWG 16	3	
LL	AWG 12	9	~ 3 µH

[*] "N x AWG.." refers to N wires wound in parallel.
[**] Teflon insulated wires are treated with etchant before winding on mandrel.

Two variants of the second stage (L3 and L4 in Fig. 9.1) were tested, one with a crowbar, S2, made of a ~ 0.2 mm thin disk with a central hole diameter about 4 mm larger than the armature diameter and another with a simple, Teflon-insulated pin as crowbar, S2 as shown in Fig. 9.2 or Fig. 9.3, respectively. Additionally, the disk had a single radial slot for avoiding induced eddy currents in the disk as depicted in Fig. 9.2. We had successfully used this type of crowbar in numerous previous experiments with our single stage generators [Neu01] and we used the same approach for the input crowbar, S1, cf. Fig. 9.1. The load inductance for the second stage was fixed at 3 µH for all tests.

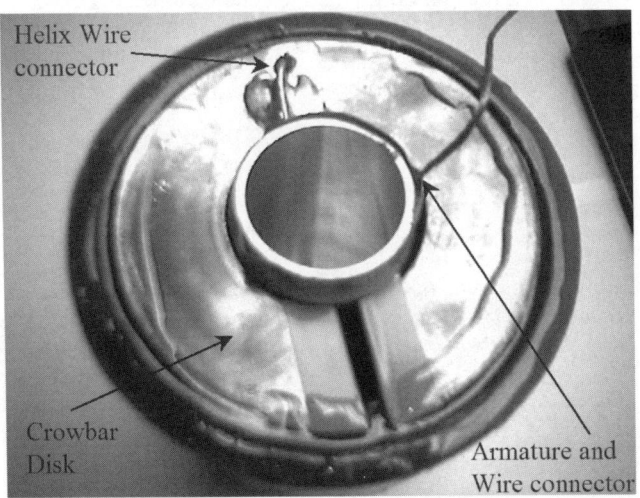

Fig. 9.2. Second stage by itself using crowbar **disk** configuration.

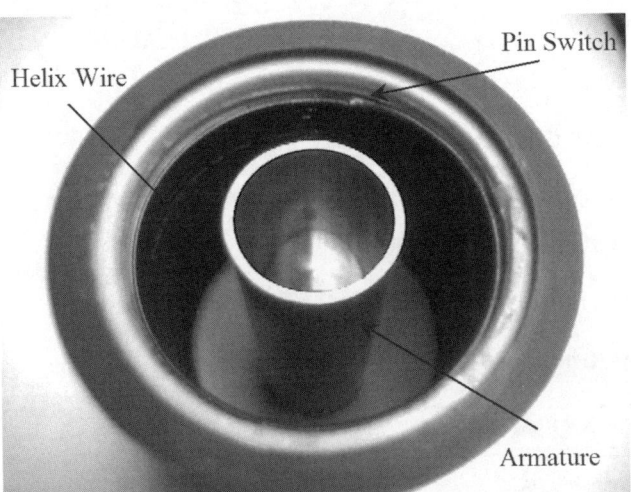

Fig. 9.3. Second stage by itself using crowbar **pin** configuration.

The conditions for current and energy multiplication, cf. Eqs. (9.3), (9.4), of the output stage are met during the entire operation of the output stage, L4, with crowbar pin configuration, see Fig. 9.4. Replacing the pin with a crowbar disk causes the amplification condition to drop below unity for the first 2 to 3 µs, thus more energy is dissipated than produced during this early phase, see Fig. 9.5. This means that ·from this point of view the crowbar disk geometry should exhibit poorer performance, with a value of $\alpha \sim 0.8$ for both configurations.

The average dL/dt of helix L2 is $\sim 8\ \Omega$, which is distinctly larger than the wire resistance of initially $\sim 1\ \Omega$ (resistance measured at 100 kHz frequency). The helix L1 is wound with 2 parallel AWG 12 magnet wires, cf. Table 9.1; all other helices are standard Teflon insulated copper wire (all wires off-the-shelf).

We used stranded AWG 12 for the ease of winding the wire onto the mandrel (the smaller wires were solid). No grooves were machined into the mandrel, and the correct pitch was adjusted utilizing custom made gages. Standard etchant was applied to the Teflon insulated wires before winding the helices, and the simple solder joints between the helices were insulated with Teflon tape and shrink tubing. The 3 µH load inductance was wound on a non-magnetic core and placed about 150 mm away from the generator's output end.

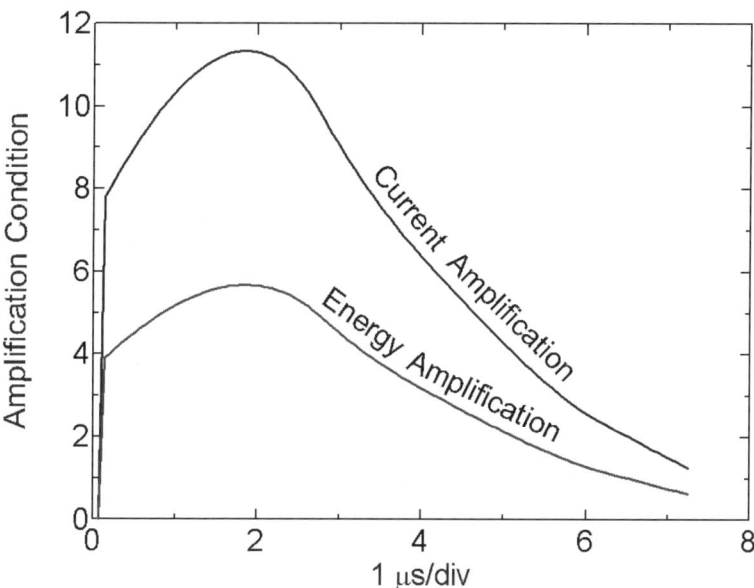

Fig. 9.4. Condition for energy and current amplification for L4 in the crowbar **pin** configuration. Upper curve – Current amplification, see Equation (9.3); bottom curve – Energy amplification, see Eq. (9.4).

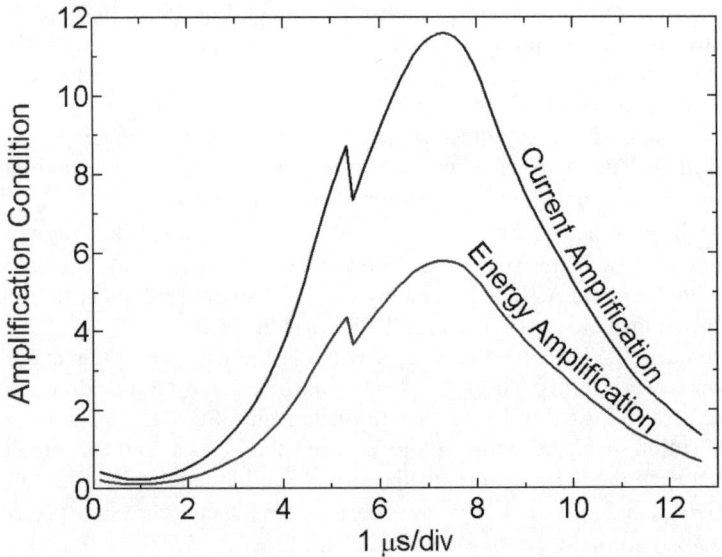

Fig. 9.5. Condition for energy and current amplification for L4 in the crowbar **disk** configuration. Upper curve – Current amplification, see Equation (9.3); bottom curve – Energy amplification, see Eq. (9.4).

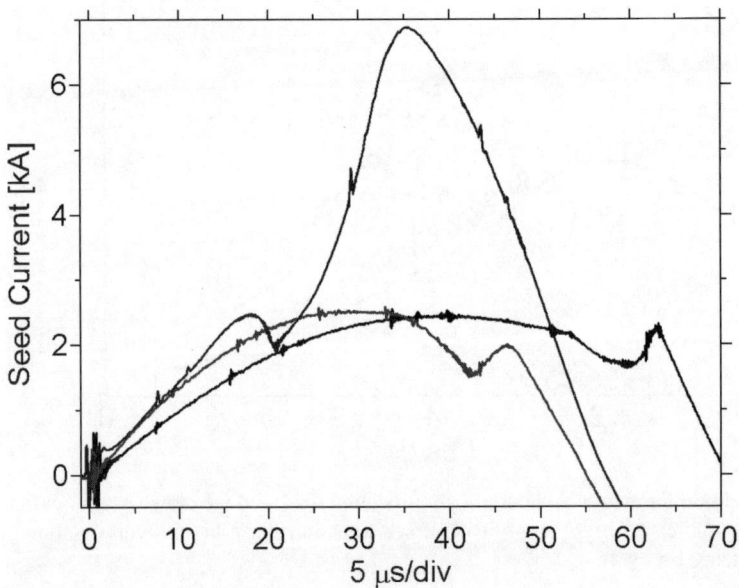

Fig. 9.6. Seed current for L1 extending solely over the first 40 turns of L2 (two lower curves with differing seed current timing) and covering all of L2 (upper curve). Seed capacitor, CS = 50 µF, 10 kV maximum DC charging voltage.

Table 9.2 Performance of the 1st stage by itself, L1 and L2, firing into L3

Seed current into L1 (10 μH)	8.4 kA
Output current into L3 (~300 nH)	222 kA
dI/dt max	15 kA/μs
Voltage max	4.5 kV
Seed Energy	352 J
Output Energy	7.4 kJ
Energy Gain, G_E	21

The experimental behavior of the first and second stage was observed separately by capacitor seeding either one of them without the other attached. These experiments revealed that the most efficient helix configuration for the first stage is achieved by limiting the helix length of field coil, L1, to the first 40 x AWG 22 turns of L2. This ensures that L1 forms a voltage step-up transformer with L2. If L1 is chosen to span the entire length of L2, a significant amount of energy is pushed back into the capacitor seed current circuit, see upper curve in Fig. 9.6. The output load for the first stage test was set to ~ 300 nH, which is close to the inductance of field coil L3, cf. Fig. 9.1.

The first stage by itself produced an energy gain of 21 when seeded with ~ 8.4 kA. (7.5 kJ output energy), see Fig. 9.7 and Table 9.2.

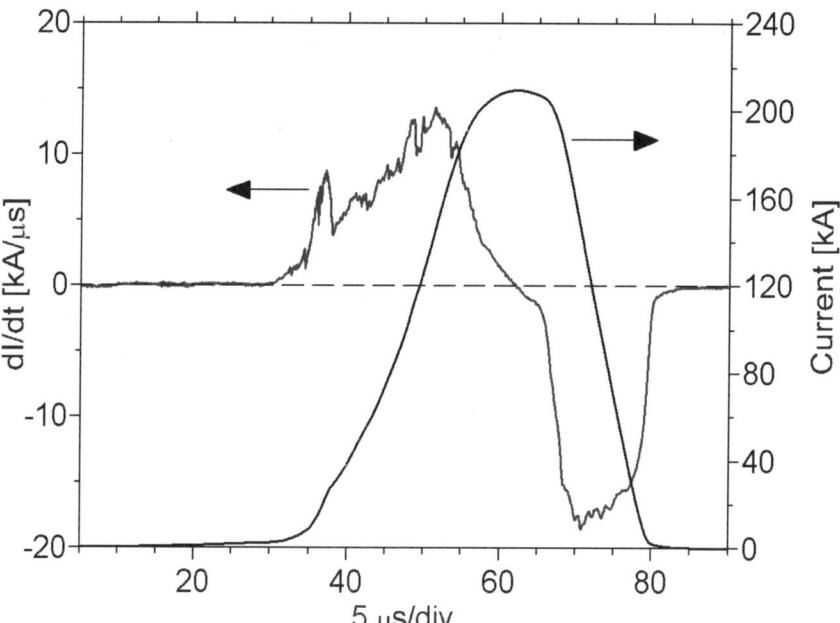

Fig. 9.7. dI/dt and current output of the first stage only, L2. Seed current was 8.4 kA and the resulting energy gain was 21.

Table 9.3 Performance of the 2nd stage by itself (crowbar **pin**), L3 and L4, firing into LL.

Seed current into L3 (0.3 µH)	25 kA
Output current into LL (3 µH)	7.5 kA
dI/dt max	2.2 kA/µs
Voltage max	6.6 kV
Seed Energy	93.8 J
Output Energy	84.4 J
Energy gain, G_E	0.9

It should be noted that we decreased the roughly ~ 400 J seed energy to ~ 200 J for the staged MFCG shots as the induced voltage in the second stage at the high seed energy level could reach ~ 100 kV or more at the second stage, likely causing breakdown of the insulation. Generally, the simple solder joints utilized for joining the wires with varying pitch as well as L2 and L3 are operating close to flawless, see Fig. 9.7.

The sharp drop in dI/dt at about 38 µs signals the moment when the armature has wiped out the first 40 turns (AWG 22) of the helix L2. About 12 µs earlier, the armature is contacting the crowbar, S1, the flux established by L1 is trapped by L2, and dI/dt exhibits a distinct positive slope when the first turns of L2 are wiped out. After about 28 µs of total runtime (at $t \sim 61$ µs in Fig. 9.7), the first stage is burned out and dI/dt goes negative. Comparing the performance of the two geometries of the 2nd stage, with crowbar pin, see Fig. 9.8 and Table 9.3, and with crowbar disk, see Fig. 9.9 and Table 9.4, reveals the inferior performance of the disk design with 30% less energy gain.

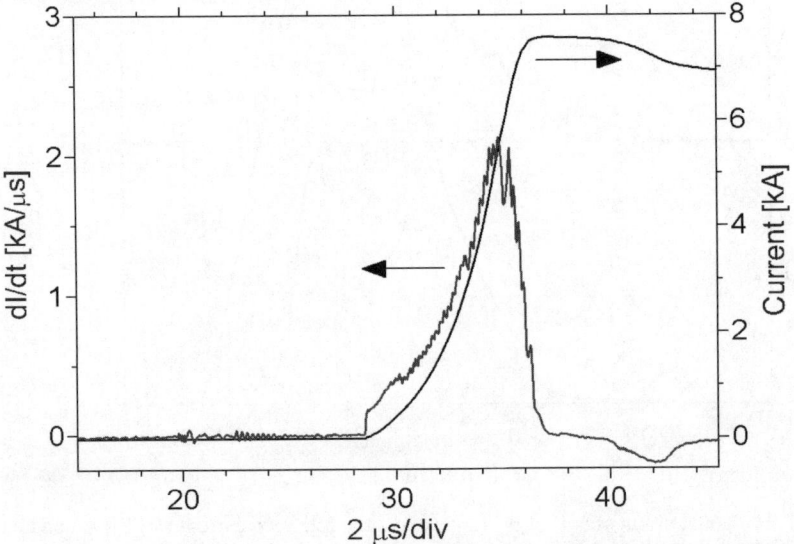

Fig. 9.8. dI/dt and current output for the second stage only with crowbar **pin**. Seed current was 25 kA into the field coil L3 = 300 nH and the resulting energy gain was 0.9.

Table 9.4 Performance of the 2nd stage by itself (crowbar **disk**), L3 and L4, firing into LL

Parameter	Value
Seed current into L3 (0.3 µH)	25 kA
Output current into LL (3 µH)	6.12 kA
dI/dt max	2 kA/µs
Voltage max	6 kV
Seed Energy	93.8 J
Output Energy	56.2 J
Energy gain, G_E	0.6

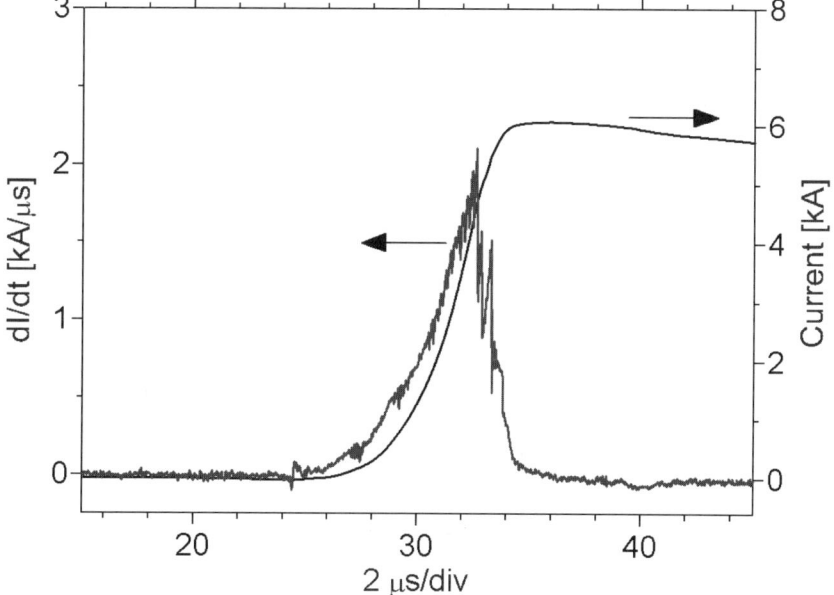

Fig. 9.9. dI/dt and current output for the second stage only with crowbar **disk**. Seed current was 25 kA into the field coil L3 = 300 nH and the resulting energy gain was 0.6.

With the methods discussed in Chap. 5.4 we quantified the contributions of intrinsic and ohmic flux losses to the combined loss, see Fig. 9.10 and Fig. 9.11. This approach requires as input the time-dependent inductance, $L(t)$, and resistance, $R(t)$, which we calculated using a 3D eddy current solver accounting for magnetic field diffusion and proximity effects.

Approximately 10% of flux is lost due to ohmic heating at the time of contact with the first turn of the crowbar disk design, Fig. 9.11, compared to the crowbar pin, that has no losses at the same point, 0 µs in Fig. 9.10, ending with 26% and 18% of flux lost due to ohmic heating, respectively, at 12.7µs and 7.3µs.

Since both crowbar disk and pin configuration have the same $L(t)$ and $R(t)$ for times after the armature contacts the first turn, the intrinsic flux losses are very similar in both cases with 29% and 31% lost, respectively, cf. Fig. 9.10 and Fig.

9.11. This is mainly due to the intrinsic loss being considered small while the armature is just sliding along the crowbar disk.

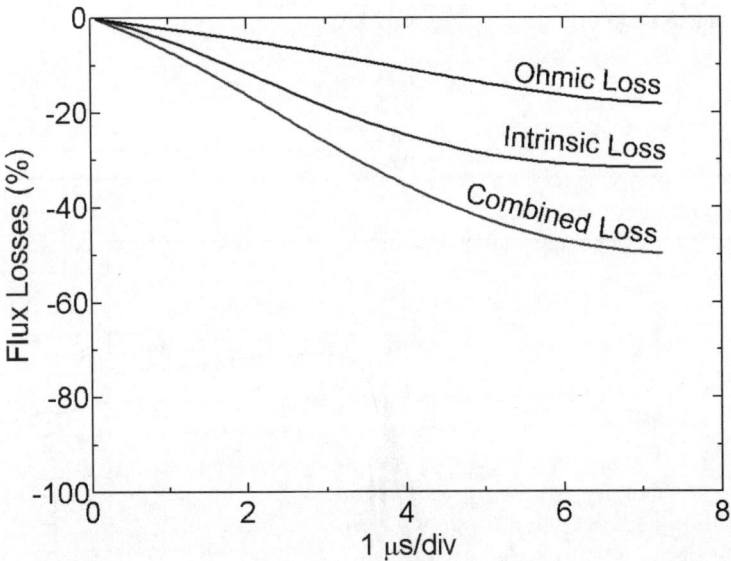

Fig. 9.10. Flux loss for the second stage MFCG into a 3 µH load with crowbar **pin**. The crowbar pin connects the second stage with the armature at the first turn at 0 µs.

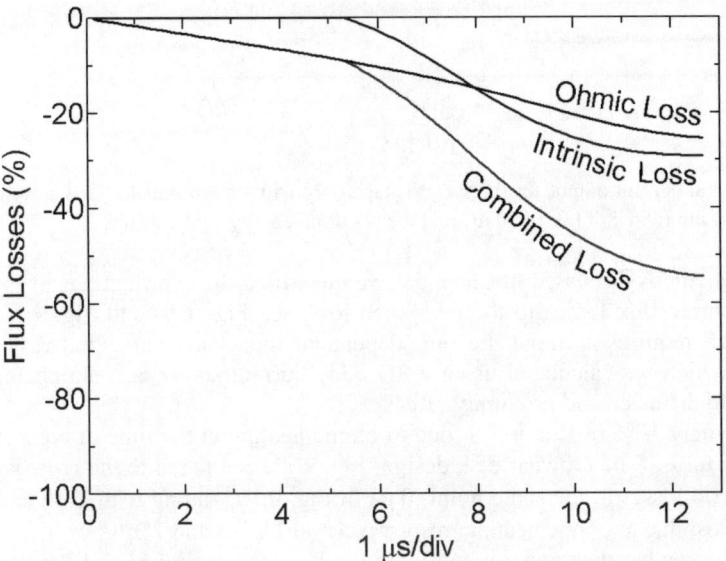

Fig. 9.11. Flux loss for the second stage into a 3 µH load with crowbar **disk**. The armature takes ~ 5.4 µs to expand from initial crowbar to the first turn.

Nevertheless, the experimental performance for the two geometries differs more than we can explain with our simple calculations. We believe that the reason for this is the heavy insulation of the crowbar disk as it is necessary to prevent early breakdown between crowbar and armature. Some of the energy will be lost into breaking down the insulation as the armature is sliding along the insulated crowbar disk. It should be noted that the secondary winding, L4, of the voltage step-up transformer formed by L3-L4 is open circuited before crowbar, thus generating a high voltage stress between the crowbar and the armature. Also, in case of the crowbar disk, the coil, L4, is switched into the circuit well before all of L2 is wiped out, thus affecting the mutual inductance and flux linkage in the system.

As desired, the energy gain, G_E, of the first stage is relatively large, whereas the second stage serves as voltage step-up with $G_E \sim 1$. The calculated figure of merit, β, in the secondary was 0.670 and 0.675 for crowbar disk and pin, respectively, which is as expected larger than 0.5 since we have to compensate for approximately 30% coupling losses between L3 and L4. While the first stage behaves more like a conventional FCG with high current and energy gains, the second stage exhibits small current amplifications of ~ 5 and comparatively large experimental flux losses of $\sim 50\%$, mainly due to the large load inductance. The effect of the large load, 3µH, connected to stage 2 is manifested in the current waveforms outputs in Fig. 9.8 and Fig. 9.9, which exhibits a large L/R time constant after generator burn out. In contrast, stage 1 in Fig. 9.7 shows a much smaller L/R constant (faster current decay), with the 300 nH load, which corresponds to the field coil, L3, of the stage 2.

The crowbar pin configuration, cf. Fig. 9.3 and Fig. 9.4, has the disadvantage that at time 0 µs the secondary winding has already lost $\sim 40\%$ of its volume for compression. Thus, the potential current and energy amplification is relatively lower than in the crowbar disk design, since the initial inductance is 75% smaller than the crowbar disk at the switching time 0 µs. However, the crowbar disk configuration, see Fig. 9.2, shows an abrupt drop in current and energy amplification at 5.4 µs, see Fig. 9.5. This drop is not visible in Fig. 9.4 with crowbar pin, since the armature makes physical contact with the first turn at $t = 0$ µs. Only at the moment of contact between armature and first turn, the intrinsic flux loss parameter is assumed to drop from $\alpha = 1$ (no intrinsic loss) to $\alpha \sim 0.8$, cf. Eq. (9.2), thus accounting for the increased flux loss throughout armature-helix contact. In addition, for the crowbar disk neither the current nor the energy is amplified in the first 2.6 µs and 3.2 µs, respectively, until the conditions for amplification from Eqs. (9.3), (9.4) are met.

The same problem arises in the first stage, which was hence designed with a high initial inductance and small pitch, 40 x AWG 22 turns of L2, to achieve a high dL/dt even during the beginnings of generator operation. After these initial 40 turns, the pitch is increased to maintain the current density and avoid excessive induced voltages throughout the compression time that might lead to breakdown.

One more disadvantage for using a crowbar disk as S2, cf. Fig. 9.1, is that the staged MFCG becomes physically somewhat longer and more difficult to align. Most of the generators using the crowbar disk configuration have shown clocking problems due to misalignment, exhibiting inferior performance compared to the crowbar pin design. We chose the crowbar pin for our most successful staged FCG design, a decision that was primarily based on the lower gain (both overall and instantaneous, see Table 9.5) and the overall unnecessarily higher complexity of the crowbar disk (for the second stage only). The first stage employed a crowbar disk, which was superior for the initial capacitive seeding.

9.1.2. Dual Stage Performance

Based on the performance of the two stages separately, one might expect the overall energy gain of the complete generator being close to the product of the individual gains ~ 18, cf. Table 9.2 and Table 9.3. However, we consider this the maximum gain, since, as mentioned earlier, coupling due to the mutual inductance between stages can cause a smaller, ~ 50%, overall gain [Leo96]. Hence, the energy gain for the staged MFCG was expected to be in the range from 9 to 18. We observed experimentally an energy gain of 13 for the complete dual-staged generator, see Fig. 9.12, which is only about 30% lower than the maximum expected gain. The only change from the single stage tests was the ~ 0.1 mm thicker Teflon insulation of the helix L3 that became necessary to avoid breakdown between the concentric helices as well as between helix and armature at higher voltage levels (In the single-stage tests, PVC insulated wire was used).

The voltage gain for the staged generator from stage 1 to stage 2 is about 14, effectively stepping up the voltage from 1 kV to 14 kV in the second stage. Of course, higher seed currents will lead to higher output energy and higher output voltage. As long as the wire is not excessively heated by the current flow, the energy gain and the voltage gain will exhibit only little decrease. So far, we achieved a maximum output energy of 3,000 J and an output voltage of 42 kV with the complete dual-stage generator running into a 3 µH load (45 kA). Fig. 9.12 shows a lower energy shot with a current output of 15.2 kA into a 3 µH load. The initial supplied current was 2.3kA into the ~ 10 µH seed coil, L1, meaning that the overall final flux of the dual stage generator, 0.05 Wb, is twice the initial flux, 0.023 Wb, having an energy gain of ~13, see Table 9.6.

Table 9.5 Conditions for amplification of current and energy.

	Overall Gain	Instantaneous Gain >1
Current	$G_I = \left(L_0/L_F\right)^\beta$	$\alpha \cdot \left|\frac{dL(t)}{dt}\right| / R(t) > 1$
Energy	$G_E = \left(L_0/L_F\right)^{2\beta - 1}$	$(2\alpha - 1) \cdot \left|\frac{dL(t)}{dt}\right| / 2R(t) > 1$

Table 9.6 Performance of the complete dual stage MFCG with 2.3 kA seed current (low energy shot).

Seed current into L1 (10 µH)	2.3 kA
Output current into LL (3 µH)	15.2 kA
dI/dt max	4.6 kA/µs
Voltage max	13.8 kV
Seed Energy	26.5 J
Output Energy	344 J
Energy gain, G_E	13

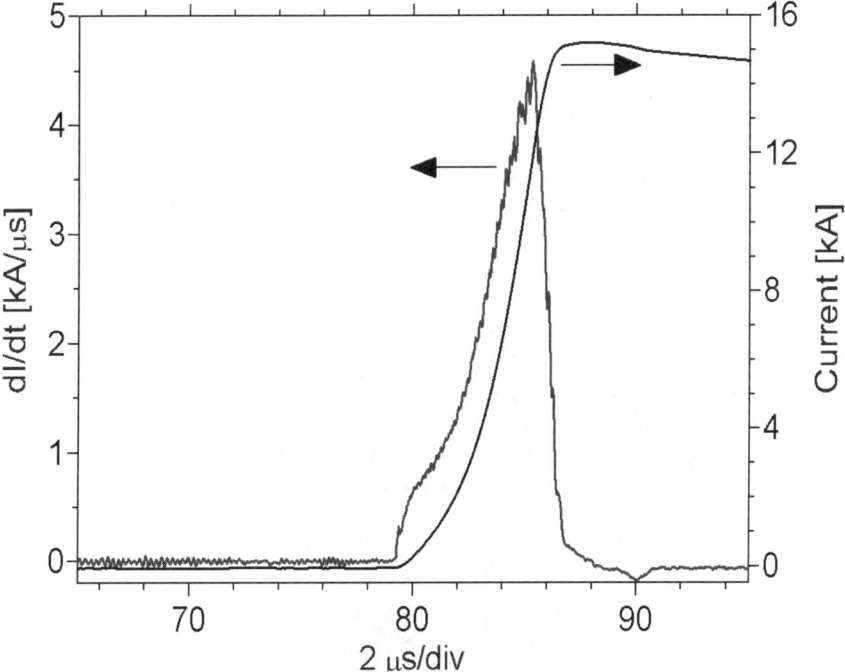

Fig. 9.12. dI/dt and current output for the complete 2 stage MFCG into a 3 µH load. Seed current was 2.3 kA and the overall energy gain was 13.

The magnetic flux density distribution during the compression of the second stage when the armature reaches the helix is depicted in Fig. 9.13. As this figure illustrates, the highest flux is produced at the contact point between the armature and the secondary winding where also most of the intrinsic flux is lost (the armature is kept static for the simulation). The magnitude of the magnetic flux density close to the secondary helix inside wall is 4 to 6 times larger at the contact point than at locations were the armature has expanded little [Neu03]. The true flux distribution in a real FCG will differ from Fig. 9.13 since the armature is moving in reality. However, to the authors' knowledge, there is presently no code available that could be used to simulate 3D magnetic fields with moving and deforming boundaries.

While we have calculated the energy gain from the experiment using the final magnetic energy and the initial magnetic energy, it should be mentioned that the demand on prime energy could be considerably higher than the initial magnetic energy. Specifically, the ohmic resistance in the capacitor-inductor seed current circuit leads to non-negligible losses in the transfer from electric field energy in the capacitor to the seed current coil's magnetic field energy. Depending on the design, as much as 50% of the initial stored energy in the capacitor is lost. As a general rule, the capacitance should be chosen as small as possible and the charging voltage as large as possible. Of course there are limits to this as electrical breakdown between crowbar S1 and armature becomes an issue due to the voltage step-up between L1 and L2 if the charging voltage of the seed capacitor becomes too large. The use of SF_6 was needed to avoid voltage-induced breakdown when stage 2 is switched. The energy loss from CS = 50 µF to L1 = 10 µH, connected via a 2 m long seed cable (~ 12 mm diameter coaxial cable), for our staged generator was about 25%, resulting in an effective energy gain of ~10 from stored energy in the capacitor to output magnetic field energy; a result that is comparable to MFCGs with somewhat larger working diameters [Leo96]. We have listed our maximum performance values in Table 9.7

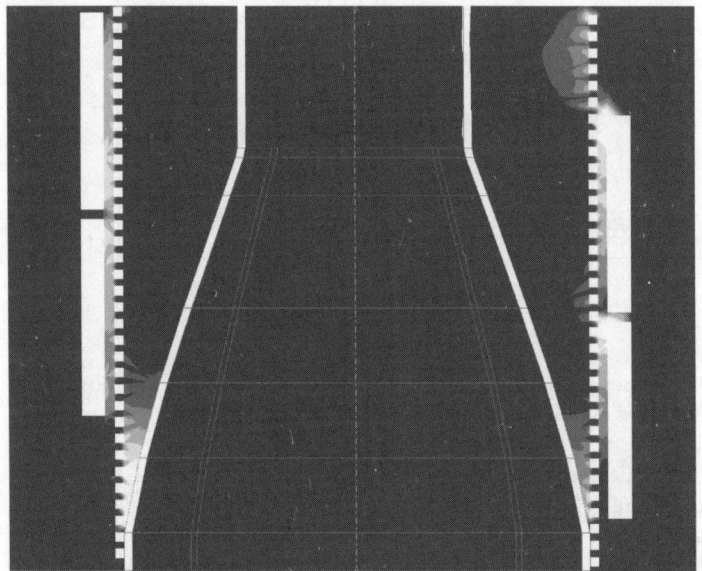

Fig. 9.13. Magnetic flux density at 100 kHz with 10kA current. ▬ - 1 Tesla, ▬ - 3.5 Tesla, ▬ -10 Tesla The primary and secondary have 34 turns and 2.5 turns, respectively. The magnetic flux in the conductors (large white rectangles represent the primary, L3) is not shown.

Table 9.7 Maximum performance achieved of the dual-stage MFCG (1st and 2nd stages together)

Parameter / Input	Output
Dual-stage FCG with **pin** crowbar into 3 µH load, 6.8 kA seed current into L1.	13.8 kA/µs, 42 kV, 45 kA
Dual-stage FCG, 231 J seed energy	3 kJ
Dual-stage FCG, Energy gain	13

9.2 Inductive Energy Storage

The 42 kV maximum voltage amplitude achieved across the 3 µH storage inductor, cf. Table 9.7, is insufficient to drive many relevant pulsed power loads, such as high power microwaves (HPM) sources. Hence, further voltage multiplication is needed to produce voltages in excess of 200 to 300 kV across a 10 to 20 Ohm load. We have chosen an approach that adds a fuse opening switch, cf. Chap. 7.1, to the dual stage FCG for commutating the inductively stored energy into a resistive load of, in our case, approx. 12 Ohm.

The presented results demonstrate the amplification of the voltage applied to a resistive type load using an inductive energy storage system and exploding wire fuse as a final switching stage. The fuse will commutate the energy stored in the inductor into the load, see Fig. 9.14. In general, the load has a much higher impedance than the fuse, and as a result the voltage across the load will be amplified. We obtained with the present system 42 kV max. voltage amplitude into a 3 µH storage inductor, cf. Table 9.7, which should produce an output voltage of > 250 kV into a 12 Ω load with optimized fuse, since typically a voltage gain of ~ 6 can be achieved with the wire fuse [Rei87].

The system performance based on experimental and simulated current/voltage waveforms and the physical limits of the relatively compact MFCG are discussed.

9.2.1 Fuse Opening Switch

The experimental setup of the exploding wire fuse used as the power conditioning system between the FCG and the load is depicted in Fig. 9.14 and Fig. 9.16. The system consists of a dual stage FCG, Fig. 9.1, with 250 mm length and 51 mm inner helix diameter, able to produce an output voltage from 100 kV to 300 kV into a 12-15 Ω ohmic load when 15-45 kA are supplied to the energy storage inductor of 3 µH using an exploding wire-opening switch with 4-16 silver parallel wires of 125 µm cross-section and 14 cm length. The same fuse length and wire type were used in all the experiments and simulations.

262   9 Practical FCG Pulsed Power System

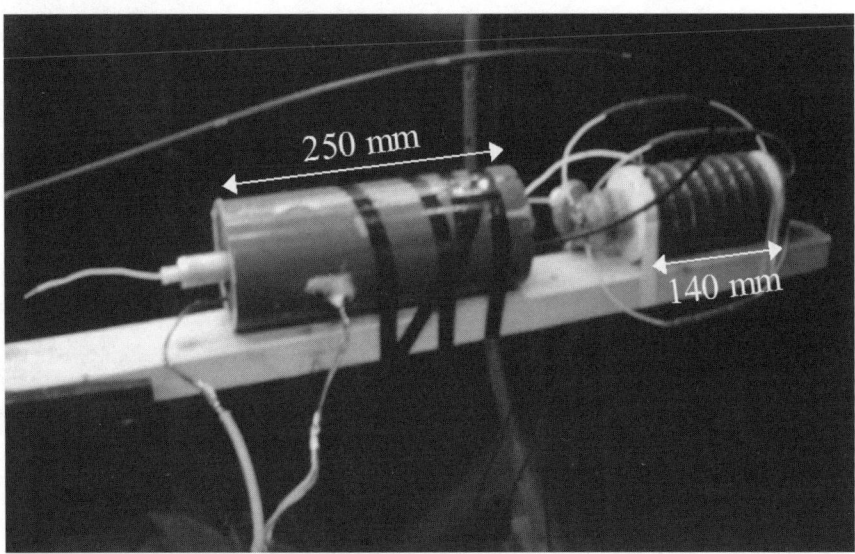

Fig. 9.14. Experimental Setup of the MFCG with the FOS. Left, MFCG (250 mm long). Right, FOS with load, cf. Fig. 9.15.

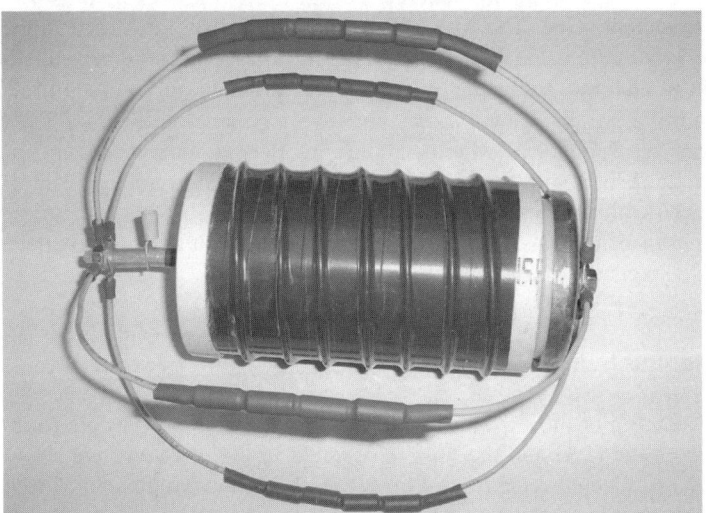

Fig. 9.15. Experimental setup of fuse opening switch (placed inside the cylinder embedded in quartz sand) with four parallel load resistors, storage inductor, and spark gap (at right side). See Fig. 9.16 for details.

9.2 Inductive Energy Storage

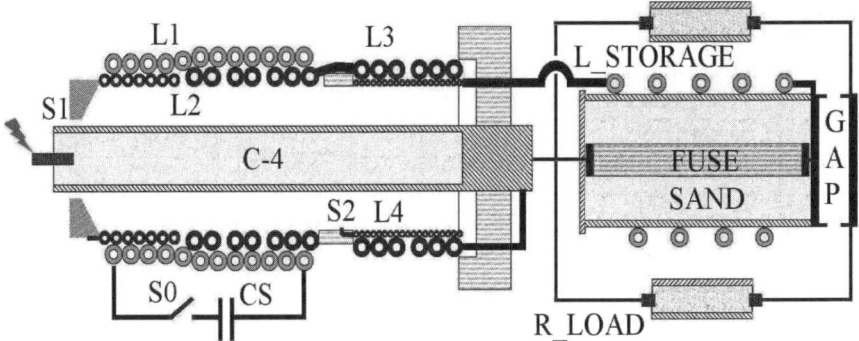

Fig. 9.16. Left, MFCG (250 mm long). Right, storage inductor, opening switch and load. Figure 4.3

The self break voltage of the spark gap used as closing switch was adjusted to close at a voltage consistent with the moment when the peak current is reached and the exploding wire fuse is opening. The enclosure of the spark-gap as well the MFCG were filled with SF_6. The fuse wires were embedded in very fine quartz sand used for sandblasting (Potters Industries, Valley Forge, PA, 2W580), which works as the quenching medium. Since the load was connected parallel to the fuse, it was built out of four parallel branches to reduce the parasitic inductance.

We used initially strings of carbon composite resistors as resistive load, which, however, quickly proved to be extremely nonlinear at the high current amplitudes. Hence, we switched to $H_2O - CuSO_4$ liquid resistors that have a higher parasitic capacitance but remain linear throughout.

To analyze the performance of the system, we measured three different currents of the system, which are MFCG output, fuse, and load current. In all cases, we derived the current by numerically integrating the measured dI/dt waveforms. Due to destruction of the current sensors with each shot, we chose to utilize self wound Rogowski coils. The Rogowski coil signals were fed through 10 m long RG-58 coaxial lines into 500 MHz analog, 2Gs/s digital oscilloscopes terminated with 50 Ω.

In addition, we also measured the output load voltage across to the load. For this purpose, we used homemade voltage dividers made with HVR ceramic resistors, model C1320A 7, connected in series with a value of 3.9 kΩ capable to hold off and stay linear up to ~18 kV each, with a total resistance of 27.3 kΩ. The shunt resistor of the divider had a value of 50 Ω. It was connected in series to the 27.3 kΩ a 2nd parallel to the 50 Ω input of the oscilloscope, resulting in a divider ratio of 1:1080. The data was captured with HP-Infinium digital storage oscilloscopes with a maximum sampling rating of 2 Gs/s. Overall, we achieve approximately 2 ns risetime with our setup.

9.2.2. Fuse Calculations

In the ideal case, the fuse will take no energy commutating the current instantaneously from the inductor to the load and it will open at the peak current stored in the 3 μH inductor. However, the fuse absorbs a considerable amount of energy during the heating explosion and turn off phases, see Chap. 7.1. To find the optimal parameters of the fuse, first we need to determine the expected amount of current produced by the generator. As outlined in Chap. 7.1, we use the experimental current to calculate the required cross sectional area of the fuse, as well the maximum voltage expected at the output load, which will determine the minimum length required to avoid breakdown or restrike in the opening phase.

The cross sectional area is very critical because if the total cross section of the wires is too large, the fuse will open after the peak current is reached in the inductor (underdriven fuse) or in the extreme case it will not open at all. If it is too small, the fuse will open too early (overdriven fuse), therefore only a small fraction of the available energy would be stored in the inductor. The more individual parallel wires are used for a given total cross sectional area, the easier it will be to interrupt the current. For this reason we used rather thin wires with 125 μm diameter, which are, however, still sufficiently strong to be easily handled. To increase the switching speed, it is important to choose a material that shows a high rate of change in resistance for a small increment of energy. This ratio should be as high as possible to yield the fastest switch. Gold shows the highest value for this ratio followed by silver, see Chap. 7.1.6. Utilizing the method from Chap. 7.1 the optimal length and cross sectional area for the opening wire fuse can be calculated, see appendix. Using silver wires of 125 μm diameter the calculations yield a length of 140 mm with 4 to 5 wires, cf. Chap. 7.1.14.

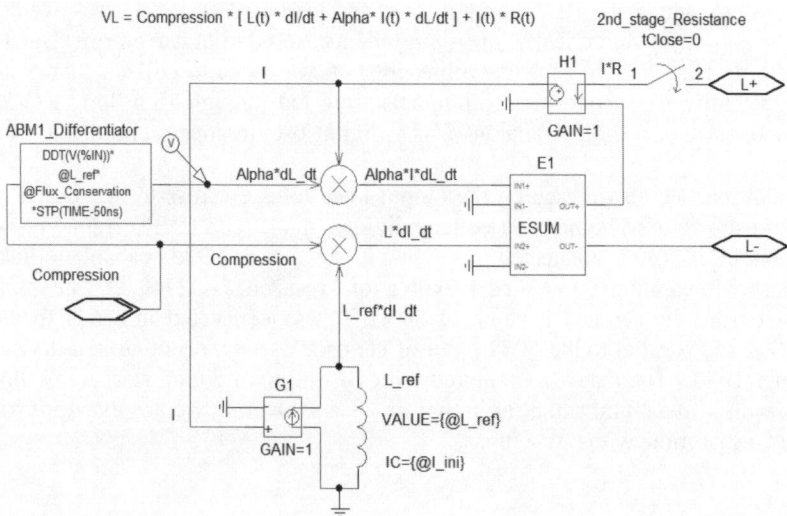

Fig. 9.17. Pspice schematic of the second stage with voltage controlled inductor.

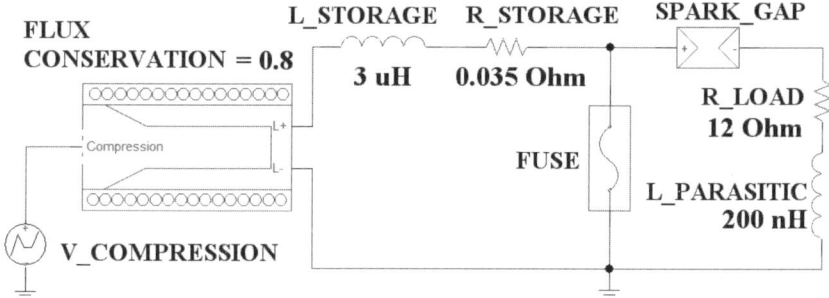

Fig. 9.18. Pspice schematic of the MFCG with Opening Exploding Wire Switch.

Utilizing the simple mathematical equations and empirical fuse material parameters from Chap. 7.1, we built a PSpice® model that predicts the dI/dt, current and load voltage of the dual stage FCG system. Since the second stage is switched into the circuit only at the time of crowbar switch, S2, closing, the dual stage FCG itself is modeled as a single stage FCG that mimics the output of the second stage only. The mutual inductance collapse between L3 and L4 is not considered in this case but rather lumped into the specific value of the intrinsic flux loss parameter ($\alpha = 0.8$ in this specific case), see Fig. 9.18. The overall model with FCG, storage inductor, fuse, spark gap, and load includes the major parasitic impedances and is depicted in Fig. 9.18.

To model a dynamically changing FCG source, the time varying inductance, $L(t)$, and resistance, $R(t)$, of the FCG have to be known. Our model senses the current that is fed back from the circuit and calculates the differential voltages at every time interval. The seed current, $I(t = 0)$ as initial condition for Eq. (9.5) is initially set to the desired current amplitude. MFCG flux losses were scaled by α going from 1, no losses at all, to 0 where all the flux is lost. Although α changes throughout the compression time, we assumed an average value, typically ~ 0.8. We modeled the resistivity of the fuse as a function of the current action integral, or deposited energy by two simple equations, Eqs. (9.7), (9.8) for the heating/melting phase and the vaporization phase, respectively, cf. Chap. 7.1.13.

$$I(t) \cdot \alpha(t) \cdot \frac{d}{dt} L(t) + L(t) \cdot \frac{d}{dt} I(t) + I(t) \cdot R(t) = 0 \qquad (9.5)$$

$$h_e = \int \left(\frac{I_{material}}{fuse_{cross-area}} \right)^2 \qquad (9.6)$$

$$\rho(h) = \rho_0 \cdot \left[1 + A \cdot \left(\frac{h}{h_e} \right)^B \right] \qquad (9.7)$$

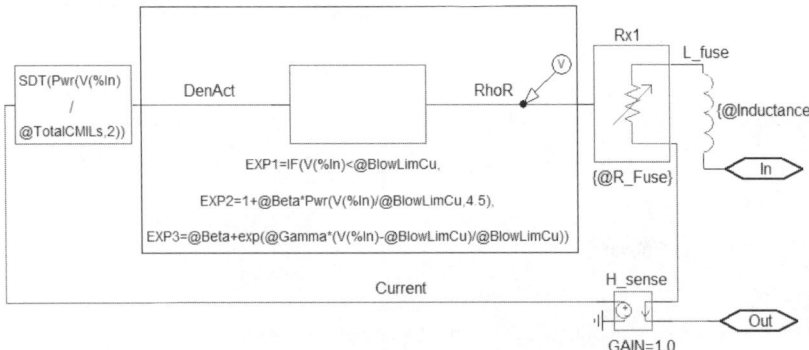

Fig. 9.19. Pspice Schematic of the Exploding Wire Switch Model, cf. Fig. 6.11

$$\rho(h) = \rho_0 \cdot \left(A + \exp\left(\frac{h - h_e}{h_e} \cdot C \right) \right) \tag{9.8}$$

The parameter h_e is the blow limit, which depends on the peak current density of the material and is calculated from Eq. (9.6) up to the explosion of the wire. This limit separates the heating/melting phase given in Eq. (9.7) from the vaporization phase given in Eq. (9.8) when the actual current action, h, that flows through the fuse is $h > h_e$. ρ_0 is the resistivity of the material at room temperature. A indicates how fast the resistivity rises for a given material. B and C are used to model the resistivity during the heating/melting phase and vaporization phase, respectively. Typical values of h_e ($10^{17} A^2 s/m^4$), A, B, and C for silver are 1.03, 36, 4.5, 100, respectively, cf. Table 7.2. The resulting PSpice® model for the fuse is depicted in Fig. 9.19.

The spark gap was modeled with an exponential resistance decrease or linear resistance slopes, see Fig. 9.20, without noticing much of a difference in the final result, since the most of the energy is transferred during the first 10% of the closing time.

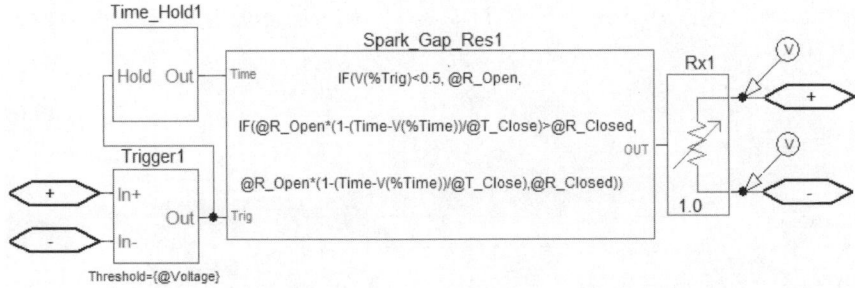

Fig. 9.20. PSpice® schematic of the spark gap model with linear $R(t)$ decay.

The calculated load voltage shape shown in Fig. 9.21 matches the experimental waveform, and the calculated fuse dI/dt shown in Fig. 9.22 is close to the experimental data as well.

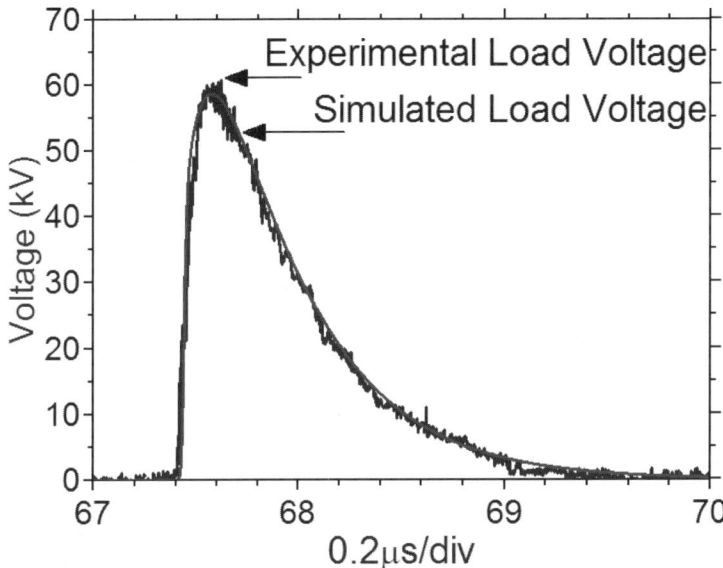

Fig. 9.21. Experimental and simulated load voltage waveforms.

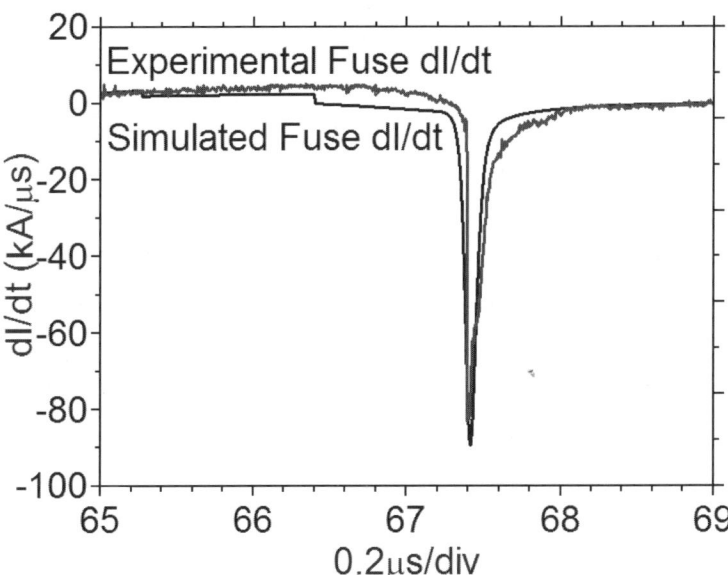

Fig. 9.22. Experimental and simulated dI/dt waveforms.

The small step in Fig. 9.22 at 66.4 µs, which corresponds to the change in slope of the current in Fig. 9.23 at ~ 5 µs, is due to the fact that we limited ourselves to only 10 inductance values as input data for the PSpice® model. We note here that the inductance calculations carried out with a 3D magnetic field solver for diffusion and proximity effects are rather time consuming. Alternatively, simpler inductance calculations are often accurate enough or a mockup of the helix can be used to directly measure the inductance as a function of the armature position. The PSpice® model linearly interpolates between the calculated inductance values. Increasing the number of inductance values, $L(t)$, would remove the step in the dI/dt waveform, however, the accuracy of the maximum current or load voltage would improve only little.

It should be noted that the dual-stage FCG itself is in limits linear with respect to seed current (energy) in and current (energy) out. However, the entire system with fuse opening switch and sparkgap is very nonlinear due to the fact that if, for instance, the seed current is chosen rather small, the fuse will be underdriven and comparatively very little energy will be commutated from the storage inductor, $L_{STORAGE}$, to the load resistance, R_{LOAD}, cf. Fig. 9.16. For a final optimization of the fuse, the dual stage FCG performance has to be repeatable within 5% to 10%. Large fluctuations in the FCG output (at constant seed input) will foil any system optimization due to the nonlinear nature of the overall explosive pulsed power system.

As depicted in Fig. 9.23, by the use of proper fuse parameters we expect to achieve 300 kV into a 13 Ω load, with 45 kA into 3 µH from the dual stage FCG.

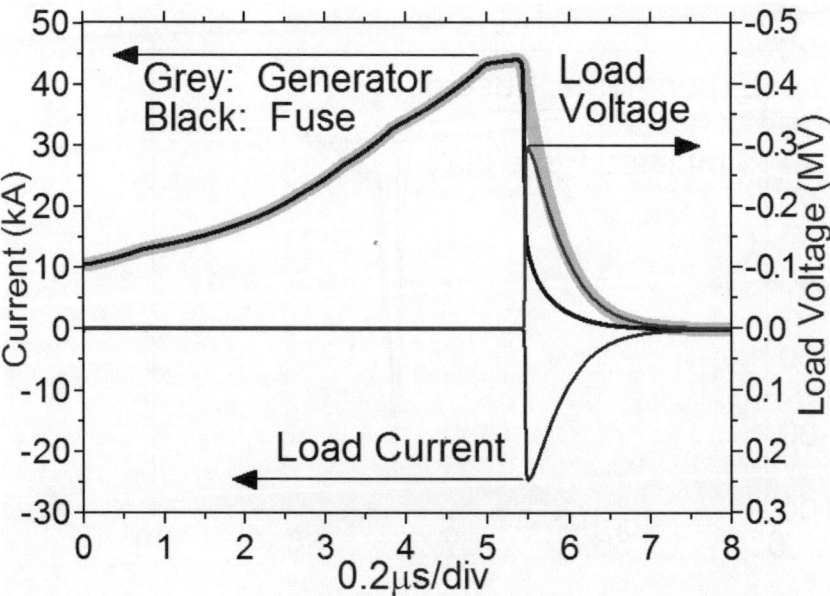

Fig. 9.23. Left axis calculated MFCG fuse and load currents. Right axis load voltage.

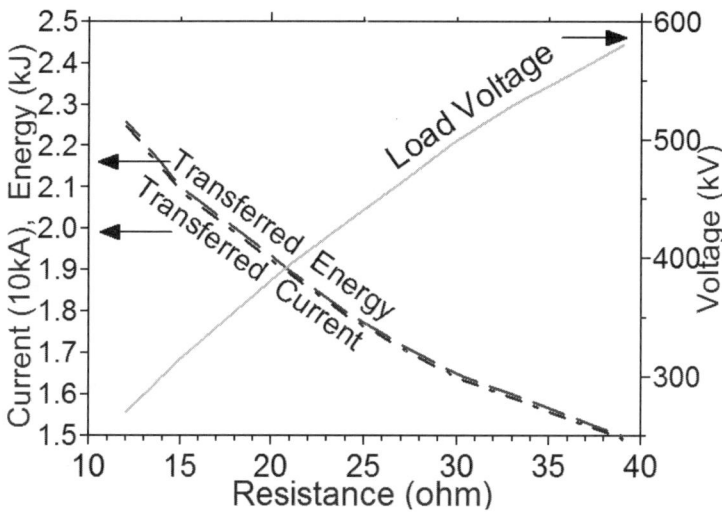

Fig. 9.24. Transferred energy, current and load voltage for silver fuse with 3 kJ maximum stored energy in the storage inductor.

Efficiency and energy transfer as well as the output voltage are depicted in Fig. 9.24 as a function of the resistive load. It is assumed that a storage inductor energy of 3 kJ (45 kA max current) is supplied by the FCG into 3 µH, cf. Table 9.7. In this case, the fuse consists of 15 silver wires of 125 µm diameter. The regime relevant to HPM sources close to and above 300 kV is of interest. A maximum of roughly 70% of the stored energy in the inductor, $L_{STORAGE}$, is transferred to a resistive load of ~ 15 Ohm.

9.3 Comparison with traditional PPS

Generating 200 kV or more with a traditional pulsed power system (PPS) supplying ten or more kA into a comparatively low impedance load (10… 20 Ohm) typically requires the use of a voltage multiplier and a pulse forming line (PFL), see Fig. 9.25. The reason that a voltage multiplier is needed lies in the fact that virtually all available capacitors are limited to 100 kV DC charging voltage. Hence, if a voltage in excess of 100 kV is needed, a voltage multiplier such as a Marx generator is utilized. We also note that capacitors are limited in their DC charging voltage simply due to the fact that it is generally easier to hold off high pulsed voltages than large DC voltages. Most PPS designers would prefer to utilize a voltage multiplier where the DC charging voltage can be kept comparably low instead of a single stage system where the pulsed voltage is merely of the same order as the charging voltage.

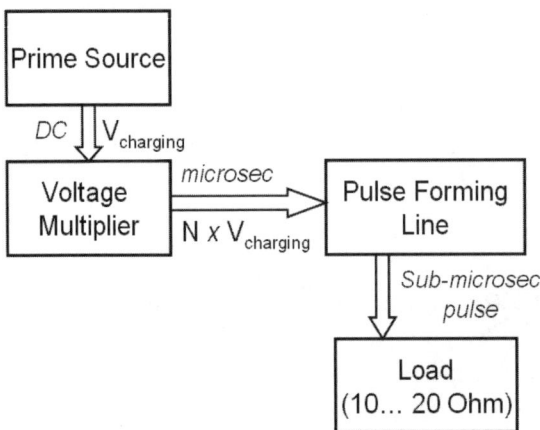

Fig. 9.25. Traditional PPS for generating sub-microsecond pulses into a 10... 20 Ohm load.

One might want to connect the load directly to the voltage multiplier, which will, however, not work in almost all cases since the output impedance of the voltage multiplier is maybe a factor 10 or more larger than the load impedance. Hence, the energy stored in the voltage multiplier is rather slowly, microsecond time scale, transferred to the pulse forming line (PFL). The PFL can be a coaxial line or strip line, however, in both cases with an impedance, $(L/C)^{1/2}$, matched to the load impedance for maximum energy transfer and fast risetime. The typical propagation speed of a pulse in an oil-filled coaxial line is of the order of 5 ns/m, which means that the PFL becomes quickly rather long if a 100 or 200 ns pulse is desired. Since the PFL also has to hold off pulsed voltages in excess of double the pulse output voltage, > 400 kV, its diameter is also rather large, so that the PFL typically takes up the most space of all 4 sub-systems depicted in Fig. 9.25. (N.B. For a simple PFL firing into a matched load, the pulsed voltage across the load will only be half of the charging voltage). A properly designed PFL will deliver a flat top pulse with fast risetime and, depending on length, sub-microsecond duration.

The length of the PFL can be significantly reduced by the use of high permittivity dielectric materials, which, however, has its challenges in that the best candidate materials are ceramics. This group of materials is difficult to control over large volumes. Water with a high permittivity exhibits high hold-off electric fields on the μs time scale, however, can only be used for intermediate energy storage as it will conduct on longer time scales. At present, research into Compact Pulsed Power (CPP), addressing the ceramic materials issue and breakdown of materials such as transformer oil, water, and liquid nitrogen are underway [Gau04].

We will discuss in the following a specific approach that uses a topology without PFL, connecting the load directly to the voltage multiplier, which we designed for low impedance. The output parameters of this more traditional, yet compact PPS are similar to the dual-stage generator presented in Chap. 9.2 so that a size comparison with the explosive PPS will be straightforward.

9.2.1. Compact Low Inductance Marx Generator

The basic idea of a Marx generator is the charging of N capacitors in parallel to the charging voltage V_{charge} and rapidly discharging the N capacitors in series, thus generating a nominal output voltage of $N \times V_{charge}$. The erected capacitance of the Marx generator is

$$C_{erected} = \frac{C_{stage}}{N} \qquad (9.9)$$

where we have used C_{stage} for the capacitance of a single stage, keeping in mind that one stage may consist of several capacitors in parallel. The erected inductance of the Marx generator increases with the number of stages:

$$L_{erected} = L_{stage} \cdot N \qquad (9.10)$$

Hence, the Marx impedance is

$$Z_{Marx} = \sqrt{L_{erected}/C_{erected}} = N \cdot \sqrt{L_{stage}/C_{stage}}. \qquad (9.11)$$

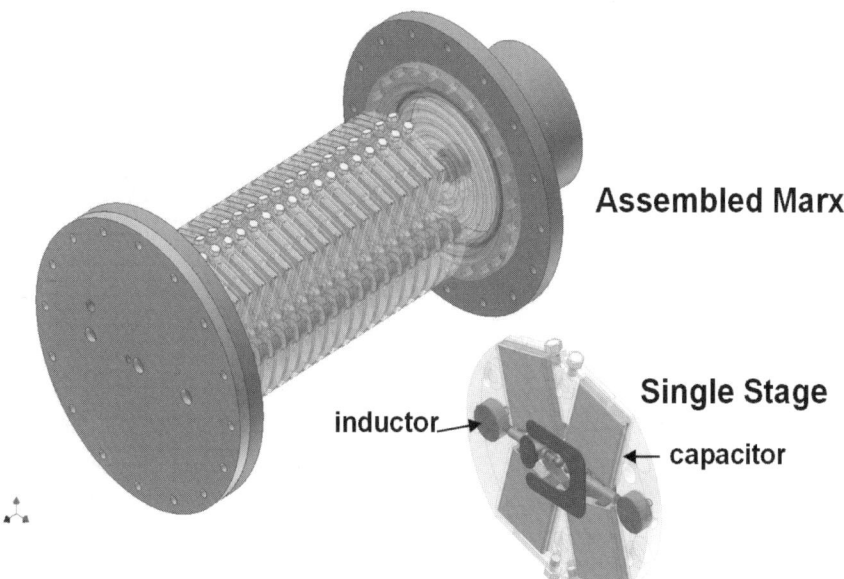

Fig. 9.26. Compact low inductance Marx generator with 25 stages (30 cm diameter enclosing steel cylinder is not shown). Length is approximately 0.7 m with 500 J stored energy, 500 kV nominal output voltage, and Marx impedance of 18 Ohms [Fmv03].

We have used 4 mica capacitors, 25 nF each, in parallel per stage, see lower right hand corner in Fig. 9.26, resulting in $C_{stage} = 100$ nF. Each of the capacitors has a self inductance of ~ 25 nH. The connections and the switching spark gaps add to this inductance, which totals to measured 1.1 µH for 20 stages or 55 nH per single stage, see Fig. 9.27. Since the erected capacitance of the 25 stage Marx generator is 4 nF, we calculate from Eq. (9.11) $Z_{Marx} = 18.5$ Ohm.

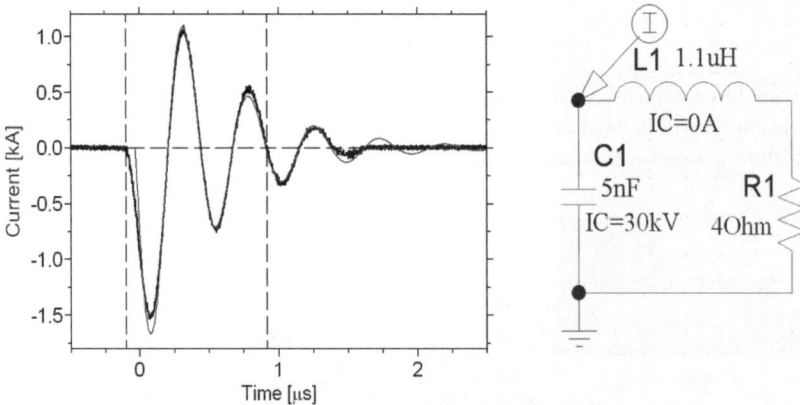

Fig. 9.27. Operation of the compact Marx generator, 20 stages only, into a short at low charging voltage, 1.5 kV, with helium in the spark gaps (left). The experimental current is distinguished by its noise amplitude from the calculated current. The lumped circuit parameters for the erected Marx (right) were adjusted to match the experimental current.

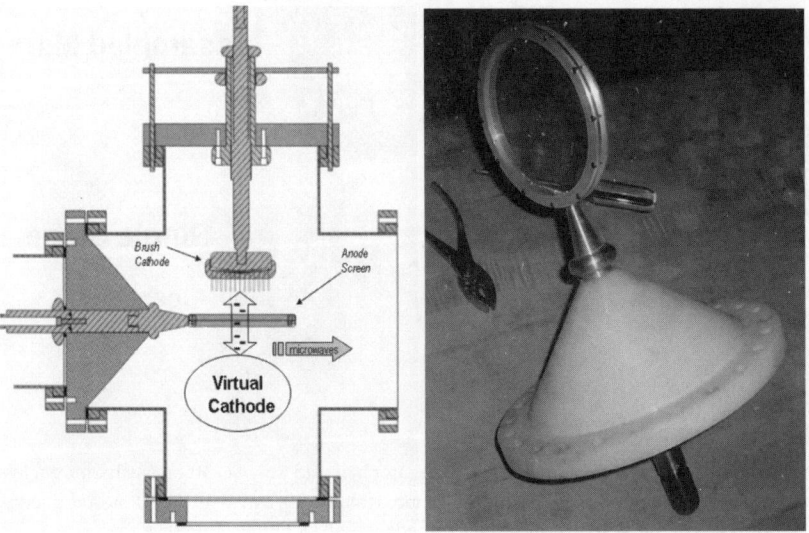

Fig. 9.28. Hundreds of MW HPM output Virtual Cathode Oscillator (Vircator), fitted inside a standard 8" vacuum cross (left) [Lar03]. Anode stalk with anode grid (right).

Neglecting resistive losses in the Marx generator itself, the expected output voltage into an 18.5 Ohm resistive load would be ½ N x V_{charge} = 250 kV. We consider this the maximum output voltage into a matched load, since resistive losses in the Marx (note the 4 Ohm parasitic resistance in Fig. 9.27), voltage sharing due to parasitic inductance in the load, as well as switching losses dissipate some of the available energy or voltage.

As discussed above, with a Marx generator impedance of merely 18.5 Ohms, connecting the load directly to the voltage multiplier becomes a viable option. However, one problem remains; the Marx output pulse is not a flat top pulse at all but basically the first half period of a damped sinusoidal oscillation of an RLC circuit. Hence, we chose as load a Virtual Cathode Oscillator (vircator), which is known to be forgiving with respect to pulse voltage fluctuations, Fig. 9.28. In addition, the vircator requires no external magnetic field to guide the electron beam [Bar01]. Thus, the vircator can be run straight off the compact Marx, without any other auxiliary power, see Fig. 9.29.

Utilizing a velvet cathode with ~ 6 cm in diameter and an anode-cathode gap of ~ 10 mm, we have presently achieved an output power of estimated ~ 50 MW with a peak voltage of 220 kV at V_{charge} = 20 kV. The ohmic resistance of the vircator is dynamically changing but is close to 18 Ω after the virtual cathode has been established with the tendency to drop to lower values towards the end of operation. The frequency of the HPM microwave radiation was around 5 GHz. Further optimization of the vircator and compact Marx is pending, but achieving at least 200 MW HPM output power appears to be realistic with the compact system design.

Fig. 9.29. Assembled Marx generator and vircator ready for testing.

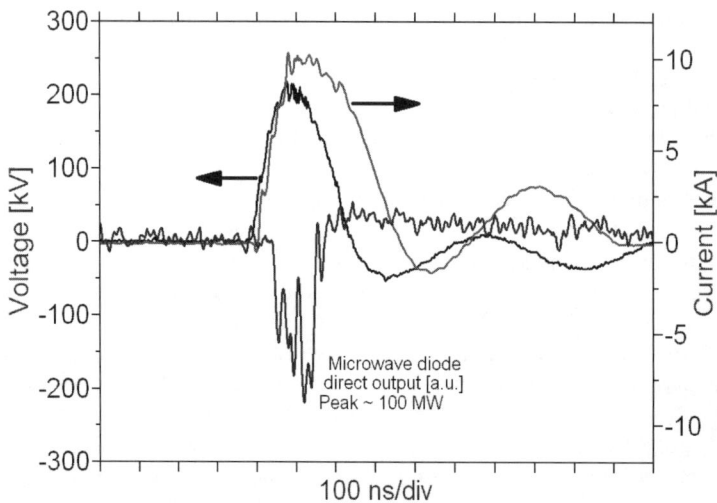

Fig. 9.30. Output voltage, current, and microwave power of the compact Marx generator, see Fig. 9.26, driving the vircator, see Fig. 9.28.

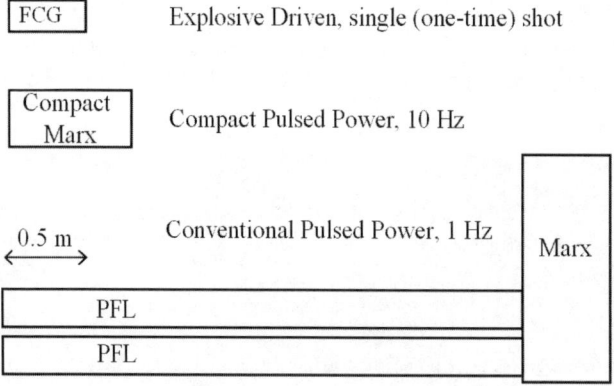

Fig. 9.31. Approximate footprints of the explosive PPS (top), compact PPS (middle), and traditional PPS (bottom), all designed to operate into a 10... 20 Ohm load with 200 to 300 kV pulse voltage.

The foot print of the compact Marx generator is several times smaller than the non-optimized traditional, everyday laboratory use, pulsed power system with PFL, see middle or bottom picture in Fig. 9.31, respectively. We would like to emphasize that, by utilizing charging inductors rather than resistors, the compact Marx can be charged within less than 100 ms from a rapid charging power supply [Gie03], thus enabling a pulse repletion rate of 10 Hz in burst mode. Continuous operation is not possible since no provisions for cooling are implemented in the present design.

Overall, the explosive PPS, which is comparable in electrical output, is much smaller than any of the two traditional PPS in Fig. 9.31. It also has the additional advantage that the required prime energy is roughly a factor 10 smaller than for the compact Marx. However, if repetitive operation or just simply more than a one-time operation is needed, the Marx generator would obviously be the better solution.

References

[Bar01] Barker BJ, Schamiloglou, E (eds) (2001) High-Power Microwave Sources and Technologies. IEEE Press and John Wiley & Sons
[Che86] Chernyshev VK, Zharinov EI, Kazakov SA, Busin VN, Vaneev VE, Korotkov MI (1986) Magnetic flux cutoffs in helical explosive magnetic generators. In: Turchi PJ (ed) Megagauss physics and technology. Plenum Press, New York, pp 455-469
[Fmv03] Compact Marx generator research supported by FMV, Sweden, 2003/2004.
[Gau04] Gaudet JA, Barker RJ, Buchenauer CJ, Christodoulou C, Dickens J, Gundersen MA, Joshi RP, Krompholz HG, Kolb JF, Kuthi A, Laroussi M, Neuber A, Nunnally W, Schamiloglu E, Schoenbach KH, Tyo JS, Vidmar RJ (2004) Research issues in developing compact pulsed power for high peak power applications on mobile platforms. Proceedings of the IEEE 92: 1144-1165
[Gie03] Giesselmann M, Heeren T, Helle T (2003) Compact, high power capacitor charger. In: Giesselmann M, Neuber A (eds) Digest of Technical Papers of the 14th IEEE International Conference, IEEE Press, Piscataway NJ, pp 707-710
[Lar03] M. Lara, J. Mankowski, J. Dickens, and M. Kristiansen (2003) Reflex-triode geometry of the virtual-cathode oscillator. In: Giesselmann M, Neuber A (eds) Digest of Technical Papers of the 14th IEEE International Conference, IEEE Press, Piscataway NJ, pp 1161-1164
[Leo96] Leontyev AA, Mintsev VB, Ushnurtsev AYe, Fortov VYe, Shurupov AV (1996) Two-staged magnetic flux compressors with flux trapping. In: Chernyshev VK, Selemir VD, Plyashkevich LN (eds) Megagauss and megaampere pulse technology and applications, part 1. RFNC-VNIIEF, Sarov, Russia, pp 322-326
[Neu01] Neuber A, Dickens J, Cornette JB, Jamison K, Parkinson R, Giesselmann M, Worsey P, Baird J, Schmidt M, Kristiansen M (2001) Electrical behavior of a simple helical flux compression generator for code benchmarking. IEEE Trans. on Plasma Science 29: 573-581
[Neu02] Neuber A, Holt T, Hernandez J, Dickens J, Kristiansen M (2002) Geometry impact on flux losses in MFCGs. In: Selemir VD, Plyashkevich LN (eds) Meggauss-9. RFNC-VNIIEF, Sarov, Russia, pp 571-577
[Neu03] Neuber AA, Hernandez JC, Dickens JC, Kristiansen M (2003) Helical MFCG for driving a high inductance load. Electromagnetic Phenomena 3: 397-404
[Rei87] Reinovsky RE (1987) Fuse opening switches for pulse power applications. In: Guenther A, Kristiansen M. (eds) Opening switches, advances in pulsed power technology, vol. 1, Plenum Press, New York, London

Index

armature
   6061 - T6 aluminum 68
   annealing 121
   brittle failure 65, 70
   cracking 70
   eccentricity 109
   end effect 67, 106
   fracturing 54
   longitudinal cracking 53
   longitudinal cracks 71
   material 8
   multi layer 94
   OFHC copper 68
   premature fracturing 61
   seamless pipe 10
   spalling 91
   structural properties 65
   surface finish 121
   wall thickness tolerance 112

boosting 22

camera
   gated image intensifier 73
   rotating mirror 73
capacitive divider 130
Chapman-Jouguet 69
   CJ plane 69
computer code
   2DL 85
   AUTODYN-2D 95
   CHARADE 60
   CHEETAH 70
   CTH 128
   LS-DYNA-3D 99
   TDL 60, 85
   TIGER 69

conductivity 36
contact
   armature/stator 97
   partial turn skipping 142
   sliding 146
   spirally moving 10
   turn skipping 109
   zipper 146
crowbar
   disk 51, 255
   glideplane 50
   moment of 4, 49
   pin 254
current
   conduction 36
   displacement 36
   linear density 41
current density 36

density
   drop in 78
   physical, C-4 69
detonation
   Rayleigh line 59
   Shock Hugoniots 59
   thermodynamic theory 58
detonator 72
diffusion
   magnetic 31
   non linear 42
   nonlinear 35, 47

efficiency
   conversion 2
   energy 23
energy
   flow 5
   output 5

prime 5
seed 2, 5
specific 1, 243
storage 2
surface factor 45
energy transfer
  transformer 223
equation
  continuity 58
  energy 58
  Eulerian, of motion 60
  Gurney 21
  Hugoniot 130
  Lagrangian, of motion 60
  Lamé's 67
  magnetic diffusion 36
  modified Gurney 120
  nonlinear magnetic diffusion 43
  of state, Jones-Wilkins-Lee 99
  of state, Steinberg 98
  thermal diffusion 37
exploding bridge wire 72

FCG
  applied 16
  double end initiated 17
  dual stage 247
  dual stage performance 258
  end initiated 17
  first stage 253
  second stage 254
  simple 16
  size 19
  volume resistance 134
FCG model
  2D 181
  PSpice 159
field coil 253
figure of merit 18, 142
finite difference 60
finite element method 97
flux
  conservation 3
  linkage 174, 176
  per unit length 147
  residual 228
flux loss
  stator geometry 139
flyer plate 127
frequency
  angular 18

HPM 273
fuse
  blow limit 167, 218
  cross sectional area 204
  equivalent action timescale 214
  example calculations 220
  foil 202
  opening switch 261
  optimal length 206
  optimal material 209
  other factors 213
  quenching medium 210
  resistivity 217
  system equivalency 216
  wire 202

gain
  current 11
  energy 3, 11
geometric optics 79
Gurney
  angle 8
  specific energy 20

helix
  current carrying 4
  stator 4
Helmholtz free energy 62

imaging
  fast optical 8
  x-ray 8
imploding liner 23
inductance
  calculations 184
  helical FCG 14
  initial 3
  load 3
  lumped 11
inductor
  collapse 160
integral of current action 205
  normalized 218

law
  Faraday 40
  second, of thermodynamics 57
Law
  Hooke's 65
loss
  avoidable 49

electric breakdown 145
electrical breakdown 134
  individual 151
  intrinsic 15, 145, 149, 173
  magnetic flux 35
  ohmic 149

Mach stem 55, 82
magnetic flux 10
  non linear loss 8
magnetic flux compression
  explosively driven 7
mandrel 247
Marx generator 271
metal fatigue 93
method of characteristics 60
Mgauss 23
mica capacitor 272
modulus
  shear 64
  Young's 63
momentum
  conservation of 57

parameter
  intrinsic flux loss 15
PFN 22, 23
  Pulse Forming Network 26
piezoelectric pin 130
plasma
  focus 25
  neon 25
power conditioning 5, 166
pulse forming line 270
pulsed power
  compact 270
  traditional 274
pulsed power system
  size 275
  traditional 269

radiation
  RF 26
rarefaction 78
rarefactions 57
resistance
  calculation 185
  shocked air 131, 137
  shocked argon 137
  shocked SF_6 132, 138
resistivity 36

  shocked metals 154
resistor
  liquid $CuSO_4$ 263

seed source 235
  capacitive 236
  example system 243
  explosively driven 239
  ferroelectric 239
  ferromagnetic 240
  permanent magnet 241
  solid state switched 238
  spark gap switched 238
  specific energy 243
shock tube 127
skin depth 12
  classical 37
  flux 45
  magnetic 37
  magnetic flux 39
stator 10
strain
  plane 66
  shear 66
strain energy 62
stress
  hoop 68
  plane 66
switch
  closing 3
  crowbar 50
  fuse, opening 201
  opening 3

transformer
  dynamic 230
  impedance 245
  passive 230
  strongly coupled 225

vircator 26, 272
voltage
  induced 3

wave
  compressive 81
  detonation 55
  shock 56, 72
  spherical detonation 81
  Taylor 69
wire

gauge number 168
insulation 144
round 30
size 29

square 30

xenon flash tube 73